Neural Networks for RF and Microwave Design

DISCLAIMER OF WARRANTY

The technical descriptions, procedures, and computer programs in this book have been developed with the greatest of care and they have been useful to the authors in a broad range of applications; however, they are provided as is, without warranty of any kind. Artech House, Inc. and the authors and editors of the book titled *Neural Networks for RF and Microwave Design* make no warranties, expressed or implied, that the equations, programs, and procedures in this book or its associated software are free of error, or are consistent with any particular standard of merchantability, or will meet your requirements for any particular application. They should not be relied upon for solving a problem whose incorrect solution could result in injury to a person or loss of property. Any use of the programs or procedures in such a manner is at the user's own risk. The editors, authors, and publisher disclaim all liability for direct, incidental, or consequent damages resulting from use of the programs or procedures in this book or the associated software.

For a listing of recent titles in the *Artech House Microwave Library*, turn to the back of this book.

Neural Networks for RF and Microwave Design

Q. J. Zhang
K. C. Gupta

Artech House
Boston • London
www.artechhouse.com

Library of Congress Cataloging-in-Publication Data
Zhang, Q. J.
  Neural networks for RF and microwave design / Q. J. Zhang, K. C. Gupta.
    p. cm. — (Artech House microwave library)
  Includes bibliographical references and index.
  ISBN 1-58053-100-8 (alk. paper)
    1. Microwave circuits—Computer-aided design. 2. Radio circuits—Computer-aided design. 3. Neural networks (Computer science). I. Gupta, K. C. II. Title. III. Series.

TK7876 .Z435 2000   00-030612
621.381'32—dc21   CIP

British Library Cataloguing in Publication Data
Zhang, Q. J. (Qi-jun)
  Neural networks for RF and microwave design. — (Artech
  House microwave library)
  1. Neural networks (Computer science) 2. Radio frequency
  3. Microwave circuits—Design and construction
  I. Title II. Gupta, K. C. (Kuldip C)
  621.3'8132

ISBN 1-58053-100-8

Cover design by Igor Valdman

© 2000 ARTECH HOUSE, INC.
685 Canton Street
Norwood, MA 02062

All rights reserved. Printed and bound in the United States of America. No part of this book or software may be reproduced or utilized in any form or by any means, electronic or mechanical, including photocopying, recording, or by any information storage and retrieval system, without permission in writing from the publisher.
  All terms mentioned in this book and software that are known to be trademarks or service marks have been appropriately capitalized. Artech House cannot attest to the accuracy of this information. Use of a term in this book and software should not be regarded as affecting the validity of any trademark or service mark.

International Standard Book Number: 1-58053-100-8
Library of Congress Catalog Card Number: 00-030612

10 9 8 7 6 5 4 3 2 1

Contents

	Preface	xv
1	**Introduction and Overview**	**1**
1.1	RF and Microwave Design	1
1.2	Artificial Neural Networks (ANNs)	3
1.3	Overview of the Book	4
	References	8
2	**Modeling and Optimization for Design**	**11**
2.1	The Design Process	11
2.1.1	Anatomy of the Design Process	11
2.1.2	Conventional Design Procedures	13
2.1.3	CAD Approach	15
2.1.4	Knowledge-Aided Design (KAD) Approach	17
2.2	RF and Microwave Circuit CAD	19
2.2.1	Modeling of Circuit Components	19
2.2.2	Computer-Aided Analysis Techniques	21
2.2.3	Circuit Optimization	36
2.3	CAD for Printed RF and Microwave Antennas	38

2.3.1	Modeling of Printed Patches and Slots	39
2.3.2	Analysis of Printed Patches and Slots	51
2.4	Role of ANNs in RF and Microwave CAD	55
2.4.1	Modeling of RF and Microwave Components	55
2.4.2	Efficient Optimization Strategies	56
2.4.3	Implementation of Knowledge-Aided Design (KAD)	56
2.5	Summary	57
	References	57
3	**Neural Network Structures**	**61**
3.1	Introduction	61
3.1.1	Generic Notation	62
3.1.2	Highlights of the Neural Network Modeling Approach	63
3.2	Multilayer Perceptrons (MLP)	64
3.2.1	MLP Structure	64
3.2.2	Information Processing by a Neuron	65
3.2.3	Activation Functions	66
3.2.4	Effect of Bias	68
3.2.5	Neural Network Feedforward	70
3.2.6	Universal Approximation Theorem	71
3.2.7	Number of Neurons	73
3.2.8	Number of Layers	74
3.3	Back Propagation (BP)	75
3.3.1	Training Process	75
3.3.2	Error Back Propagation	76
3.4	Radial Basis Function Networks (RBF)	77
3.4.1	RBF Network Structure	78
3.4.2	Feedforward Computation	79
3.4.3	Universal Approximation Theorem	81
3.4.4	Two-Step Training of RBF Networks	81

3.5	Comparison of MLP and RBF Neural Networks	81
3.6	Wavelet Neural Networks	83
3.6.1	Wavelet Transform	83
3.6.2	Wavelet Networks and Feedforward Computation	84
3.6.3	Wavelet Neural Network with Direct Feedforward From Input to Output	86
3.6.4	Wavelet Network Training	87
3.6.5	Initialization of Wavelets	87
3.7	Arbitrary Structures	88
3.8	Clustering Algorithms and Self-Organizing Maps	90
3.8.1	Basic Concept of the Clustering Problem	91
3.8.2	k-Means Algorithm	94
3.8.3	Self-Organizing Map (SOM)	94
3.8.4	SOM Training	95
3.8.5	Using a Trained SOM	96
3.9	Recurrent Neural Networks	97
3.10	Summary	100
	References	101
4	**Training of Neural Networks**	**105**
4.1	Microwave Neural Modeling: Problem Statement	105
4.2	Key Issues in Neural Model Development	106
4.2.1	Data Generation	107
4.2.2	Range and Distribution of Samples in Model Input Parameter Space	108
4.2.3	Data Splitting	114
4.2.4	Data Scaling	116
4.2.5	Initialization of Neural Model Weight Parameters	119
4.2.6	Overlearning and Underlearning	120

4.2.7	Quality Measures for a Neural Model	126
4.3	Neural Network Training	128
4.3.1	Categorization of Training Techniques	129
4.3.2	Gradient-Based Methods	130
4.3.3	Line Minimization	131
4.3.4	Local Minimum and Global Minimum	132
4.4	Back Propagation Algorithm and Its Variants	133
4.5	Training Algorithms Using Gradient-Based Optimization Techniques	137
4.5.1	Conjugate Gradient Training Method	137
4.5.2	Quasi-Newton Training Method	139
4.5.3	Levenberg-Marquardt and Gauss-Newton Training Methods	140
4.6	Nongradient-Based Training: Simplex Method	141
4.7	Training With Global Optimization Methods	143
4.7.1	Genetic Algorithms	143
4.7.2	Simulated Annealing (SA) Algorithms	145
4.8	Training Algorithms Utilizing Decomposed Optimization	146
4.9	Comparisons of Different Training Techniques	147
4.10	Feedforward Neural Network Training: Examples	148
	References	151
5	**Models for RF and Microwave Components**	**155**
5.1	Modeling Procedure	155
5.1.1	Selection of Model Inputs and Outputs	156
5.1.2	Training Data Generation	156
5.1.3	Error Measures	157

5.1.4	Integration of EM-ANN Models with Circuit Simulators	157
5.2	Models for Vias and Multilayer Interconnects	158
5.2.1	Microstrip Transmission Line Model	159
5.2.2	Broadband GaAs One-Port Microstrip Via	160
5.2.3	Broadband GaAs Two-Port Microstrip Via	162
5.2.4	Stripline-to-Stripline Multilayer Interconnect	163
5.2.5	Microstrip-to-Microstrip Multilayer Interconnect	165
5.2.6	Integration of EM-ANN Models with a Network Simulator	166
5.3	EM-ANN Models for CPW Components	168
5.3.1	EM-ANN Modeling of CPW Transmission Lines	169
5.3.2	Modeling of CPW Bends	169
5.3.3	EM-ANN Models for CPW Opens and Shorts	174
5.3.4	EM-ANN Modeling of CPW Step-in-Width	175
5.3.5	EM-ANN Modeling of CPW Symmetric T-Junctions	176
5.4	Other Passive Components' Models	178
5.4.1	Spiral Inductors	179
5.4.2	Multiconductor Transmission Lines	179
5.4.3	Microstrip Patch Antennas	187
5.4.4	Waveguide Filter Components	188
	References	190
6	**Modeling of High-Speed IC Interconnects**	**195**
6.1	Introduction	195

6.2	High-Speed Interconnect Modeling and Signal Integrity Analysis	197
6.2.1	Traditional Techniques	197
6.2.2	Neural Network Approach	200
6.3	Application Examples	203
6.3.1	Example A: Three Parallel Coupled Interconnects	203
6.3.2	Example B: Two Asymmetric Interconnects	206
6.3.3	Example C: An Eight-Bit Digital Bus Configuration	207
6.3.4	Example D: Interconnect Circuit With Nonlinear Terminations	209
6.3.5	Example E: Signal Integrity Optimization	210
6.3.6	Example F: Neural Networks for Interconnects on a Printed Circuit Board	213
6.4	Discussion	216
6.4.1	Run-Time Comparison	216
6.4.2	Performance Evaluation	218
6.5	Conclusions	222
	References	223

7	**Active Component Modeling Using Neural Networks**	**227**
7.1	Introduction	227
7.2	Direct Modeling Approach	228
7.2.1	Transistor DC Model	229
7.2.2	Small-Signal Models	230
7.2.3	Large-Signal Models	233
7.2.4	Time-Varying Volterra Kernel-Based Model	237
7.3	Indirect Modeling Approach Through a Known Equivalent Circuit Model	239

7.4	Discussion	245
	References	246
8	**Design Analysis and Optimization**	**249**
8.1	Design and Optimization Using ANN Models	249
8.2	Optimization of Component Structure	250
8.3	Circuit Optimization Using ANN Models	251
8.3.1	CPW Folded Double-Stub Filter	251
8.3.2	CPW Power Divider	252
8.4	Multilayer Circuit Design and Optimization Using ANN Models	255
8.5	CPW Patch Antenna Design and Optimization	265
8.5.1	Transmission Line Model for CPW Antennas	265
8.5.2	CPW Patch Antenna Design Using EM-ANN Models	269
8.5.3	CPW Patch Antenna Design Optimization Using EM-ANN Models	269
8.6	Yield Optimization of a Three-Stage MMIC Amplifier	272
8.7	Remarks	278
	References	281
9	**Knowledge-Based ANN Models**	**283**
9.1	Introduction	283
9.1.1	Motivation	283
9.1.2	Rule-Based Knowledge Networks	284
9.1.3	Microwave-Oriented Knowledge Structures	285
9.2	Knowledge-Based Neural Networks (KBNN)	285
9.2.1	Model Structure	286

9.2.2	Neural Network Training	289
9.2.3	Finished Model for User and Discussion	289
9.2.4	KBNN Examples	290
9.3	Source Difference Method	303
9.3.1	Model Structure	308
9.3.2	Preprocessing	308
9.3.3	Neural Network Training	309
9.3.4	Finished Model for User	309
9.4	Prior Knowledge Input Method (PKI)	310
9.4.1	Model Structure	310
9.4.2	Preprocessing	310
9.4.3	Neural Network Training	311
9.4.4	Finished Model for User	311
9.5	Space-Mapped Neural Networks	312
9.5.1	Model Structure	312
9.5.2	Space-Mapping Concept	313
9.5.3	Space-Mapped Neuromodeling	313
9.5.4	Frequency in Neuromapping	314
9.6	Hierarchical Neural Networks and Neural Model Library Development	315
9.6.1	Development of a Library of Models: Problem Statement	316
9.6.2	Hierarchical Neural Network Structure	316
9.6.3	Base Models and Their Training	317
9.6.4	Hierarchical Neural Model and Its Training	319
9.6.5	Finished Model for User	321
9.6.6	Algorithm for Overall Library Development	322
9.6.7	Discussion	323
9.6.8	Hierarchical Neural Model Examples	323
9.7	Summary	333
	References	334

10	**Concluding Remarks and Emerging Trends**		**337**
10.1	Summary of the Book		337
10.2	Impact of Neural Nets on RF and Microwave Design		340
10.2.1	Insertion in Design Tools		340
10.2.2	ANN Models Linked to Design Software		340
10.2.3	Efficient Use of EM Simulators		341
10.2.4	Development of Efficient Optimization Strategies		341
10.2.5	Implementation of Knowledge-Aided Design		342
10.3	Trends and Challenges		342
	References		346
	Appendix A: NeuroModeler Introductory Version		**347**
A.1	System Requirements		347
A.2	How to Install the Software		347
A.3	Quick Start the Program Using an Example		348
A.4	User Interactions		349
A.4.1	Minimum User Interactions		349
A.4.2	Extra User Control		350
A.5	Highlights of the Introductory Version		350
A.6	Information on Upgrade to the Standard Version		350
	About the Authors		**353**
	Index		**357**

Preface

Neural networks, also called artificial neural networks (ANN), are information processing systems inspired by the ability of the human brain to learn from observations and to generalize by abstraction. The fact that neural networks can be trained for totally different applications, has resulted in their use in diverse fields such as pattern recognition, speech processing, control, medical applications, and more. The recent introduction of neural networks to the RF and microwave field marks the birth of an unconventional alternative to modeling and design problems in RF and microwave CAD. Neural networks can learn and generalize from data allowing model development even when component formulas are unavailable. Neural network models are universal approximators allowing reuse of the same modeling technology for both linear and nonlinear problems and at both device and circuit levels. Yet neural network models are simple and model evaluation is very fast. Recent works have led to their use for modeling microstrip lines, vias, CPW discontinuities, spiral inductors, FETs, and VLSI interconnects; for speeding up harmonic balance simulations and optimizations; and for Smith chart representation and automatic impedance matching. These pioneering works herald a brand new opportunity to conquer some of the toughest RF and microwave CAD problems today and tomorrow.

An increasing number of RF and microwave engineers as well as researchers have begun to take a serious interest in this emerging technology. Although a number of research papers on this topic have appeared in the literature recently, there is no book describing this technology from the RF/microwave engineer's perspective. All the existing books on neural networks—written mostly for signal processing, pattern recognition, process control, and so on—do not address RF and microwave modeling and design problems. This book has been prepared with RF/microwave designers, researchers, and graduate students as its primary audience. The subject of neural networks will be described from the point of view of, and in the language of, RF and microwave engineers. The issues, challenges, formulations, and solutions important to the RF and microwave areas are described uniquely in this book.

Following an introduction to RF/microwave design and neural networks in Chapter 1, we describe the RF and microwave process and problems and relate them to neural networks in Chapter 2. Chapters 3 and 4 describe various kinds of neural network structures and methods for training neural networks, and thus provide the ANN background needed for the following chapters. ANN modeling of various types of RF and microwave components is discussed in Chapter 5. Examples such as transmission line structures, coplanar waveguide circuit design, and microstrip patch antennas are included. These component models can be linked to commercially available microwave network simulators. Use of ANN models for optimizing the geometry of some of these components is also discussed. ANN modeling of interconnects used in high-speed digital circuits—as well as in RF and microwave circuits—is the topic of Chapter 6. The development of small-signal and large-signal models for various active devices like FETs, HBTs, and HMETs is the topic of Chapter 7. This is followed by a chapter that describes, specifically for RF and microwave designers, how to incorporate neural network models in circuit simulation and design. Examples of optimization of CPW circuits, patch antennas, multilayer filters, and amplifiers are included. Chapter 9 highlights a unique and exciting technical area, combining prior RF and microwave knowledge with neural networks leading to a methodology for knowledge-based design for RF and microwave circuits. The book concludes with a chapter summarizing the concepts presented in previous chapters and discussing some of the emerging trends for future research and development in this exciting field. To further help readers in putting ANN concepts into practice, an Introductory Version of NeuroModeler software is attached. Appendix A provides a brief instruction for using the software.

We thank Mark Walsh, Managing Editor of Artech House, for his enthusiasm in this book project. We also thank our former Ph.D. students, Dr. Fang Wang of IBM (supervised by Q. J. Zhang), Dr. Paul Watson of Air Force Research Lab (supervised by K. C. Gupta), and Dr. Choonsik Cho of Hyundai Electronics (supervised by K. C. Gupta) for their excellent work in this field. Several important parts of the book are based upon their Ph.D. theses. Q. J. Zhang also wishes to thank his colleagues M. S. Nakhla and J. W. Bandler for their research collaborations, his former students A. H. Zaabab and A. Veluswami for their research contributions, which are also used in the book, and his students V. Devabhaktuni, F. Wang, L. Lin, C. Xi, Y. Fang, G. Chahl and S. Thurairasa, and Drs. M. Yagoub and R. Achar of Carleton University for their help during the preparation of this manuscript. K. C. Gupta is also thankful to his colleague Professor Roop Mahajan for introducing him to the exciting world of artificial neural networks. The support and cooperation of the staff at Artech House—including Sean Flannagan, Tina Kolb, and Barbara Levenwirth—are appreciated. And finally we thank our families for their love and support throughout this project.

Q. J. Zhang
K. C. Gupta

1

Introduction and Overview

1.1 RF and Microwave Design

The rapid development of commercial markets for wireless communication products over the past decade has led to an explosion of interest in improved circuit design approaches in the radio frequencies (RF) and microwave areas. This new market for high frequency expertise is replacing the older discipline of Department of Defense (DOD)-oriented RF/microwave electronics—with its emphasis on performance at any price—after the defense build-down in the late 1980s and early 1990s. Modern wireless communication systems require a thorough understanding of RF and microwave circuit design techniques in addition to a background in digital communication techniques and familiarity with existing and emerging wireless communication protocol standards. The wireless industry's emphases on time-to-market and low cost, resulting from first-pass-design, are placing enhanced demands on computer-aided design (CAD) tools for RF/microwave circuits, antennas, and systems.

RF and microwave circuit design has progressed considerably and has achieved a certain level of maturity in recent years. Electromagnetic (EM) simulation techniques for high-frequency structures developed over the past decade have helped to bring the CAD for hybrid RF/microwave circuits and monolithic RF/microwave integrated circuits (MMIC) to its current state of the art. The key contribution of EM simulation techniques to RF/microwave CAD has been in the domain of accurate models for microwave components [1]. As a result, the microstripline-based RF/microwave-circuit design has been brought to almost a first-pass success level. Modeling still remains a major bottleneck for CAD of certain classes of RF/microwave circuits (such as coplanar waveguide (CPW) circuits, multiple-layered circuits, and integrated circuit-antenna modules) and for most of the millimeter-wave (mm-wave, above about 40 GHz) circuits. The efficient use of EM simulation techniques in the development of accurate RF/microwave CAD tools is still a topic of research. The use of EM simulators for accurate and practical design of RF/microwave circuits can be made possible by innovative developments in diakoptics methods for frequency- and time-domain analyses. The diakoptics (or decomposition or segmentation) approach consists of partitioning the circuits into smaller parts, carrying out the characterization of each of the smaller parts, and combining these characterizations (by network theory-based methods) to yield the response of the overall circuit. Since the EM analysis is carried out only for a smaller circuit portion at any time, the computational requirement can become manageable and practical for design purposes.

Another tendency in RF and microwave design is the increasing need for optimization-based design automation. Stringent design specifications on the circuits need to be met using optimization-based computer algorithms that require iterative circuit evaluations. Furthermore, the drive for time-to-market in the wireless industry demands optimizations to include effects such as manufacturing tolerances, process variations, and so on. Statistical analysis such as the Monte Carlo analysis and yield-driven optimization become necessary. This also leads to a highly repetitive computational process in which circuit simulations are performed again and again. Added to this challenge is the increasing size and complexities of the circuits that need to be optimized. The overall task could become computationally very prohibitive. However, the pursuit of computation speed and accuracy has often been a trade-off process where models with better accuracy usually come with more computation and fast models have often sacrificed accuracy. New modeling techniques that allow very fast model evaluation and at the same time do not sacrifice accuracy are needed in order to allow massive and highly repetitive model evaluations in optimization and yield-driven design.

One of the recent developments that may lead to an efficient usage of EM simulation for RF and microwave CAD (and is related to the theme of this book) is the use of artificial neural network (ANN) models trained by full-wave EM simulators [2–4]. In this methodology, EM simulation is used to obtain S-parameters for all the components to be modeled over the ranges of designable parameters for which these models are expected to be used. Training ANN configurations using the data obtained from EM simulations develops an ANN model for each one of the components. Such ANN models have been shown to retain the accuracy obtainable from EM simulators and at the same time exhibit the efficiency (in terms of computer time required) that is obtained from lumped network models normally implemented in commercially available microwave network simulators (like HP's MDS). Similar

ANN models need to be developed for several novel components, subcircuits, and prototype circuits in RF/microwave frequency wireless communication systems. ANNs are also well suited for modeling active devices (including thermal effects) and for circuit optimization and statistical design [5].

Another aspect of RF/microwave design that should receive attention in the near future is the possibility of employing knowledge-based tools for initial design. The overall design process consists of several steps starting from:

1. Problem identification and moving through;
2. Specifications generation;
3. Concept generation;
4. Analysis;
5. Evaluation;
6. Initial design;
7. Detailed design.

Currently available RF/microwave CAD tools address only the last step, i.e., transformation from an initial design to an optimized detailed design. It has been pointed out [6] that knowledge-based tools are needed for earlier stages of the design process. As pointed out later in this book, ANNs can be used for embedding the knowledge needed for the initial stages of the design process.

1.2 Artificial Neural Networks (ANNs)

Artificial neural networks (ANNs) have emerged as a powerful technique for modeling general input/output relationships. In the past, ANNs have been used for many complex tasks. Applications have been reported in areas such as control [7], telecommunications [8], biomedical [9], remote sensing [10], pattern recognition [11], and manufacturing [12], just to name a few. However, in recent years, ANNs are being used more and more in the area of RF/microwave design [13]. Applications reported in the literature include automatic impedance matching [14], microstrip circuit design [15], microwave circuit analysis and optimization [16, 17], active device modeling [17–19], modeling of passive components [2–4, 20–26], and modeling for electro/opto interconnections [27]. As a recent development, the pioneering work on straight applications of standard neural network techniques to microwave problems has now led to advanced work in RF and microwave-oriented neural network structures [28–29], training algorithms [29–30], knowledge based networks [6, 26, 31–32], and methodologies for libraries of microwave neural models development [33].

Artificial neural network models can be more accurate than polynomial regression models [34–39], allow more dimensions than look-up table models [39], and allow multiple outputs for a single model. Models using ANNs are developed by providing sufficient training data (i.e., EM simulation or measured data) from which it learns the underlying input/output mapping. Several valuable characteristics are offered by ANNs [40]. First, no prior knowledge about the input/output mapping is required for model development. Unknown relationships are inferred from the data provided for training. Therefore, with an ANN, the fitted function is represented by the network and does not have to be explicitly defined. Second, ANNs can generalize, meaning they can respond correctly to new data that has not been used for model development. Third, ANNs have the ability to model highly nonlinear as well as linear input/output mappings. In fact, it has been shown that ANNs are capable of forming an arbitrarily close approximation to any continuous nonlinear mapping [41]. ANNs provide a general methodology for the development of accurate and efficient electromagnetically trained ANN (EM-ANN) models for use in CAD of RF/microwave circuits, antennas, and systems. These models are capable of providing EM simulation accuracy within a microwave circuit simulator environment and thus lead to an accurate and efficient CAD.

The primary objective of this book is to provide the required background and expose the current state of the art in the applications of ANN technology to RF/microwave design.

1.3 Overview of the Book

As the book is addressed to RF/microwave designers, students, and researchers, the background reviews of modeling and optimization for design and neural network structures are provided in Chapters 2 and 3, respectively. Methods for training—the most important step in the development of neural nets—is the topic of Chapter 4. Materials in these background chapters converge to Chapter 5 on ANN modeling for RF/microwave design. Specific examples of ANN modeling in different domains of RF/microwave design are presented in the next three chapters. ANNs for high speed interconnects are discussed in Chapter 6. This is followed by active device models in Chapter 7 and several examples of circuit analysis and optimization in Chapter 8. The emerging area of knowledge-based ANN models is the topic of Chapter 9. The book is then drawn together in a concluding chapter. Figure 1.1 depicts the conceptual flow graph used for the organization of the book. The following notes give some introduction to each of the remaining chapters.

1.3.1 Chapter 2—Modeling and Optimization for Design

This chapter discusses the anatomy of the overall design process. Conventional design and computer-aided design methodologies are presented. Concepts of knowledge-aided design are introduced. The importance of modeling and optimization in the design process is brought out. It is in these two areas that ANNs are likely to play very crucial roles. CAD methods for RF and microwave circuits and printed antennas are briefly reviewed. The role of ANNs in RF and microwave CAD is discussed.

1.3.2 Chapter 3—Neural Network Structures

The starting point for introducing neural networks is a discussion of structures (or architectures) used for constructing neural networks. Most commonly used feedforward neural network configurations and the back propagation method used for training such ANNs are described. Other ANN structures discussed in this chapter include radial basis function (RBF) network, wavelet neural network, and self-organizing maps (SOM). Brief reviews of arbitrary and recurrent ANNs are also included.

1.3.3 Chapter 4—Training of Neural Networks

Training algorithms are integral and the most important part of neural network development. The first part of this chapter documents key issues related to training; namely data generation, range and distribution of samples in the input parameter space, data scaling, initialization of weight parameters, and quality measures for the trained neural network. The actual training process—algorithms for finding values of weights associated with various neurons—can be viewed as an optimization process. Thus, various well-known optimization techniques can be used for training ANN structures. Methods discussed in this chapter include the back propagation algorithm based on the steepest descent principle, the conjugate gradient algorithm, the quasi-Newton algorithm, Levenberg-Marquardt and Gauss-Newton algorithms, the use of decomposed optimization, the simplex method, genetic algorithms, and simulated annealing algorithms.

6   Neural Networks for RF and Microwave Design

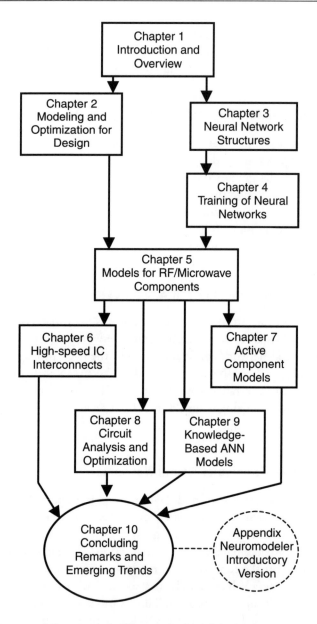

Figure 1.1 Conceptual flow graph for the organization of the book.

1.3.4 Chapter 5—Models for RF and Microwave Components

This chapter includes several examples of the techniques described in earlier chapters to the development of CAD models for RF/microwave components.

Component examples include one-port and two-port microstrip vias, vertical interconnects used in multilayer circuits, CPW lines, CPW bends, and other discontinuities occurring in CPW circuits. Also included in this chapter is the ANN modeling of some other passive components such as spiral inductors, multiconductor transmission lines, microstrip patch antennas, and wave guide filters.

1.3.5 Chapter 6—High-Speed IC Interconnects

High-speed interconnect constitutes an important enabling technology item in high-speed computer circuits. ANN macromodeling techniques allow high-speed interconnect optimization to be carried out. This chapter presents a detailed description of this important application of neural networks.

1.3.6 Chapter 7—Active Component Models

Active device modeling is an important and critical area of microwave CAD, and it is an area in which ANNs can play a significant role. This chapter describes the direct modeling of the device's external behavior (both for small-signal and nonlinear models) as well as indirect modeling through a known equivalent circuit model. The discussion also includes a time-varying Volterra kernel-based model and circuit representation of the neural network models.

1.3.7 Chapter 8—Circuit Analysis and Optimization

This chapter presents examples of ANN-model-based analysis and the optimization of RF/microwave circuits. Two examples of CPW circuits (a power divider and a folded double-stub filter), the optimization of a CPW patch antenna, the optimization of multilayer band pass filters, and the optimization of a three-stage MMIC MESFET amplifier are discussed.

1.3.8 Chapter 9—Knowledge-Based ANN Models

This chapter addresses the important topic of embedding knowledge in ANN models as well as the use of prior knowledge in reducing the training time of ANN structures. The effect of adding knowledge is illustrated through a knowledge-based neural network structure. Several methods for combining circuit knowledge with neural networks are described, including the source-difference method, the prior-knowledge input method, space mapped neural models, and hierarchical neural networks. Examples of these techniques are included.

1.3.9 Chapter 10—Concluding Remarks and Emerging Trends

A short chapter in which some conclusions are drawn completes the book. The current status of ANN applications to RF/microwave design is summarized from the earlier chapters. Some thoughts on the likely impact of the neural network technology on RF and microwave design are articulated. Current trends and challenges facing this area are then given.

References

[1] Wu, D., et al., "Accurate Numerical Modeling of Microstrip Junctions and Discontinuities," *Int. J. Microwave mm-Wave Computer-Aided Eng.*, Vol. 1, No. 1, 1991, pp. 48–58.

[2] Watson, P. M., and K. C. Gupta, "EM-ANN Models for Microstrip Vias and Interconnects in Multilayer Circuits," *IEEE Trans. Microwave Theory and Techniques*, Vol. 44, Dec. 1996, pp. 2495–2503.

[3] Creech, G. L., et al., "Artificial Neural Networks for Fast and Accurate EM-CAD of Microwave Circuits," *IEEE Trans. Microwave Theory and Techniques*, Vol. 45, May 1997, pp. 794–802.

[4] Veluswami, A., M. S. Nakhla, and Q. J. Zhang, "The Application of Neural Networks to EM-Based Simulation and Optimization of Interconnects in High-Speed VLSI Circuits," *IEEE Trans. on Microwave Theory and Techniques*, Vol. 45, May 1997, pp. 712–723.

[5] Zaabab, A. H., Q. J. Zhang, and M. Nakhla, "A Neural Network Modeling Approach to Circuit Optimization and Statistical Design," *IEEE Trans. on Microwave Theory and Techniques*, Vol. 43, June 1995, pp. 1349–1358.

[6] Gupta, K. C., "ANN and Knowledge-Based Approaches for Microwave Design," in *Directions for the Next Generation of MMIC Devices and Systems*, N. K. Das and H. L. Bertoni, Eds., NY: Plenum, 1996, pp. 389–396.

[7] Balakrishnan, S. N., and R. D. Weil, "Neurocontrol: A Literature Survey," *Mathematical and Computer Modeling*, Vol. 23, No. 1–2, 1996, pp. 101–117.

[8] Cooper, B. S., "Selected Applications of Neural Networks in Telecommunication Systems," *Australian Telecommunication Research*, Vol. 28, No. 2, 1994, pp. 9–29.

[9] Alvager, T., T. J. Smith, and F. Vijai, "The Use of Artificial Neural Networks in Biomedical Technologies: An Introduction," *Biomedical Instrumentation and Technology*, Vol. 28, No. 4, 1994, pp. 315–322.

[10] Goita, K., et al., "Literature Review of Artificial Neural Networks and Knowledge Based Systems for Image Analysis and Interpretation of Data in Remote Sensing," *Canadian Journal of Electrical and Computer Engineering*, Vol. 19, No. 2, 1994, pp. 53–61.

[11] Smetanin, Y. G., "Neural Networks as Systems for Pattern Recognition: A Review," *Pattern Recognition and Image Analysis*, Vol. 5, No. 2, 1995, pp. 254–293.

[12] Nunmaker, J. F. Jr., and R. H. Sprague Jr., "Applications of Neural Networks in Manufacturing," *Proceedings of the Twenty-ninth Hawaii International Conference on System Sciences*, Vol. 2, 1996, pp. 447–453.

[13] Zhang, Q. J., and G. L. Creech (Guest Editors), *International Journal of RF and Microwave Computer-Aided Engineering*, Special Issue on Applications of Artificial Neural Networks to RF and Microwave Design, Vol. 9, NY: Wiley, 1999.

[14] Vai, M., and S. Prasad, "Automatic Impedance Matching with a Neural Network," *IEEE Microwave and Guided Wave Letters*, Vol. 3, No. 10, Oct. 1993, pp. 353–354.

[15] Horng, T., C. Wang, and N. G. Alexopoulos, "Microstrip Circuit Design Using Neural Networks," *MTT-S Int. Microwave Symp. Dig.*, 1993, pp. 413–416.

[16] Zaabab, A. H., Q. J. Zhang, and M. Nakhla, "Analysis and Optimization of Microwave Circuits and Devices Using Neural Network Models," *MTT-S Int. Microwave Symp. Dig.*, 1994, pp. 393–396.

[17] Litovski, V. B., et. al., "MOS Transistor Modeling Using Neural Network," *Electronics Letters*, Vol. 28, No. 18, 1992, pp. 1766–1768.

[18] Gunes, F., F. Gurgen, and H. Torpi, "Signal-Noise Neural Network Model for Active Microwave Devices," *IEE Proc.-Circuits, Devices, Syst.*, Vol. 143, No. 1, Feb. 1996, pp. 1–8.

[19] Shirakawa K., et. al., "A Large-Signal Characterization of an HEMT Using a Multilayered Neural Network," *IEEE Trans. on Microwave Theory and Techniques*, Vol. 45, No. 9, Sept. 1997, pp. 1630–1633.

[20] Creech, G. L., et al., "Artificial Neural Networks for Accurate Microwave CAD Applications," *IEEE MTT-S Int. Microwave Symp. Dig.*, 1996, pp. 733–736.

[21] Watson, P., and K. C. Gupta, "EM-ANN Models for Via Interconnects in Microstrip Circuits," *IEEE MTT-S Int. Microwave Symp. Dig.*, 1996, pp. 1819–1822.

[22] Watson, P. M., C. Cho, and K. C. Gupta, "EM-ANN Model Synthesis of Physical Dimensions for Multilayer Asymmetric Coupled Transmission Line Structures," *International Journal of RF and Microwave Computer-Aided Engineering*, Vol. 9, No. 3, 1999, pp. 175–186.

[23] Watson, P. M., and K. C. Gupta, "EM-ANN Modeling and Optimal Chamfering of 90° CPW Bends with Air-Bridges," *IEEE MTT-S Int. Microwave Symp. Dig.*, June 1997, pp. 1603–1606.

[24] Watson, P. M., and K. C. Gupta, "Design and Optimization of CPW Circuits Using EM-ANN Models for CPW Components," *IEEE Trans. on Microwave Theory and Techniques*, Vol. 45, No. 12, Dec. 1997, pp. 2515–2523.

[25] Gupta, K. C., and P. M. Watson, "Transmission Line Model for CPW Antennas Using ANN Modeling Approach," *IEEE AP-S and URSI Radio Science Meeting Digest*, Montreal, Canada, July 1997, p. 212.

[26] Watson, P. M., K. C. Gupta, and R. L. Mahajan, "Development of Knowledge Based Artificial Neural Network Models for Microwave Components," *IEEE MTT-S Int. Microwave Symp.*, 1998, Digest, pp. 9–12.

[27] Zhang, Q. J., et al. "Ultra Fast Neural Models for Analysis of Electro/Opto Interconnects," *IEEE Electronic Components and Technology Conf.*, San Jose, CA, May 1997, pp. 1134–1137.

[28] Zhang, Q. J., F. Wang, and V. Devabhaktuni, "Neural Network Structures for RF and Microwave Applications," *IEEE AP-S Antennas and Propagations International Symp.*, (Orlando, FL), July 1999, pp. 2576–2579.

[29] Wang, F., et al., "Neural Network Structures and Training Algorithms for Microwave Applications," *International Journal of RF and Microwave CAE*, Special Issue on Applications of Artificial Neural Networks to RF and Microwave Design, Vol. 9, 1999, pp. 216–240.

[30] Devabhaktuni, V., C. Xi, F. Wang, and Q. J. Zhang, "Robust Training of Microwave Neural Models," *IEEE MTT-S International Microwave Symp.*, (Anaheim, CA), June 1999, Digest, pp. 145–148.

[31] Wang, F., and Q. J. Zhang, "Incorporating Functional Knowledge into Neural Networks," *IEEE Int. Conf. Neural Networks*, (Houston, TX), June 1997, pp. 266–269.

[32] Wang, F., and Q. J. Zhang, "Knowledge-Based Neural Models for Microwave Design," *IEEE Trans. on Microwave Theory and Techniques*, Vol. 45, Dec. 1997, pp. 1349–1358.

[33] Wang, F., V. Devabhaktuni, and Q. J. Zhang, "A Hierarchical Neural Network Approach to the Development of Library of Neural Models for Microwave Design," *IEEE Trans. on Microwave Theory and Techniques*, Vol. 46, Dec. 1998, pp. 2391–2403.

[34] Smith, A. E., and A. K. Mason, "Cost Estimation Predictive Modeling: Regression versus Neural Network," *Engineering Economist*, Vol. 42, No. 2, Feb. 1997, pp. 137–160.

[35] Finnie, G. R., G. E. Wittig, and J. M. Desharnais, "A Comparison of Software Effort Estimation Techniques: Using Function Points with Neural Networks, Case-Based Reasoning, and Regression Models," *Journal of Systems and Software*, Vol. 39, No. 3, Mar. 1997, pp. 281–290.

[36] Tsintikidis, D., et. al., "A Neural Network Approach to Estimating Rainfall from Spaceborne Microwave Data," *IEEE Trans. on Geoscience and Remote Sensing*, Vol. 35, No. 5, May 1997, pp. 1079–1093.

[37] Hierlemann, A., et. al., "Environmental Chemical Sensing Using Quartz Microbalance Sensor Arrays: Applications of Multicomponent Analysis Techniques," *Sensors and Materials*, Vol. 7, No. 3, Mar. 1995, pp. 179–189.

[38] Danon, Y., and M. J. Embrechts, "Least Squares Fitting Using Artificial Neural Networks," in *Intelligent Engineering Systems Through Artificial Neural Networks*, Vol. 2, NY: ASME Press, 1992.

[39] Zhang, Q. J., F. Wang, and M. S. Nakhla, "Optimization of High-Speed VLSI Interconnects: A Review," *Int. J. of Microwave and Millimeter-Wave CAE*, Vol. 7, 1997, pp. 83–107.

[40] Bishop, C. M., *Neural Networks for Pattern Recognition*, New York: Oxford University Press Inc., 1996.

[41] Hornik, K., M. Stinchcombe, and H. White, "Multi-Layered Feed-Forward Neural Networks are Universal Approximations," *Neural Networks*, Vol. 2, 1990, pp. 259–266.

2

Modeling and Optimization for Design

This chapter discusses the anatomy of the overall design process. Conventional design and computer-aided design methodologies are presented. Concepts of knowledge-aided design are introduced. Current status of CAD techniques for RF and microwave circuits and printed antennas is reviewed. The importance of modeling and optimization in the design process is brought out. It is in these two areas that ANNs are likely to play very crucial roles. The anticipated role of neural network techniques in knowledge-aided design is pointed out.

2.1 The Design Process

2.1.1 Anatomy of the Design Process

The sequence of various steps in a typical RF and microwave design process (actually for any design!) [1] is shown in Figure 2.1. One starts with problem identification. This phase is concerned with determining the need for a product. A product is identified, resources allocated, and end-users are targeted. The next step is drawing up the product design specification (PDS), which describes the requirements and performance specifications of the product. This is followed by a concept generation stage, where preliminary design decisions are made. Several alternatives will normally be considered. Decisions made at this stage determine the general configuration of the product and thus have enormous implications for the remainder of the design process. At each of these design stages, there is usually a need for feedback to earlier stages and reworking of the previous steps. The analysis and evaluation of the conceptual design lead to concept refinement, for example, by placing values on numerical attributes.

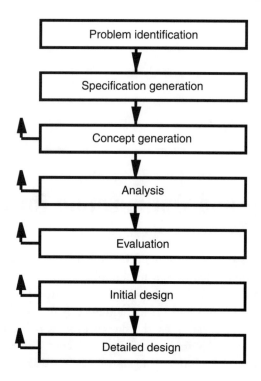

Figure 2.1 Sequence of steps in a typical design process.

The performance of the conceptual design is tested for its response to external inputs and its consistency with the design specifications. These steps lead to an initial design.

The step from initial design to the final detailed design involves modeling, computer-aided analysis, and optimization. CAD tools currently available to us for RF and microwave design primarily address this step only.

A review of the sequence of steps in Figure 2.1 points out that the analysis of the design of an RF component, circuit, or antenna is needed at two different stages of the design process.

Once the concept embodying the configuration for the design is arrived at, there is a need for analysis to evaluate the potential performance of the tentative design. At this stage, approximate analysis methods (such as those based on very approximate models) provide a sufficiently accurate approach. The second place in the design process where an analysis is needed is in the last step that converts the initial design into an optimized detailed design. Accuracy of the analysis process here gets translated directly into a match between the design performance and the design specifications. At this stage we need an accurate computer-aided analysis based on most accurate models for various components.

The design process outlined above can be considered to consist of two segments. Initial steps starting from the product identification to the initial design may be termed *design-in-the-large* [2]. The second segment that leads from an initial design to the detailed design has been called *design-in-the-small*. It is for this second segment that the most current RF and microwave CAD tools have been developed.

It is in the design-in-the-large segment that important and expensive design decisions are made. Here, the previous experience of the designers plays a significant role and a knowledge-based system is the most likely candidate technology that could help designers. Understanding this part of the design process is a prerequisite for developing successful design tools for any RF and microwave product. An extensive discussion on knowledge-based design and related topics is available in a three volume treatise on artificial intelligence in engineering design [3].

There are three design philosophies applicable to the design of RF and microwave circuits, antennas, and systems. This discussion is applicable to other design domains also. These are:

- Conventional design procedure;
- CAD approach;
- Knowledge-aided design (KAD) approach.

2.1.2 Conventional Design Procedures

The conventional design process is the methodology that designers used before the CAD methods and software were developed. A flow diagram depicting the conventional design procedure is shown in Figure 2.2.

One starts with the desired design specifications and arrives at an initial configuration for the circuit or antenna to be designed. Available design data and previous experience are helpful in selecting this initial configuration. Analysis and synthesis procedures are used for deciding values of various parameters of the design. A laboratory model is constructed for the initial design and measurements are carried out for evaluating its characteristics. Performance achieved is compared with the desired specifications and if the given specifications are not met, the design is modified. Adjustment, tuning, and trimming mechanisms incorporated into the design are used for carrying out these modifications. Measurements are carried out again and the results are compared with the desired specifications. The sequence of modifications, measurements, and comparison is carried out iteratively until the desired specifications are achieved. At times the specifications are compromised in view of the practically feasible performance of the module. The final design configuration thus obtained is sent for prototype fabrication. This procedure had been used for the design

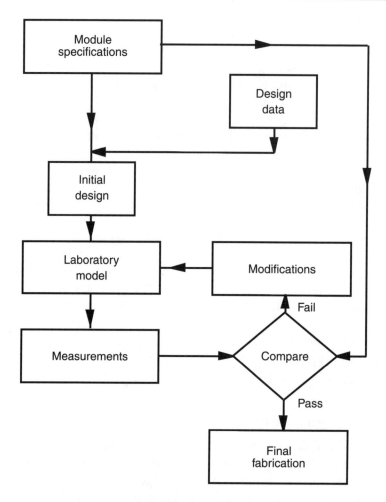

Figure 2.2 The conventional design procedure that was used for RF and microwave designs before CAD methods were developed.

of RF and microwave circuits and antennas for quite some time. However, it has become increasingly difficult to use this iterative experimental method successfully because of the following considerations:

- Increased complexity of modern systems demands more precise and accurate design of RF and microwave subsystems. Consequently, the effect of tolerances in the design has become increasingly important.
- A larger variety of active and passive components are now available for achieving a given RF and microwave design function. The choice

of the appropriate RF or microwave structure becomes difficult if the iterative experimental approach is used.

- It is very difficult to incorporate any modifications in the module fabricated by integrated circuit technology.

The method developed for dealing with this situation is known as computer-aided design (CAD). Computer-aided design, in its strict interpretation, may be taken to mean any design process where the computer is used as a tool. However, usually the word CAD implies that without the computer as a tool, that particular design process would have been impossible or much more difficult, more expensive, more time consuming, less reliable, and more than likely would have resulted in an inferior product.

2.1.3 CAD Approach

A typical flow diagram for a CAD procedure is shown in Figure 2.3.

As before, one starts with a given set of specifications. Synthesis methods and available design data (at times prestored in computer memory) help to arrive at the initial design. The performance of this initial module is evaluated by a computer-aided analysis. Numerical models for various components (passive and active) used in the product to be designed are needed for the analysis. These are called from the library of subroutines developed for this purpose. The performance characteristics of the design obtained as results of the analysis are compared with the given specifications. If the results fail to satisfy the desired specifications, the designable parameters of the design are altered in a systematic manner. This constitutes the key step in the optimization. Several optimization strategies include sensitivity analysis of the design performance for calculating changes in the designable parameters. The sequence of the design performance analysis, comparison with the desired performance, and parameter modification is performed iteratively until the specifications are met or the optimum performance of the design (within the given constraints) is achieved. The design is now fabricated and the experimental measurements are carried out. Some modifications may still be required if the modeling and/or analysis are not accurate enough. However, these modifications are hopefully very small. The aim of the CAD method is to minimize the experimental iterations as far as practicable.

The process of CAD, as outlined above, consists of three important segments, namely:

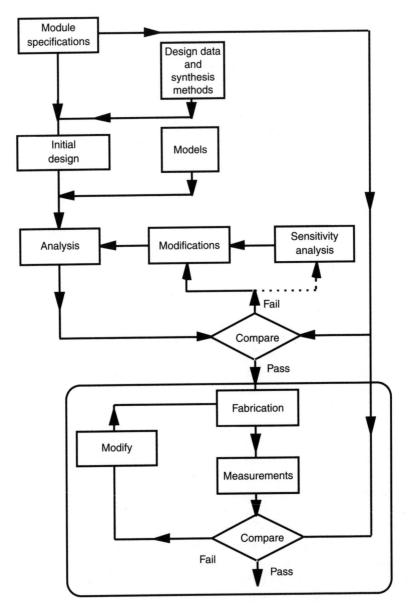

Figure 2.3 Computer-aided design methodology suitable for RF and microwave design.

1. Modeling;
2. Analysis;
3. Optimization.

Modeling involves characterization of various active and passive components to the extent of providing a numerical model that can be handled by the computer. In the case of RF or microwave designs, one comes across a variety of active and passive elements. Semiconductor devices used include bipolar and MESFET transistors, point contact and Schottky barrier detectors, varactor and PIN diodes, and also transferred electron and avalanche devices. Passive elements used in microwave modules include sections of various transmission structures, lumped components, dielectric resonators, nonreciprocal components and planar (two-dimensional) elements, and a variety of radiating element configurations. Transmission structures could be coaxial line, waveguide, stripline, microstrip line, coplanar line, slot line, or a combination of these. Not only do these transmission structures need to be characterized fully for impedance, phase velocity, attenuation, and so forth, it also becomes necessary to model the parasitic reactances caused by geometrical discontinuities in these transmission lines.

Modeling of components in microwave circuits had been the main difficulty in successful implementation of CAD techniques at microwave frequency. However, the development of electromagnetic (EM) simulation techniques developed over the last decade has helped to construct adequate models and bring microwave hybrid and monolithic circuit CAD software to a level of maturity. Modeling still remains the major bottleneck for CAD of certain classes of microwave circuits such as coplanar waveguide CPW circuits, multilayered circuits, and for integrated circuit-antenna modules. Current research in efficient use of EM simulation techniques [4] and in use of artificial neural network models [5–7] will lead to further improvement in CAD tools for microwave circuits and antennas.

2.1.4 Knowledge-Aided Design (KAD) Approach

Development of knowledge aids may be based on developing a task structure [8] for the design process. A generic task-oriented methodology involves:

- A description of the tasks;
- Proposed methods for it;
- Decomposition of the task into subtasks;
- Methods available for various subtasks;
- The knowledge base required for implementing various methods;
- Any control strategies for these methods.

A method for accomplishing a generic design task is known as the propose-critique-modify (PCM) approach, shown in Figure 2.4.

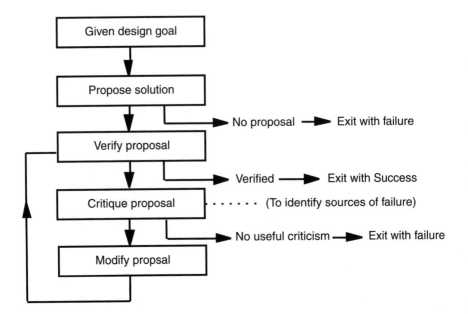

Figure 2.4 Propose-critique-modify (PCM) approach for arriving at an initial design.

This approach consists of the following:

- Proposal of partial or complete design solutions;
- Verification of proposed solutions;
- Critiquing the proposal by identifying causes of failure, if any;
- Modification of proposals to satisfy design goals.

The use of knowledge-based approaches to the initial stages of RF and microwave circuit and antenna design is an area that needs to be explored. We currently heavily depend upon the accumulated experience of senior designers for executing these design steps. Recognizing the significant contribution of these steps to the final design efforts in developing technology aids for this purpose would be worthwhile.

Knowledge-based systems developed for initial design of RF and microwave circuits and other subsystems would also be very helpful for instruction or training of design engineers. For example, a system that can present all the relevant options for, say, designs of high power amplifier at RF and microwave frequencies, could educate the designer about the relative merits of various configurations as well as lead to a design for meeting a particular set of specifications.

2.2 RF and Microwave Circuit CAD

As discussed in Section 2.1.3 on "CAD Approach," three important segments of CAD are:

1. Modeling;
2. Analysis;
3. Optimization.

This section is an overview of these three aspects as applicable to RF and microwave circuit CAD. Details of RF and microwave circuit CAD methodology are available in several books [10–12]. This section deals with RF and microwave circuit CAD techniques based on network analysis. This approach is used extensively for practical design of RF and microwave hybrid and integrated circuits. Field simulation and analysis techniques based on frequency domain and time domain electromagnetic simulation methods are also applicable to RF component and circuit analyses. Electromagnetic solvers on big computers are also now fast enough to carry out optimization.

2.2.1 Modeling of Circuit Components

An accurate and reliable characterization of RF and microwave circuit components is one of the basic prerequisites of successful CAD. The degree of accuracy to which the performance of microwave integrated circuits can be predicted depends on the accuracy of characterization and modeling of components. Kinds of elements and passive devices that need to be characterized depend upon the transmission medium used for circuit design. Most research efforts aimed at characterization for high frequency CAD purposes have been reported for microstrip lines [13, 14]. Some results are available for slot lines and fin lines also [13, 15, 16]. However, for coplanar lines and suspended substrate transmission structures, modeling techniques are still in a state of infancy.

In addition to the transmission media, implementation of CAD requires characterization of various junctions and discontinuities in transmission structures. Effects of these junctions and discontinuities become more and more significant as one moves to the high microwave frequency range. At higher frequencies, radiation associated with discontinuities needs to be considered. Variations of effects of two typical discontinuities with frequency are illustrated in Figure 2.5 (from [17]).

Figure 2.5(a) shows the input VSWR introduced by right-angled bends in 25-Ω and 75-Ω microstrip lines (on a 0.079-cm-thick substrate with

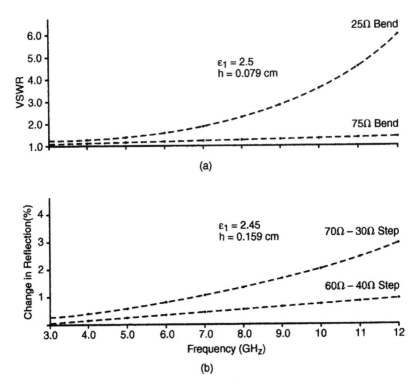

Figure 2.5 (a) Frequency variations of VSWR introduced by right-angled bends in 25-Ω and 75-Ω microstrip lines; (b) frequency variations of the percentage changes in reflection coefficients introduced by two-step discontinuities in microwave circuits (from [17], Bahl, I., and P. Bhartia, *Microwave Solid State Circuit Design*, © 1988 John Wiley and Sons. Reprinted with permission from John Wiley and Sons, Inc.).

ϵ_r = 2.5). We note that the input VSWR caused by the 90 degree bend increases monotonically with frequency and is larger for low impedance (wider) lines, the specific values being 3.45 for a 25-Ω line at 10 GHz as compared to 1.30 for a 75-Ω line at the same frequency. Figure 2.5(b) shows the effect of discontinuity reactance on the behavior of a step (change-in-width) discontinuity. The reflection coefficient increases from its nominal value. At 12 GHz, this increase is about 1% for a 40-Ω to 60-Ω step, and about 3% for a 30-Ω to 70-Ω impedance change. These results are based on quasistatic characterization of these discontinuities [10] and are included to show that the discontinuity reactances should not be ignored for design at higher frequencies.

Quasistatic results for microstrip discontinuity characterization have been available since the late 1970s [10, 13, 18] and have been used in some of the commercially available CAD packages. A more accurate analysis of bends,

T-junctions, and crossings (based on field matching using equivalent parallel-plate waveguide models of discontinuity configurations) became available later [19]. Vigorous hybrid-mode frequency-dependent characterizations of microstrip open ends and gaps were reported around 1981–82 [20–22]. These results are based on Galerkin's method in spectral transform domain. This technique has been extended to other types of discontinuities also [23, 24]. The most powerful tool for modeling of transmission line discontinuities in planar microwave circuits is electromagnetic simulation based on moment-method solution of integral equations for current distribution in microstrip or CPW circuits [25–28]. Models based on this approach are used extensively in commercially available microwave CAD software.

Even when accurate numerical results are available, efficient transfer of these results for CAD is not straightforward. Lumped element models with closed form expressions for values of various parameters [10] has been the most commonly used approach. Many of these closed-form expressions suffer from limited validity with respect to various parameters (impedance, frequency, geometry, and so forth) and limit the accuracy of CAD. The substantial progress in state of the art microwave CAD that has taken place over the last decade is due to improvements in transmission line component models in this software.

Recent applications of artificial neural network (ANN) computing to modeling of microwave components [5–7, 29, 30] (discussed in this book) is an attractive alternative for CAD oriented models. In this approach, electromagnetic simulation is used to obtain S-parameters for all the components to be modeled over the ranges of designable parameters for which these models are expected to be used. An ANN model for each one of the components is developed by training an ANN configuration using the data obtained from EM simulations. Such ANN models have been shown to retain the accuracy obtainable from EM simulators and at the same time exhibit the efficiency (in terms of computer time required) that is obtained from lumped network models normally implemented in commercially available microwave network simulators (like the HP-EEsof's MDS, ADS).

In addition to the modeling of transmission lines and their discontinuities, implementation of the CAD procedure requires accurate models for various active devices like GaAs MESFETs, varactors, and pin diodes. Equivalent circuit models for small signal and large signal behavior of active devices are needed for CAD. As discussed later in Chapter 7 of this book, the ANN modeling approach is also well suited for the modeling of active devices (including thermal effects) [32, 33].

2.2.2 Computer-Aided Analysis Techniques

As discussed in Section 2.1 and shown in Figures 2.2 and 2.3, computer-aided analysis constitutes the key step in the CAD procedure. Since the analysis forms a part of the optimization loop, the analysis subprogram is executed again and again (typically 100 times or more) for a specific circuit design. For this reason, an efficient analysis algorithm constitutes the backbone of any CAD package.

Any general microwave circuit can be viewed as an arbitrarily connected ensemble of multiport components as illustrated in Figure 2.6.

The circuit shown here consists of eight components (A, B, C, D, E, F, G, H), of which G is a one-port component; A, E, and C are two-port components; D, F, and H are three-port components; and B is a four-port component. Different ports of these components are connected together. There are two external ports (1 and 2). The analysis problem in this case may be stated as follows: characterization of components A through H being known (say in terms of individual S-matrices), find the S-matrix of the overall combination with reference to the two external ports 1 and 2. The outcome of this analysis will be a 2 × 2 S-matrix. The process of evaluating the circuit performance from the known characterizations of its constituents is termed *circuit analysis*. It requires two inputs: component characterizations and the topology of interconnections. The analysis process is depicted symbolically in Figure 2.7.

The two approaches used for frequency-domain analysis of linear microwave circuits are: nodal admittance matrix analysis and scattering matrix analysis.

Nodal Admittance Matrix Method

In this method, the voltages V at all the nodes of the circuit to be analyzed are assumed as unknowns. A set of equations is constructed based on the fact that the sum of the currents flowing into any node in the circuit is zero. In

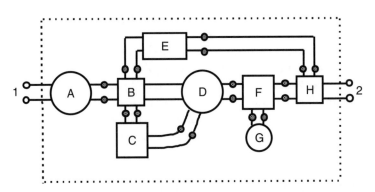

Figure 2.6 A general microwave circuit viewed as an ensemble of multiport components.

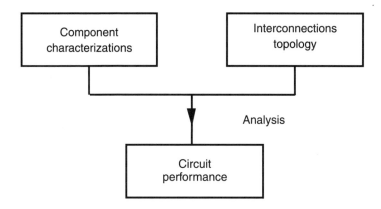

Figure 2.7 Block diagram representation of the circuit analysis process.

microwave circuits the nodal equations are usually set up in the definite form; that is, the voltage of each node is taken with respect to a designated node of the circuit, which is called the *reference node*. The equations are easily set up in the form of a matrix expression [11]:

$$YV = I_0 \qquad (2.1)$$

where

Y = a square nodal admittance matrix (a degree of this matrix equals the number of the nodes in the analyzed circuit);

V = a vector of node voltages;

I_0 = a vector of terminal currents of the independent current sources connected between the nodes and the reference node of the circuit.

The structure of the nodal Y-matrix can be illustrated by considering a circuit example shown in Figure 2.8.

The circuit equations are derived from the application of Kirchhoff's current law at each node of the circuit.

Node 1:

$$Y_1 V_1 + Y_2(V_1 - V_2) + Y_5(V_1 - V_3) = I_1 \qquad (2.2)$$

Node 2:

$$Y_3 V_2 + Y_2(V_2 - V_1) + Y_4(V_2 - V_3) + y_{11} V_2 + y_{12} V_3 = 0 \qquad (2.3)$$

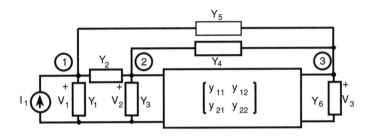

Figure 2.8 A network example for illustrating the nodal admittance matrix method.

Node 3:

$$Y_6 V_3 + Y_4(V_3 - V_2) + Y_5(V_3 - V_1) + y_{21} V_2 + y_{22} V_3 = 0 \quad (2.4)$$

These equations may be written in the matrix form as

$$\begin{array}{c} \text{Node} \\ 1 \\ 2 \\ 3 \end{array} \begin{bmatrix} Y_1 + Y_2 + Y_5 & -Y_2 & -Y_5 \\ -Y_2 & Y_2 + Y_3 + Y_4 + y_{11} & -Y_4 + y_{12} \\ -Y_5 & -Y_4 + y_{21} & Y_4 + Y_5 + Y_6 + y_{22} \end{bmatrix} \begin{bmatrix} V_1 \\ V_2 \\ V_3 \end{bmatrix} \quad (2.5)$$

$$= \begin{bmatrix} I_1 \\ 0 \\ 0 \end{bmatrix}$$

Equation (2.5) is an example of (2.1) formulated for the circuit shown in Figure 2.8.

In a general case, the nodal admittance matrix of a circuit may be derived using the rules developed for various circuit components [11] such as an admittance Y, independent current and voltage sources, controlled current and voltage sources, and multiterminal elements including sections of transmission lines, coupled line sections, MESFETs, bipolar transistors, and so forth. It can be shown that a multiterminal device (element) does not have to be replaced by its equivalent circuit composed of two-terminal elements.

When a multiterminal device is described by its definite admittance matrix $\mathbf{y}_d$, the elements of $\mathbf{y}_d$ may be entered into the matrix $\mathbf{Y}$ of the whole circuit only if the reference node of the device agrees with the reference node of the whole circuit.

Figure 2.9 presents a three-terminal device as a two-port network, with the terminal 3 taken as the reference node.

Modeling and Optimization for Design

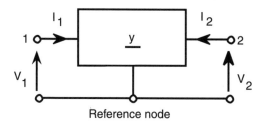

Figure 2.9 Three-terminal device as a two-port network with one terminal taken as its reference node.

The definite admittance matrix of the device is given by the following equation:

$$\begin{bmatrix} I_1 \\ I_2 \end{bmatrix} = \begin{bmatrix} y_{11} & y_{12} \\ y_{21} & y_{22} \end{bmatrix} \begin{bmatrix} V_1 \\ V_2 \end{bmatrix} \quad (2.6)$$

If the reference node of the three-terminal device agrees with the reference node of the whole circuit, and if node 1 of the terminal device is connected to the node i of the circuit and node 2 to the node j, the terms of the definite admittance matrix y_d of the device are added to the nodal admittance matrix Y of the whole circuit as:

$$\begin{matrix} & (i) & (j) \\ (i) & \begin{bmatrix} \vdots & \vdots \\ \cdots & y_{11} & y_{12} & \cdots \\ \vdots & \vdots \\ \cdots & y_{21} & y_{22} & \cdots \\ \vdots & \vdots \end{bmatrix} \end{matrix} \quad (2.7)$$

The conversion of a multiterminal element admittance matrix from indefinite form (with the reference node being an undefined node external to the circuit) into definite form, and vice versa, can be performed easily because the indefinite admittance matrix of an n terminal circuit satisfies the following relations:

$$\sum_{i=1}^{n} y_{ij} = 0, \quad \text{for } j = 1, 2, \ldots, n \quad (2.8a)$$

$$\sum_{j=1}^{n} y_{ij} = 0, \quad \text{for } i = 1, 2, \ldots, n \quad (2.8b)$$

The nodal admittance matrix Y and the right-hand vector I_0 of (2.1) are constructed in accordance with the rules developed for various components [11]. The resulting matrix Y and the vector I_0 are computed at the frequency of interest. The entries Y_{ij} of the nodal admittance matrix may become very small or very large as the circuit function (response) is computed for different values of frequency. The most important are values of the main diagonal entries Y_{ii} of the nodal admittance matrix. For example, if an inductor and a capacitor are connected to the node i, at a series resonant frequency of these elements, the entry Y_{ii} becomes undefined. Also, the nodal admittance matrix of a circuit with dependent sources may become singular and the set of circuit equations cannot be solved. In microwave circuits, we assume very often a lossy circuit. The determinant of the nodal admittance matrix of such a circuit is a polynomial in the frequency variable ω. The matrix Y becomes singular at the zeros of the polynomial, and the system of circuit equations has no solution.

Computer-aided analysis of microwave circuits in the frequency domain based on the nodal admittance matrix requires multiple computation of the solution of a system of linear equations with complex coefficients. An important problem is the numerical instability of a solution process performed by computer, which may occur if the so-called condition number of the coefficient matrix of the linear equation system is too large.

The nodal admittance matrix in general is a sparse matrix, which means that many entries of the matrix are equal to zero, particularly for large circuits. In conventional numerical procedures used to solve a system of linear equations, the arithmetic operations are performed on all nonzero and zero valued entries of the coefficient matrix. The whole solution procedure involves $N_c^3/3 + N_c^2 - N_c/3$ complex number multiplications and divisions (N_c = the order of the coefficient matrix). To save computation time and minimize storage requirements, the sparsity of the coefficient matrix must be taken into account in the solution procedure.

Computation of Circuit Functions

The solution of the nodal matrix expression (2.1) provides us with the numerical values of all node voltages of a circuit. In most practical cases, as for example filter or amplifier design, there is one source of excitation in the circuit and we are interested in one or two voltages in the circuit.

For the sake of clarity in the considerations to follow, we assume that an independent source of one ampere has been connected between node 1 and the reference node and that the output node is node N_c. The node voltage equations are now

$$YV = I_0 = \begin{bmatrix} 1 \\ 0 \\ \vdots \\ 0 \end{bmatrix} \qquad (2.9)$$

Let e_i be a vector of zero entries except the ith, which is one. The postmultiplication of a matrix by e_i results in extracting the ith column from the matrix. If matrix Z is the inverse of Y, then $YZ = I$, where I is an identity matrix of order N_c. By postmultiplication of YZ by e_1, we have

$$YZe_1 = \begin{bmatrix} 1 \\ 0 \\ \vdots \\ 0 \end{bmatrix} \qquad (2.10)$$

From (2.9) and (2.10), we can conclude

$$V = Ze_1 \qquad (2.11)$$

We define a transfer function $H(j\omega)$ as the ratio of the output voltage V_k to the input current I_1 of the independent current source connected between node 1 and the reference node:

$$H(j\omega) = \frac{V_k}{I_1} = |H(j\omega)|e^{j\phi(\omega)} \qquad (2.12)$$

The magnitude of $H(j\omega)$ is called the gain of the circuit and $\phi(\omega)$ is the phase. The term $|H(j\omega)|$ is directly related to the available power gain:

$$G_T = \frac{P_L}{P_{SA}} = 4\,\mathrm{Re}\{Y_S\}\mathrm{Re}\{Y_L\}|H(j\omega)| \qquad (2.13)$$

In (2.13), P_L is the active power dissipated in the load admittance Y_L connected between the output node and the reference node, and P_{SA} is the available power of the current source with the internal admittance Y_S.

In filter design problems, the quantity insertion loss (IL) is more commonly used:

$$IL = -20\log(|Y_S + Y_L| \cdot |H(j\omega)|) \qquad (2.14)$$

Other circuit functions are also important. We have the input admittance

$$Y_{in} = \frac{I_1}{V_1} - Y_S \qquad (2.15)$$

or, equivalently, the reflection coefficient

$$\Gamma_{in} = \frac{Y_{in} - Y_S}{Y_{in} + Y_S} = 1 - 2\frac{V_1}{I_1}Y_S \qquad (2.16)$$

These network functions can be computed, once the node voltages V_1 and V_k are determined.

Scattering-Matrix Analysis

Scattering-matrix analysis is applicable to any general microwave circuit configuration when all the circuit components are modeled in terms of their scattering parameters. In this method of analysis [10], the circuit diagram is configured such that there are no unconnected ports. That is, sources (along with their source impedances) are connected to the input ports, and all output ports are terminated in respective loads. For circuit analysis, it is adequate to consider matched sources and matched loads. Consider the example of an impedance matching circuit and its equivalent representation shown in Figure 2.10(a) and (b).

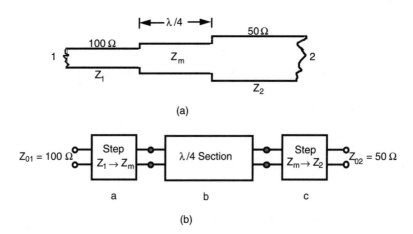

Figure 2.10 (a) A quarter-wave matching network; (b) interconnected multiport representation of the circuit in part (a).

For implementing this method, this circuit may be depicted as shown in Figure 2.11(a). Independent sources (Figure 2.11(b)) in such a representation may be described by the relation

$$b_g = s_g a_g + c_g \qquad (2.17)$$

where c_g is the wave impressed by the generator. For sources that are matched (or isolated) $s_g = 0$ and $b_g = c_g$. All other components in the circuit are described by

$$b_i = S_i a_i \qquad (2.18)$$

where a_i and b_i are incoming and outgoing wave variables, respectively, for the ith component with n_i ports and S_i is the scattering matrix.

The governing relation for all components (say the total number is N) in the circuit can be put together as

$$b = Sa + c \qquad (2.19)$$

where

$$b = \begin{bmatrix} b_1 \\ b_2 \\ \vdots \\ b_N \end{bmatrix}, \quad a = \begin{bmatrix} a_1 \\ a_2 \\ \vdots \\ a_N \end{bmatrix}, \quad \text{and } c = \begin{bmatrix} c_1 \\ c_2 \\ \vdots \\ c_N \end{bmatrix} \qquad (2.20)$$

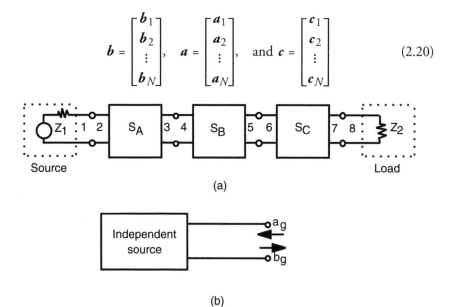

Figure 2.11 (a) Network of Figure 2.10(b) rewritten for implementing the generalized scattering-matrix analysis procedure; (b) independent source introduced for scattering-matrix analysis.

Here, b_1, a_1, ..., are themselves vectors with the number of elements equal to the number of ports of that particular component. The size of vector b, a, or c is equal to the total number of ports of all the components in the modified network representation. For the network shown in Figure 2.11(a), this number is 8; counting one port for the source, two ports each for the three components A, B, and C; and one port for the load. The vector c will have nonzero values only for the output ports of independent sources in the network. For the network of Figure 2.11(a),

$$c = [c_1, 0, 0, 0, 0, 0, 0, 0]^T \qquad (2.21)$$

The output port of the source has been numbered 1, and the superscript T indicates transpose of a vector or matrix.

The matrix S is a block diagonal matrix whose submatrices along the diagonal are the scattering-matrices of various components. In the general case, the matrix S may be written as

$$S = \begin{bmatrix} (S_1) & 0 & — & — & 0 \\ 0 & (S_2) & — & — & 0 \\ — & — & (—) & — & — \\ — & — & — & (—) & — \\ 0 & 0 & — & — & (S_N) \end{bmatrix} \qquad (2.22)$$

where 0's represent null matrices. For the network of Figure 2.11(a), the S-matrix will look like

$$S = \begin{bmatrix} [S_{11}^S] & 0 & 0 & 0 & 0 \\ 0 & \begin{bmatrix} S_{22}^A & S_{23}^A \\ S_{32}^A & S_{33}^A \end{bmatrix} & 0 & 0 & 0 \\ 0 & 0 & \begin{bmatrix} S_{44}^B & S_{45}^B \\ S_{54}^B & S_{55}^B \end{bmatrix} & 0 & 0 \\ 0 & 0 & 0 & \begin{bmatrix} S_{66}^C & S_{67}^C \\ S_{76}^C & S_{77}^C \end{bmatrix} & 0 \\ 0 & 0 & 0 & 0 & [S_{88}^L] \end{bmatrix} \qquad (2.23)$$

where S_{11}^S and S_{88}^L are one-port S-parameter characterizations for the source and the load, respectively. When both the source and the load are matched,

$S_{11}^S = S_{88}^L = 0$. The other three (2 × 2) S-matrices characterize components A, B, and C, respectively.

Equation 2.19 contains the characterizations of individual components but does not take into account the constraints imposed by interconnections. For a pair of connected ports, the outgoing wave variable at one port must equal the incoming wave variable at the other (assuming that wave variables at two ports are normalized with respect to the same impedance level). For example, if port j of one component is connected to port k of the other component, as shown in Figure 2.12, the incoming and outgoing wave variables are related as

$$a_j = b_k \text{ and } a_k = b_j \qquad (2.24)$$

or

$$\begin{bmatrix} b_j \\ b_k \end{bmatrix} = \begin{bmatrix} 0 & 1 \\ 1 & 0 \end{bmatrix} \begin{bmatrix} a_j \\ a_k \end{bmatrix} = [\Gamma]_{jk} \begin{bmatrix} a_j \\ a_k \end{bmatrix} \qquad (2.25)$$

We can extend $[\Gamma]_{jk}$ to write an interconnection matrix describing all the connections in the circuit. We express

$$\boldsymbol{b} = \Gamma \boldsymbol{a} \qquad (2.26)$$

where Γ is a square matrix of the same size as that of $\boldsymbol{S}$. The size of the Γ-matrix is given by the sum total of the ports in all the components in the circuit. The size is 8 × 8 in the example of Figure 2.11(a). For the example of Figure 2.11(a), the Γ matrix may be written as

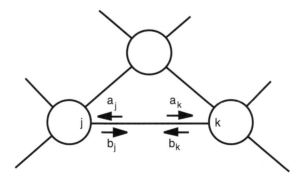

Figure 2.12 Interconnection of ports j and k.

$$\begin{bmatrix} b_1 \\ b_2 \\ b_3 \\ b_4 \\ b_5 \\ b_6 \\ b_7 \\ b_8 \end{bmatrix} = \begin{bmatrix} 0 & 1 & 0 & 0 & 0 & 0 & 0 & 0 \\ 1 & 0 & 0 & 0 & 0 & 0 & 0 & 0 \\ 0 & 0 & 0 & 1 & 0 & 0 & 0 & 0 \\ 0 & 0 & 1 & 0 & 0 & 0 & 0 & 0 \\ 0 & 0 & 0 & 0 & 0 & 1 & 0 & 0 \\ 0 & 0 & 0 & 0 & 1 & 0 & 0 & 0 \\ 0 & 0 & 0 & 0 & 0 & 0 & 0 & 1 \\ 0 & 0 & 0 & 0 & 0 & 0 & 1 & 0 \end{bmatrix} \begin{bmatrix} a_1 \\ a_2 \\ a_3 \\ a_4 \\ a_5 \\ a_6 \\ a_7 \\ a_8 \end{bmatrix} \quad (2.27)$$

Note that the matrix is symmetrical and there is only a single 1 in any row or any column. The latter signifies that any port is connected only to one other port. This excludes the possibility of three ports being connected at one single point. If such a junction exists in a circuit, it must be considered a three-port component, as shown in Figure 2.13.

The ports a', b', and c' of this new component depicting the junction j are connected to the ports a, b, and c, respectively, in a one-to-one manner and the nature of the Γ-matrix is similar to the one indicated in (2.27).

Equations (2.19) and (2.26) describe the circuit completely and may be combined into a single equation as

$$\Gamma a = Sa + c \quad (2.28)$$

or $(\Gamma - S)a = c$.

Setting $(\Gamma - S) = W$, we have

$$a = W^{-1}c = Mc \quad (2.29)$$

W is called the connection-scattering matrix. For our example of Figure 2.11(a), W may be written as

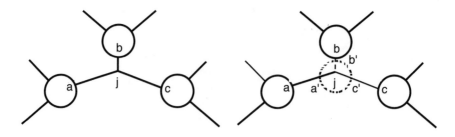

Figure 2.13 A three-way connection at j treated as a three-port component.

$$\begin{bmatrix}
 & 1 & 2 & 3 & 4 & 5 & 6 & 7 & 8 \\
1 & -S_{11}^S & 1 & 0 & 0 & 0 & 0 & 0 & 0 \\
2 & 1 & -S_{11}^A & -S_{12}^A & 0 & 0 & 0 & 0 & 0 \\
3 & 0 & -S_{21}^A & -S_{22}^A & 1 & 0 & 0 & 0 & 0 \\
4 & 0 & 0 & 1 & -S_{11}^B & -S_{12}^B & 0 & 0 & 0 \\
5 & 0 & 0 & 0 & -S_{21}^B & -S_{22}^B & 1 & 0 & 0 \\
6 & 0 & 0 & 0 & 0 & 1 & -S_{11}^C & -S_{12}^C & 0 \\
7 & 0 & 0 & 0 & 0 & 0 & -S_{21}^C & -S_{22}^C & 1 \\
8 & 0 & 0 & 0 & 0 & 0 & 0 & 1 & -S_{11}^L
\end{bmatrix}$$

(2.30)

We note that the main diagonal elements in **W** are the negative of the reflection coefficients at the various component ports. The other (nondiagonal) elements of **W** are negative of the transmission coefficients between different ports of the individual components. All other elements are zero except for those corresponding to the two ports connected together (the Γ-matrix elements) which are 1's. The zero/nonzero pattern in the **W**-matrix depends only on the topology of the circuit and does not change with component characterizations or frequency of operation. The tri-diagonal pattern of (2.30) is a characteristic of a chain of two-port components cascaded together.

The circuit analysis involves solution of the matrix equation (2.29) to find components of the vector $\mathbf{a}$. Variables $\mathbf{a}$'s corresponding to loads at various external ports are found by choosing the vector $\mathbf{c}$ to consist of a unity at one of the external ports and zeros elsewhere. This leads to the evaluation of a column of the **S**-matrix of the circuit. Say, for $c_j = 1$ and all other $c_{i, i \neq j} = 0$, we find values of vector $\mathbf{a}$. Then

$$S_{ij} = \frac{b_l}{a_j} = a_{l'} \quad \left(a_i = \begin{cases} 1, & \text{for } i = j \\ 0, & \text{otherwise} \end{cases} \right) \qquad (2.31)$$

where l' is the load port corresponding to the lth port of the circuit and j is the other circuit port (for which the impressed wave variable is unity). Equation (2.31) yields the jth column of the scattering matrix. Considering the example of Figure 2.11(a), if we make $c_1 = 1$, other c's being zero, then the first column of the matrix for the circuit in Figure 2.10 (i.e., cascade of three components A, B, and C) is given by

$$S_{11} = a_1 = M_{11}, \quad S_{21} = a_8 = M_{81} \qquad (2.32)$$

Here subscripts of S refer to circuit ports of Figure 2.10 and subscripts of a correspond to component ports of Figure 2.11(a). For obtaining the second column of the S-matrix, we take $c_8 = 1$ and other c's to be zero and find a_1 and a_8 again to yield

$$S_{12} = a_1 = M_{18}, \quad S_{22} = a_8 = M_{88} \tag{2.33}$$

It may be noted that (2.29) yields much more information (namely wave variables at all the internal connected ports) than what is needed for calculation of the S-matrix of the circuit. Knowledge of the wave variables at internal ports is needed for carrying out a sensitivity analysis of the circuit using the adjoint network method. For this reason, the present method is suitable for CAD purposes.

Analysis of Nonlinear Circuits

The computer-aided circuit analysis techniques discussed above are applicable to linear circuits only. For design of nonlinear circuits, such as oscillators, frequency multipliers, and mixers, it becomes necessary to modify the analysis technique. A general analysis procedure for nonlinear microwave circuits is discussed in this section.

Linear and Nonlinear Subnetworks

In many nonlinear microwave circuits, the nonlinearity is restricted to a single device (a MESFET or Schottky diode, for example) operating in the nonlinear region. Since the rest of the circuit is linear, it is desirable to separate the circuit into linear and nonlinear subnetworks that may be treated separately. Nonlinear and linear subnetworks are shown schematically in Figure 2.14.

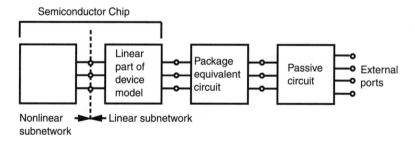

Figure 2.14 Separation of a nonlinear microwave circuit in linear and nonlinear subnetworks.

It may be noted that the linear part of the device model and the package equivalent circuit are also included in the linear subnetwork.

The nonlinear subnetwork consists of nonlinear circuit elements in the device model (nonlinear controlled current sources, diodes, voltage-dependent capacitors, current-dependent resistors, and so forth). These nonlinear components (and hence the resulting nonlinear subnetwork) are usually best simulated in terms of time-domain voltage and current vectors $v(t)$ and $i(t)$. A general time-domain representation of the nonlinear subnetwork could be of the form [34, 35]

$$i(t) = f\left\{i(t), \frac{di}{dt}, v(t), \frac{dv}{dt}\right\} \quad (2.34)$$

where vector f is a nonlinear function of various currents, voltages, and their time derivatives. Higher order derivatives may also appear. For the present discussion, it is assumed that the vector function f is known (hopefully analytically) from the nonlinear modeling of the device.

The linear subnetwork, on the other hand, can be easily analyzed by a frequency-domain circuit analysis program and characterized in terms of an admittance matrix as

$$I(\omega) = Y(\omega)V(\omega) + J(\omega) \quad (2.35)$$

where V, I are the vectors of voltage and current phasors at the subnetwork ports, Y represents its admittance matrix, and J is a vector of Norton equivalent current sources.

Analysis of the overall nonlinear circuit (linear subnetwork plus nonlinear subnetwork) involves continuity of currents given by (2.34) and (2.35) at the interface between linear and nonlinear subnetworks.

Harmonic Balance Method

In principle, the solution to the nonlinear circuit problem can always be found by integrating the differential equations that describe the system. However, in most microwave nonlinear circuits we are interested in the steady state response with periodic excitations and periodic responses with a limited number of significant harmonics. In such a situation, the harmonic balance technique is much more efficient.

In the harmonic balance technique [36], the nonlinear network is decomposed into a minimum possible number of linear and nonlinear subnetworks. The frequency-domain analysis of the linear subnetwork is carried out at a frequency ω_0 and its harmonics. The characterization in (2.35) is thus generalized as:

$$I_k(k\omega_0) = Y(k\omega_0)V_k(k\omega_0) + J_k(k\omega_0) \qquad (2.36)$$

where $k = 1, \ldots, N_h$, with N_h being the number of significant harmonics considered. The nonlinear subnetwork is analyzed in the time domain and the response obtained is in the form of (2.34). A Fourier expansion of the currents yields

$$i(t) = \left\{ \mathrm{Re} \sum_{k=0}^{N_h} F_k(k\omega_0)\exp(jk\omega_0 t) \right\} \qquad (2.37)$$

where coefficients F_k are obtained by a FFT algorithm. The piecewise harmonic-balance technique involves a comparison of (2.36) and (2.37) to yield a system of equations as

$$F_k(k\omega_0) - Y(k\omega_0)V_k(k\omega_0) - J_k(k\omega_0) = 0, \quad \text{for } k = 0, 1, \ldots, N_h \qquad (2.38)$$

The solution of this system of equations yields the response of the circuit in terms of voltage harmonics V_k. Numerically, the solution of (2.38) is obtained by minimizing the harmonic balance error

$$\Delta\epsilon_b(V) = \left\{ \sum_{k=0}^{N_h} |F_k(k\omega_0) - Y(k\omega_0)V(k\omega_0) - J_k(k\omega_0)|^2 \right\}^{1/2} \qquad (2.39)$$

Optimization techniques used for circuit optimization are used for solving Equation (2.38) and thus obtaining steady-state periodic solution of nonlinear circuits.

In most of the circuit-antenna module applications, only the steady-state response is needed and the harmonic balance method discussed above is appropriate for this purpose. However, when transient response is also desired (as for evaluating certain modulation aspects of grid oscillators), it becomes necessary to use time-domain techniques.

2.2.3 Circuit Optimization

Optimization is an important step in the CAD process and, as depicted in Figure 2.15, converts an initial (and quite often unacceptable) design into an optimized final design meeting the given specifications.

Modeling and Optimization for Design

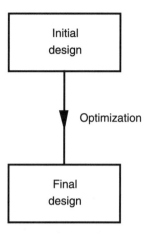

Figure 2.15 Role of the optimization process.

Optimization procedures involve iterative modifications of the initial design, followed by circuit analysis and comparison with the specified performance.

Let $\boldsymbol{\phi}$ be a vector of designable parameters in the circuit. Let the weighted error between kth circuit response $F_k(\boldsymbol{\phi}, \omega)$ and the corresponding upper design specifications $S_k^U(\omega)$ at ith frequency point $\omega = \omega_i$ be defined as,

$$W_{ki}^U(F_k(\boldsymbol{\phi}, \omega_i) - S_k(\omega_i)), \qquad (2.40)$$

where W_{ki}^U is a positive weighting factor. For lower design specifications $S_k^L(\omega)$ imposed on the circuit response $F_k(\boldsymbol{\phi}, \omega)$ with a weighting factor of W_{ki}^L, the error function is given by,

$$W_{ki}^L(S_k^L(\omega_i) - F_k(\boldsymbol{\phi}, \omega_i)) \qquad (2.41)$$

Let $\boldsymbol{e}$ be a vector containing all error functions of the form of (2.40) and (2.41) for all k's and i's. Let

$$U(\boldsymbol{\phi}) = \|e\|_p \qquad (2.42)$$

be the pth norm of $\boldsymbol{e}$. The optimization problem is to find

$$\min_{\boldsymbol{\phi}} U(\boldsymbol{\phi}) \qquad (2.43)$$

subject to the constraints on ϕ. If $p = \infty$ the optimization problem becomes mini-max optimization, which is one of the popular formulations of circuit optimization [37, 38].

There are two different ways of carrying out the modification of designable parameters in an optimization process. These are known as *gradient methods* and *direct search methods* of optimization. Gradient methods use information about the derivatives of the performance functions (with respect to designable parameters) for arriving at the modified set of parameters. This information is obtained from the sensitivity analysis. On the other hand, the direct search methods do not use gradient information, and parameter modifications are carried out by searching for the optimum in a systematic manner. A flow chart for the optimization process is shown in Figure 2.16.

The optimization methods useful for RF and microwave circuits and antennas are identical to those used in other disciplines and are well documented in the literature [10–12].

2.3 CAD for Printed RF and Microwave Antennas

CAD techniques for RF and microwave antennas are not as well developed as those for RF and microwave circuits. One of the factors responsible for this

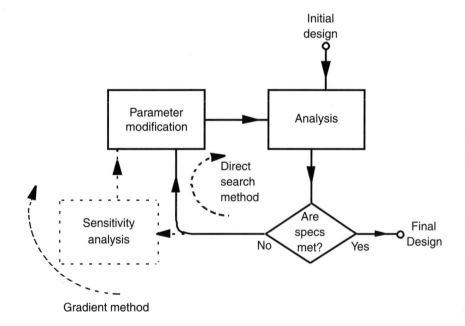

Figure 2.16 Flow charts for direct search and gradient methods of optimization.

situation is the variety of different configurations used for radiators at RF and microwave frequencies. Design techniques used for, say, reflector antennas, are different from those needed for wire antennas (dipoles, Yagi arrays, and so forth) which are quite different from those needed for slotted waveguide array or printed microstrip antennas, for that matter. The class of antennas most appropriate for integration with circuits are printed microstrip patches and slot radiators. CAD considerations for this group of antennas are discussed in this section.

2.3.1 Modeling of Printed Patches and Slots

General CAD methodology depicted in Figure 2.3 is appropriate for microstrip patches and slot radiators also. Design of microstrip patches has been studied more widely (for example, see [39]) than that of slot radiators. Two kinds of modeling approaches are used for microstrip patches:

1. Equivalent network models;
2. Numerical models based on EM simulation methods.

Network Models for Microstrip Patches

The similarity of microstrip patch radiators to microstrip resonators has led to the development of three kinds of network models for microstrip patches. These are:

1. Transmission line model [39, Ch. 10];
2. Cavity model [40];
3. Multiport network model [39, Ch. 9].

Transmission Line Model

The transmission line model is the simplest of the network models for microstrip patch antennas. In this model, a rectangular microstrip antenna patch is viewed as a resonant section of a microstrip transmission line. A detailed description of the transmission-line model is available in literature [39, 41–43]. The basic concept is shown in Figure 2.17 which illustrates the transmission line models for an unloaded rectangular patch; a rectangular patch with a feed line along the radiating edge; and a rectangular patch with a feed line along the nonradiating edge.

Z_{0p} is the characteristic impedance of a microstrip line of width W_p, and ϵ_{rep} is the corresponding effective dielectric constant. B_e and G_e are capacitive

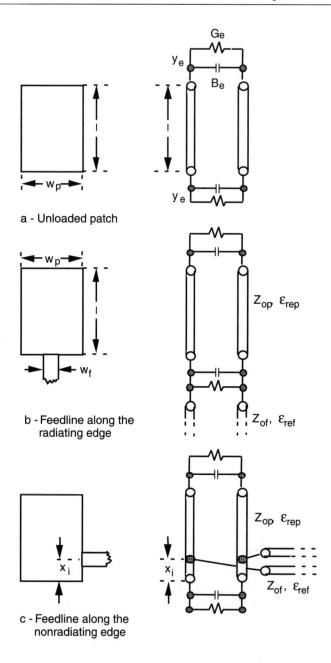

Figure 2.17 Transmission line models for three rectangular microstrip patch configurations.

and conductive components of the edge admittance Y_e. The susceptance B_e accounts for the fringing field associated with the radiating edge of the width W_p, and G_e is the conductance contributed by the radiation field associated with each edge. Power carried away by the surface wave(s) excited along the slab may also be represented by a lumped loss and added to G_e. In Figure 2.17(b) and (c), Z_{0f} and ϵ_{ref} are the characteristic impedance and the effective dielectric constant for the feeding microstrip line of width W_f. In both of these cases, the parasitic reactances, associated with the junction between the line and the patch, have not been taken into account.

Transmission line models may also be developed for two-port rectangular microstrip patches [44, 45]. These configurations are used in the design of series-fed linear (or planar) arrays. Models for two types of two-port rectangular microstrip patches are shown in Figure 2.18.

Figure 2.18(a) illustrates the equivalent transmission line network when the two ports are located along the radiating edges, and Figure 2.18(b) shows the transmission line model [45] when the two ports are along the nonradiating edges. It has been shown [44, 45] that when the two ports are located along the nonradiating edges, transmission from port 1 to port 2 can be controlled by suitable choices of distances x_1 and x_2. Again, the two models shown in Figures 2.18(a) and (b) do not incorporate the parasitic reactances associated with the feed-line-patch junctions.

There are several limitations inherent to the concept of the transmission line model for microstrip antennas. The basic assumptions include: fields are uniform along the width W_p of the patch, and there are no currents transverse to the length l of the patch. Detailed analysis of the rectangular patches [46] has shown that, even at a frequency close to the resonance, field distribution along the radiating edge is not always uniform. Also, the transverse currents are caused by the feeding mechanism and are invariably present. Moreover, the circularly polarized rectangular microstrip antennas (whose operation depends upon the excitation of two orthogonal modes) cannot be represented by the transmission line model discussed above. Clearly, a more accurate method for modeling of microstrip antennas is needed.

Cavity Model

A planar two-dimensional cavity model for microstrip patch antennas [47, 48] offers considerable improvement over the one-dimensional transmission line model discussed in the previous section. In this method of modeling, the microstrip patch is considered as a two-dimensional resonator surrounded by a perfect magnetic wall around the periphery. The fields underneath the patch are expanded in terms of the resonant modes of the two-dimensional resonator.

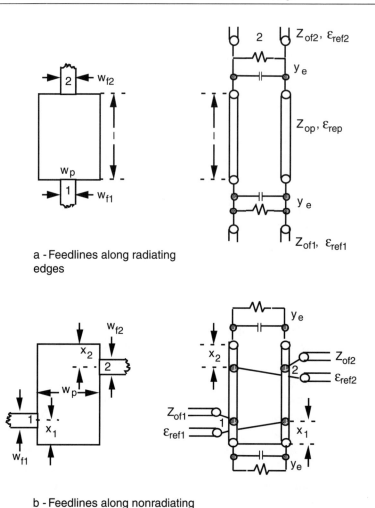

a - Feedlines along radiating edges

b - Feedlines along nonradiating edges

Figure 2.18 Transmission line modes for two-port rectangular microstrip patch antennas.

This approach is applicable to a variety of patch geometries. These geometries, the corresponding modal variations denoted by Ψ_{mn}, and the resonant wave numbers k_{mn} are shown in Table 2.1.

E and H fields are related to Ψ_{mn} by

$$E_{mn} = \Psi_{mn}\hat{z} \qquad (2.44)$$

$$H_{mn} = \hat{z} \times \nabla_t \Psi_{mn} / j\omega\mu \qquad (2.45)$$

Table 2.1
Variation of Modal Fields (Ψ_{mn}) and Resonant Wave Numbers (k_{mn}) for Various Patch Geometries Analyzed by the Cavity Method (Reproduced with permission from [47], © IEEE, 1979)

Geometry	Expressions
Rectangle	$\Psi_{mn} = \cos\dfrac{m\pi}{a}x \cos\dfrac{n\pi}{b}y$ $k_{mn} = \sqrt{\left(\dfrac{m\pi}{a}\right)^2 + \left(\dfrac{n\pi}{b}\right)^2}$
Circle (disk)	$\Psi_{mn} = J_n(k_{mn}\rho)e^{jn\phi}$ $J_n'(k_{mn}a) = 0$
Circular ring	$\Psi_{mn} = [N_n'(k_{mn}a)J_n(k_{mn}\rho) - J_n'(k_{mn}a)N_n(k_{mn}\rho)]e^{jn\phi}$ $\dfrac{J_n'(k_{mn}a)}{N_n'(k_{mn}a)} = \dfrac{J_n'(k_{mn}b)}{N_n'(k_{mn}b)}$
Circular segment	$\Psi_{m\nu} = J_\nu(k_{m\nu}\rho)\cos\nu\phi$ $\nu = n\pi/\alpha,\ J_\nu'(k_{m\nu}a) = 0$
Circular ring segment	$\Psi_{mn} = [N_\nu'(k_{m\nu}a)J_\nu(k_{m\nu}\rho) - J_\nu'(k_{m\nu}a)N_\nu(k_{m\nu}\rho)]\cos\nu\phi$ $\nu = n\pi/\alpha$ $\dfrac{J_\nu'(k_{m\nu}a)}{N_\nu'(k_{m\nu}a)} = \dfrac{J_\nu'(k_{m\nu}b)}{N_\nu'(k_{m\nu}b)}$
Ellipse	Even modes: $\Psi_{mn} = \text{Re}_m(\xi, \chi e_n)Se_m(\eta, \chi e_n)$ $\text{Re}_m(a, \chi e_n) = 0,\ \chi e_n = kq$ major axis $= 2q\cosh a$ minor axis $= 2q\sinh a$ Odd modes: Replacing e by o in the above
Disk with slot	$\Psi_{mn} = J_{n/2}(k_{mn}\rho)\cos(n\phi/2)$ $J_{n/2}'(k_{mn}a) = 0$ $\alpha \approx 2\pi,\ \nu = n/2$

Table 2.1 (continued)
Variation of Modal Fields (Ψ_{mn}) and Resonant Wave Numbers (k_{mn}) for Various Patch Geometries Analyzed by the Cavity Method (Reproduced with Permission from [47], © IEEE, 1979)

Right isosceles

(a) $\Psi_m = \cos\dfrac{m\pi}{a}x - \cos\dfrac{m\pi}{a}y$

$k_m = \sqrt{2}\dfrac{m\pi}{a}$

(b) $\Psi_m = \cos\dfrac{m\pi}{a}x \cos\dfrac{m\pi}{a}y$

$k_m = \sqrt{2}\dfrac{m\pi}{a}$

Equilateral triangle

$\Psi_{mn} = \cos\dfrac{2\pi l}{3b}\left(\dfrac{u}{2}+b\right)\cos\left(\dfrac{\pi(m+n)(v-w)}{9b}\right)$

$+ \cos\dfrac{2\pi m}{3b}\left(\dfrac{u}{2}+b\right)\cos\left(\dfrac{\pi(n-l)(v-w)}{9b}\right)$

$+ \cos\dfrac{2\pi n}{3b}\left(\dfrac{u}{2}+b\right)\cos\left(\dfrac{\pi(l-m)(v-w)}{9b}\right)$

$l = -(m+n),\ u = -\dfrac{\sqrt{3}}{2}x + \dfrac{1}{2}y$

$v - w = -\dfrac{\sqrt{3}}{2}x + \dfrac{3}{2}y$

$b = a/2\sqrt{3}$

$k_{mn}^2 = \left(\dfrac{4\pi}{3a}\right)^2 (m^2 + n^2 + mn)$

where $\hat{z}$ is a unit vector normal to the plane of the patch. Resonant wave numbers k_{mn} are solutions of

$$(\nabla_t^2 + k_{mn}^2)\Psi_{mn} = 0 \qquad (2.46)$$

with

$$\frac{\partial \Psi_{mn}}{\partial p} = 0 \qquad (2.47)$$

on the magnetic wall (periphery of the patch). ∇_t is the transverse part of the Laplacian operator and p is perpendicular to the magnetic wall.

The fringing fields at the edges are accounted for by extending the patch boundary outwards and considering the effective dimensions to be somewhat larger than the physical dimensions of the patch. The radiation is accounted for by considering the effective loss tangent of the dielectric to be larger than the actual value. If the radiated power is estimated to be P_r, the effective loss tangent δ_e may be written as

$$\delta_e = \frac{P_r + P_d}{P_d}\delta_d \qquad (2.48)$$

where P_d is the power dissipated in the dielectric substrate and δ_d is the loss tangent for the dielectric medium. The effective loss tangent given by (2.48) can be further modified to incorporate the conductor loss P_c. The modified loss tangent δ_e is given by

$$\delta_e = \frac{P_r + P_d + P_c}{P_d}\delta_d \qquad (2.49)$$

The input impedance of the antenna is calculated by finding the power dissipated in the patch for a unit voltage at the feed port, and is given by

$$Z_m = \frac{|V|^2}{[P + 2j\omega(W_E - W_M)]} \qquad (2.50)$$

where $P = P_d + P_c + P_{SW} + P_r$, P_{SW} is the power carried away by the surface wave, W_E is the time-averaged electric stored energy, and W_M is the time-averaged magnetic energy. The voltage V equals $E_z d$ averaged over the feed-strip width (d is the substrate thickness). The far-zone field and radiated power are computed by replacing the equivalent magnetic-current ribbon on the patch's perimeter by a magnetic line current of magnitude Kd on the ground plane (xy plane). The magnetic current source is given by

$$K(x, y) = \hat{n} \times \hat{z}E(x, y) \qquad (2.51)$$

where $\hat{n}$ is a unit vector normal to the patch's perimeter and $\hat{z}E(x, y)$ is the component of the electric field perpendicular to the ground plane.

The limitations of the cavity model arise from the very approximate modeling of the fields outside the patch. The radiation field is accounted for by artificially increasing the dielectric loss inside the cavity by introducing the concept of effective loss tangent. Thus the radiated power needs to be estimated a priori before the voltage at the edges of the patch can be computed.

A cavity model for microstrip patch antennas may also be formulated by considering a planar two-dimensional resonator with an impedance boundary wall all around the edges of the patch. A direct form of network analog (DFNA) method for the analysis of such a cavity model has been discussed in [49]. In this version of the cavity model approach, the fringing field and the radiated power are not included inside the cavity, but are localized at the edges of the cavity resonator. This procedure is conceptually more accurate, but the solution for fields in a cavity with complex admittance walls is much more difficult to evaluate. Numerical methods become necessary for implementing this version of the cavity model.

The cavity model approach (including the version with impedance walls) does not allow any external mutual coupling among microstrip patches to be modeled and accounted for.

The multiport network modeling approach mentioned below overcomes these limitations, and is the most versatile of the network modeling approaches for microstrip patches.

Multiport Network Model (MNM)

In the multiport network modeling approach, the fields underneath the patch and those outside the patch are modeled separately. In this respect, the MNM approach differs significantly from the cavity model discussed earlier. However, the characterization of fields underneath the patch is similar to that in the cavity model (except for the equivalent loss tangent concept used in the cavity model to account for radiation). Thus, the MNM approach may be considered to be an extension of the cavity model method.

Details of the MNM approach for microstrip patches on single-layer and two-layer substrates are discussed in [50, Ch. 4].

EM Simulation-Based Numerical Models

An alternative to the use of network models for microstrip patch antennas is to employ general purpose electromagnetic (EM) simulation software for the evaluation of currents or field distribution associated with microstrip patch antennas. Various techniques for electromagnetic field simulation [51–53] have been developed to the level of user friendly commercially available software during the last decade.

Integral-Equation–Based Full-Wave Analysis

The most commonly used method of electromagnetic simulation of microstrip patch antennas is based on the moment-method solution of the current distribution on the conducting patch and the associated feed structure. This approach has also been used in several electromagnetic simulation softwares (like HP's Momentum, Sonnet Software's Em, Ansoft's Strata, and Zeland Software's IE3D) developed for the design of microwave integrated circuits and printed antennas.

The integral-equation–based full-wave analysis approach, when applied to microstrip antennas, is comprised of the following basic steps:

1. Formulation of an integral equation in terms of electric current distribution of the patch and the associated feed structure;
2. Solution for the current distribution using the method of moments;
3. Evaluation of the network parameters (S_{11}, Z_{in}, and so forth) at the input port or multiport parameters for antennas with two (or more) ports;
4. Evaluation of the radiation characteristics from the electric current distribution on the patch, and so forth.

In this approach (as well as in other EM simulation-based methods), modeling and analysis of microstrip antennas are closely integrated into a single procedure. The computer-aided design procedure shown in Figure 2.3 gets modified, as shown in Figure 2.19.

Two difficult steps in this procedure are:

1. The choice of the initial design (a transmission line modeling approach is usually used);
2. Design parameter modifications.

Optimization techniques similar to those used for microwave circuits can be used for this purpose. Because of the computer-intensive nature of EM

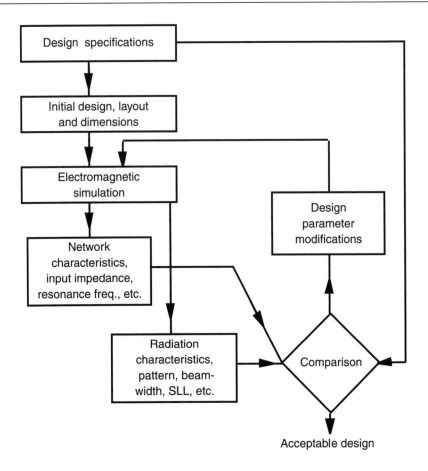

Figure 2.19 Microstrip antenna design using EM simulators.

simulation software, antenna optimization using EM simulation becomes a very slow process. Nevertheless, use of EM simulators for circuit and antenna optimization has gained some acceptance in recent years [54]. Specialized techniques like space-mapping and decomposition [55]—as well as the use of artificial neural networks [7, 30]—have been developed for making efficient use of EM simulation in optimization and design, and these are applicable to microstrip antenna design also.

Finite-Difference Time-Domain Simulation of Microstrip Patches

In addition to the integral equation formulation for full-wave analysis of microstrip patches, finite-difference time-domain (FDTD) simulation is also used for these antennas. Basic formulation of the FDTD method (as a central difference discretization of Maxwell's curl equations in both time and space) is well known [51–53], and is identical for circuit and antenna problems. However, details of the implementation by various researchers differ with respect to excitation treatment, boundary conditions, and postprocessing of results to obtain parameters of interest in frequency domain.

Modeling of Slot Antennas

Slot antennas may be considered to have evolved from slot transmission lines [13] in a manner similar to the evolution of microstrip patch antennas from microstrip transmission lines. Based on this analogy, a transmission line model can be used for rectangular slot resonators, as shown in Figure 2.20, and extended to the transmission line model for a slot antenna, as shown in Figure 2.21.

The conductance G_r represents the radiation conductance of the slot. Note that G_r is connected at the midpoint of the half-wavelength resonant slot, which is the location of the maximum voltage across (and hence that of the maximum magnetic current along) the slot. This conductance G_r can be evaluated by equating the power radiated by the slot to the equivalent power dissipated in G_r when a voltage V_s exists across the conductance. That is,

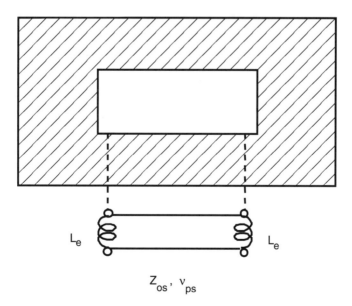

Figure 2.20 A transmission line resonator model of a resonant slot.

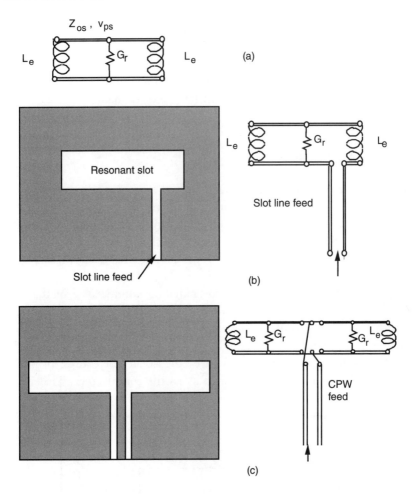

Figure 2.21 A simple equivalent transmission line model of a rectangular half-wave resonant slot; (a) feedline not shown; (b) resonant slot ($\lambda/2$) fed by an off-centered slot line; and (c) one-wavelength resonant slot fed by a CPW line.

$$G_r = \frac{2P_{rad}}{V_s^2} \quad (2.52)$$

Inductors L_e represent the inductance due to slot-line short-circuited terminations at the two ends of the resonant slot. The value for L_e is obtained from the characterization of short-circuited ends of a slot-line [13]. The equivalent model shown in Figure 2.21(a) does not include representation of the feedline. A slot-line–fed half-wave resonant slot can be modeled by the equivalent network shown in Figure 2.21(b). The discontinuity reactance contributed by the junction between the feeding slot-line and the resonant slot is not shown in Figure 2.21(b) and needs to be included for accurate modeling and analysis of the resonant slot antenna.

Figure 2.21(c) depicts the implementation of the equivalent transmission line model for a one-wavelength resonant slot fed at the midpoint by a coplanar-waveguide (CPW) feed. Note that in this case, two radiation conductances G_r are included at the two points of the maximum voltage across the slot. A network for representing the external mutual coupling between the two halves of the antenna needs to be incorporated and is not shown in Figure 2.21(c). Also, as for the model shown in Figure 2.21(b), the discontinuity reactance at the junction of the feedline and the slot is not included in Figure 2.21(c).

The basic idea of the equivalent transmission line modeling of rectangular slot antennas has been verified [56] by comparison with full-wave electromagnetic simulation. Just like transmission line models for microstrip patches, this modeling approach is expected to play a significant role in the initial first-order design of slot antennas and integrated circuit-antenna modules incorporating slot antennas.

In addition to the network modeling of slot antennas described in this section, numerical modeling based on electromagnetic simulation (discussed earlier for microstrip patches) is equally valid and employed for slot antennas as well. In the integral equation formulation for slot structures, it is appropriate to solve for equivalent magnetic current distribution in the slot regions in place of electric current distribution over the conductor areas (as commonly carried out for microstrip structures).

2.3.2 Analysis of Printed Patches and Slots

The computer-aided analysis of printed antennas is generally carried out in two broad steps:

1. Evaluation of the aperture field in terms of equivalent electric and/or magnetic current distribution on the top surface of the substrate;
2. Evaluation of the far field from the aperture field distribution.

Usually the network characteristics of the antenna (such as resonance frequency, input impedance, bandwidth, and so forth) are computed in step 1; and radiation characteristics (such as beam width and direction, side-lobe level, polarization, and so forth) become parts of computations in step 2.

When the network modeling approach is used for a microstrip patch antenna, network solution techniques are used for finding network parameters, namely, input impedance for single-port radiators and S-parameters for two (or multiple) port radiators. In addition, solution of the network model yields voltage distribution along the periphery of the patches. This voltage distribution is written in terms of the equivalent magnetic current distribution, which is used for the calculation of the radiation field in the step 2 in the procedure outlined above.

When integral-equation–based EM simulation is used for modeling and analysis of microstrip patch antennas, the first part of the computation results in electric current over the conducting patch and the associated feed structure. This electric current distribution can be used for finding the radiation field in the far zone of the antenna. The Green's function expressing the field produced by an infinitesimal electric current element on the ground dielectric substrate is already known. In fact, this Green's function (see [50], Ch. 5) is the starting point for the integral equation formulation. An integral of this Green's function weighted by the current distribution yields the total field, and, when evaluated with the far field approximation, provides results for the radiation field.

When simulating slot antennas by the moment-method-based integral equation approach, we solve for the equivalent magnetic current distribution in the slots (the nonconducting portion on the upper surface of the substrate). In this case, the far field is evaluated by using a Green's function for fields produced by an elementary magnetic current source.

When FDTD simulation is used, it becomes necessary to limit the computational domain by enclosing the antenna in a virtual box. The next step is the evaluation of the equivalent electric and magnetic current densities, J_s and M_s, which are related to the field components on the surface of the box by $J_s = \hat{n} \times H$ and $M_s = -\hat{n} \times E$ with $\hat{n}$ being a normal outward unit vector on the surface of the box. Since the field components on the surface of the virtual box calculated by FDTD are in the time domain, one makes use of the discrete Fourier transform (DFT) to obtain the corresponding frequency domain phasor quantities.

Common to all of the above methods of evaluating the aperture field in terms of electromagnetic currents is the near-field to far-field transformation for evaluating the antenna characteristics. This transformation is based upon the surface equivalence theorem (for details, see texts like [57]). Far-field evaluation starts with the computation of two vector potentials A and F as:

$$\vec{A} = \mu \iint_{S'} \frac{\vec{J}_S e^{-jkr''}}{4\pi r''} dS' \tag{2.53}$$

$$\vec{F} = \epsilon \iint_{S'} \frac{\vec{M}_S e^{-jkr''}}{4\pi r''} dS' \tag{2.54}$$

The electromagnetic fields at a far-field point, **P**, shown in Figure 2.22 can be expressed as follows

$$\vec{E} = -j\omega\vec{A} - \frac{j\omega}{k^2}\nabla(\nabla \cdot \vec{A}) - \frac{1}{\epsilon}\nabla \times \vec{F} \qquad (2.55)$$

$$\vec{H} = -j\omega\vec{F} - \frac{j\omega}{k^2}\nabla(\nabla \cdot \vec{F}) - \frac{1}{\mu}\nabla \times \vec{A} \qquad (2.56)$$

Since $r' \ll r$ for the far-field point **P**, we can assume

$$r'' = \sqrt{r^2 + r'^2 - 2rr'\cos\Psi} \cong r - r'\cos\Psi \qquad (2.57)$$

and the two vector potentials in (2.53) and (2.54) can be rewritten as

$$\vec{A} = \mu \frac{e^{-jkr}}{4\pi r} \iint_{S'} \vec{J}_S e^{jkr'\cos\psi} dS' \qquad (2.58)$$

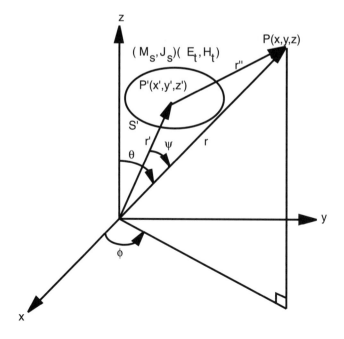

Figure 2.22 Coordinate system used for discussing near-field to far-field transformation.

$$\vec{F} = \epsilon \frac{e^{-jkr}}{4\pi r} \iint_{S'} \vec{M}_S e^{jkr'\cos\psi} dS' \qquad (2.59)$$

where

$$r'\cos\psi = r' \cdot \hat{r} = x'\sin\theta\cos\phi + y'\sin\theta\sin\phi + z'\cos\theta \qquad (2.60)$$

Since the integrands in (2.58) and (2.59) are functions of $\vec{J}_S$, $\vec{M}_S$, r' and Ψ only, we can define two new radiating vectors, $\vec{N}$ and $\vec{L}$, as shown below:

$$\vec{N} = \iint_{S'} \vec{J}_S e^{jkr'\cos\psi} dS' \qquad (2.61)$$

$$\vec{L} = \iint_{S'} \vec{M}_S e^{jkr'\cos\psi} dS' \qquad (2.62)$$

which are related to the vector potentials $\vec{A}$ and $\vec{F}$ by

$$\vec{A} = \mu \frac{e^{-jkr}}{4\pi r} \vec{N} \qquad (2.63)$$

$$\vec{F} = \epsilon \frac{e^{-jkr}}{4\pi r} \vec{L} \qquad (2.64)$$

Now, if we substitute (2.63) and (2.64) in (2.55) and (2.56), and drop the terms which decay faster than $1/r$, we can obtain the radiating fields at the far-field point as shown below:

$$E_\theta = \eta H_\phi = -j\frac{e^{-jkr}}{2\lambda r}(\eta N_\theta + L_\phi) \qquad (2.65)$$

$$E_\phi = -\eta H_\theta = j\frac{e^{-jkr}}{2\lambda r}(-\eta N_\phi + L_\theta) \qquad (2.66)$$

Consequently, we can calculate the time-averaged Poynting vector (average power density) at a far-field point (r, θ, ϕ) as follows:

$$W_r = \frac{1}{2}\text{Re}[\vec{E} \times \vec{H}^*] = \frac{1}{2}\text{Re}[E_\theta H_\phi^* - E_\phi H_\theta^*] \quad (2.67)$$

$$= \frac{\eta}{8\lambda^2 r^2}\left[\left|N_\theta + \frac{L_\phi}{\eta}\right|^2 + \left|N_\phi - \frac{L_\theta}{\eta}\right|^2\right]$$

By simply multiplying the above power density by the square of the distance, r^2, we obtain the radiation intensity of the antenna at a certain direction (θ, ϕ) as follows:

$$U = \frac{\eta}{8\lambda^2}\left[\left|N_\theta + \frac{L_\phi}{\eta}\right|^2 + \left|N_\phi - \frac{L_\theta}{\eta}\right|^2\right] \quad (2.68)$$

Thus we note that the second broad step mentioned in the beginning of this section on computer-aided analysis is a straightforward implementation of the equivalence theorem in electromagnetics.

2.4 Role of ANNs in RF and Microwave CAD

2.4.1 Modeling of RF and Microwave Components

As pointed out in Section 2.1.3, accurate and efficient modeling of RF and microwave components is a very critical step in the CAD process. It has been well-recognized [4] that the accuracy of modeling has been the main bottleneck in implementing CAD tools for first pass design of monolithic RF and microwave circuits. Significant improvements in RF and microwave models have been made in recent years by the use of electromagnetic simulation tools. However, electromagnetic computation by itself is too computationally intensive to be used as a part of an interactive CAD brochure. Closed form expressions, fitted lumped element models, and numerical models derived from electromagnetic simulation or measurements are generally used. Most of these models have inherent limitations of accuracy and validity over a restricted range of parameter values. Improvements in modeling are still needed for higher microwave and millimeter-wave frequencies, two-layer and multiple layer microwave circuits, coplanar-waveguide (CPW) circuits, and integrated circuit-antenna modules.

As mentioned in Chapter 1, use of artificial neural networks provides a powerful approach for developing CAD models. This approach is discussed in several following chapters of this book. Chapter 5 describes ANN models developed for RF and microwave components. Component examples include one-port and two-port microstrip vias, vertical interconnects used in multilayer circuits, CPW lines, CPW bends, and other discontinuities occurring in CPW circuits. High-speed interconnects constitute an important enabling technology item in high-speed computer circuits. As discussed in Chapter 6, ANN macromodeling techniques allow high-speed interconnect optimization to be carried out. Chapter 7 is an overview of ANN techniques used for active RF and microwave device models. Active device modeling is an important and critical area of microwave CAD, and it is an area where ANNs can play a significant role. This chapter describes direct modeling of the device's external behavior (for both small signal and nonlinear models) as well as indirect modeling through a known equivalent circuit model. Discussion also includes time-varying Volterra kernel-based model and circuit representation of the neural network models.

The approaches discussed in these three chapters are applicable to other RF and microwave components as well.

2.4.2 Efficient Optimization Strategies

The optimization process, as illustrated in Figures 2.15 and 2.16, is another important and time-consuming part of the CAD process. Repeated analysis of the design that is needed for the optimization process can be accelerated by using ANN macromodels of the design to be optimized. Chapters 6 and 8 of this book illustrate this approach. In Chapter 6, ANN models are used for yield optimization of high-speed IC interconnects. Chapter 8 provides examples of optimization of multilayer circuits, CPW circuits, CPW fed slot antennas, and MESFET amplifiers, using the ANN modeling approach.

2.4.3 Implementation of Knowledge-Aided Design (KAD)

The use of knowledge-based approaches to the initial stages of RF and microwave circuit and antenna design is an area that needs to be explored. Suitably trained ANN macromodels could become a useful vehicle for incorporating knowledge into the design process. The discussion on knowledge-based ANN models included in Chapter 9 of this book is a pointer in that direction. As pointed out in this chapter, prior knowledge about a component embedded in the ANN model helps in reducing the training time needed for developing an accurate ANN model. This knowledge-based design approach can be extended to the overall design process, say for high efficiency RF amplifiers. Based on the design rules developed for various classes of high-efficiency amplifiers, an ANN model can be trained to provide the designer with an appropriate configuration for a given set of specifications. An ANN macromodel for this configuration can be used to modify an embedded typical design into a design that meets the given specifications.

Recent efforts in developing effective ways to incorporate RF and microwave information in the neural network structures has led to creation of several unique RF/microwave oriented neural network approaches [5, 6, 58, 59]. We can look forward to increased research activity in neural network-based knowledge-aided design.

2.5 Summary

This chapter has discussed the anatomy of the overall design process. Conventional design and computer-aided design methodologies have been presented. Concepts of knowledge-aided design are introduced. Current status of CAD techniques for RF and microwave circuits and printed antennas is reviewed. The importance of modeling and optimization in the design process is brought out. It is in these two areas that ANNs are likely to play very crucial roles. The anticipated role of neural network techniques in knowledge-aided design is pointed out.

References

[1] Hopgood, A. A., "Systems for design and selection," *Knowledge-Based Systems for Engineers and Scientists*, Boca Raton, FL: CRC Press, 1993, Ch. 8.

[2] Green, M., "Conceptions and Misconceptions of Knowledge-Aided Design," *Knowledge-Aided Design*, New York: Academic Press, 1992, Ch. 1.

[3] Tong, C., and D. Sriram, *Artificial Intelligence in Engineering Design*, New York: Academic Press, 1992.

[4] Gupta, K. C., "Emerging Trends in Millimeter-Wave CAD," *IEEE Trans. Microwave Theory and Techniques*, Vol. MTT-46, June 1998, pp. 747–755.

[5] Gupta, K. C., "ANN and Knowledge-Based Approaches for Microwave Design," in *Directions for the Next Generation of MMIC Devices and Systems*, N. K. Das and H. L. Bertoni, eds., New York, NY: Plenum, 1996, pp. 389–396.

[6] Watson, P. M., K. C. Gupta, and R. L. Mahajan, "Development of Knowledge-Based Artificial Neural Network Models for Microwave Components," *1998 IEEE International Microwave Symposium Digest*, Baltimore MD, June 1998, pp. 9–12.

[7] Veluswami, A., M. S. Nakhla, and Q. J. Zhang, "The Application of Neural Networks to EM-Based Simulation and Optimization of Interconnects in High-Speed VLSI Circuits," *IEEE Trans. Microwave Theory and Techniques*, Vol. 45, No. 5, May 1997, pp. 712–723.

[8] Chandrasekaran, B., "Generic Tasks in Knowledge-Based Reasoning—High-Level Building Blocks for Expert System Design," *Expert*, Vol. 1, No. 23, 1986.

[9] Chandrasekaran, B., "Design Problem Solving: A Task Analysis," *Knowledge-Aided Design*, M. Green, ed., New York: Academic Press, 1992, Ch. 2.

[10] Gupta, K. C., R. Garg, and R. Chadha, *Computer Aided Design of Microwave Circuits*, Norwood MA: Artech House, 1981.

[11] Dobrowolski, J. A., *Introduction to Computer Methods for Microwave Circuit Analysis and Design*, Norwood, MA: Artech House, 1991.

[12] Dobrowolski, J. A., *Computer-Aided Analysis, Modeling and Design of Microwave Networks—The Wave Approach*, Norwood, MA: Artech House, 1996.

[13] Gupta, K. C., et al., *Microstrip Lines and Slotlines*, 2nd Ed., Norwood, MA: Artech House, 1996.

[14] Edwards, T. C., *Foundations for Microstrip Circuit Design*, 2nd ed. New York: John Wiley and Sons, 1991.

[15] Bhartia, P., and P. Pramanick, "Fin-Line Characteristics and Circuits," *Infrared and Millimeter Waves*, vol. 17, New York: Academic Press, 1988.

[16] Hoffman, R. K., *Handbook of Microwave Integrated Circuits*, Norwood, MA: Artech House, 1987.

[17] Bahl, I. J., and P. Bhartia, *Microwave Solid State Circuit Design*, New York: John Wiley and Sons, 1988, p. 757.

[18] Hammerstad, E. O., and F. Bekkadal, "Microstrip Handbook," ELAB Report STF44 A74169, University of Trondheim, Norway, 1975.

[19] Mehran, R., "Frequency Dependent Equivalent Circuits for Microstrip Right Angle Bends, T-Junctions and Crossings," *Arch. Elek. Ubertragug*, Vol. 30, 1976, pp. 80–82.

[20] Jansen, R. H., and N. H. L. Koster, "A Unified CAD Basis for Frequency Dependent Characterization of Strip, Slot and Coplanar MIC Components," *Proc. 11th European Microwave Conf.* (Amsterdam, the Netherlands), 1981, pp. 682–687.

[21] Jansen, R. H., "Hybrid Mode Analysis of End Effects of Planar Microwave and Millimeter-Wave Transmission Line," *Proc. IEE*, Part H, Vol. 128, 1981, pp. 77–86.

[22] Koster, N. H. L., and R. H. Jansen, "The Equivalent Circuits of the Asymmetrical Series Gap in Microstrip and Suspended Substrate Lines," *IEEE Trans. Microwave Theory and Techniques*, Vol. MTT-30, Aug. 1982, pp. 1273–1279.

[23] Jansen, R. H., "The Spectral Domain Approach for Microwave Integrated Circuits," *IEEE Trans. Microwave Theory and Techniques*, Vol. MTT-33, Oct. 1985, pp. 1043–1056.

[24] Koster, N. H. L., and R. H. Jansen, "The Microstrip Step Discontinuity: A Revised Description," *IEEE Trans. Microwave Theory and Techniques*, Feb. 1986, pp. 213–223.

[25] Katehi, P. B., and N. G. Alexopoulos, "Frequency Dependent Characteristics of Microstrip Discontinuities in Millimeter-Wave Integrated Circuits," *IEEE Trans. Microwave Theory and Techniques*, Vol. MTT-33, Oct. 1985, pp. 1029–1035.

[26] Jackson, R. W., and D. M. Pozar, "Full-Wave Analysis of Microstrip Open-End and Gap Discontinuities," *IEEE Trans. Microwave Theory and Techniques*, Oct. 1985, pp. 1036–1042.

[27] Mosig, J. R., and F. E. Gardiol, "Analytical and Numerical Techniques in the Green's Function Treatment of Microstrip Antennas and Scatterers," *Proc. IEE*, Part H, Vol. 130, 1983, pp. 175–182.

[28] Wu, C. I., et al., "Accurate Numerical Modeling of Microstrip Junctions and Discontinuities," *Int. J. Microwave MM-Wave Computer-Aided Eng.*, Vol. 1, No. 1, 1991, pp. 48–58.

[29] Watson, P. M., and K. C. Gupta, "EM-ANN Models for Microstrip Vias and Interconnects in Multilayer Circuits," *IEEE Trans. Microwave Theory and Techniques*, Vol. 44, Dec. 1996, pp. 2495–2503.

[30] Watson, P. M., and K. C. Gupta, "Design and Optimization of CPW Circuits using EM-ANN Models for CPW Components," *IEEE Trans. Microwave Theory and Techniques*, Vol. 45, Dec. 1997, pp. 2515–2523.

[31] Mahajan, R. L., and K. C. Gupta, "Physical-Neural Network Modeling for Electronic Packaging Applications," *Proc. Wireless Commun. Conf.*, Boulder, CO: August 1996, pp. 157–162.

[32] Rousset, J., et al., "An Accurate Neural Network Model of FET for Intermodulation and Power Analysis," 26th European Microwave Conference, Czech Republic, Sept. 1996.

[33] Zaabab, H., Q. J. Zhang, and M. S. Nakhla, "A Neural Network Modeling Approach to Circuit Optimization and Statistical Design," *IEEE Trans. Microwave Theory and Techniques*, Vol. 43, 1995, pp. 1349–1358.

[34] Chua, L. O., *Introduction to Non-Linear Network Theory*, New York: McGraw-Hill, 1969.

[35] Chua, L. O., and P. M. Lin, *Computer-Aided Analysis of Electronic Circuits*, Chapters 5, 7, 10, and 11, Englewood Cliffs, NJ: Prentice-Hall, 1975.

[36] Nakhla, M. S., and J. Vlach, "A Piecewise Harmonic Balance Technique for Determination of Periodic Response of Nonlinear Systems," *IEEE Trans. Circuit Systems*, Vol. CAS-23, 1976, pp. 85–91.

[37] Zhang, Q. J., F. Wang, and M. S. Nakhla, "Optimization of High Speed VLSI Interconnects: A Review," *Int. Journal on Microwave and Millimeter-Wave Computer-Aided Engineering*, Special Issue on Optimization oriented Microwave CAD, Vol. 7, 1997 (invited), pp. 83–107.

[38] Bandler, J. W., and Q. J. Zhang, "Next Generation Optimization Methodologies for Wireless and Microwave Circuit Design," (plenary session invited paper), *IEEE MTT-S Int. Topical Symp. on Technologies for Wireless Applications Dig.*, Vancouver, BC: Feb. 1999, pp. 5–8.

[39] James, J. R., and P. S. Hall eds., *Handbook of Microstrip Antenna*, Vols. I and II, London U. K.: Peter Peregrinus Ltd., 1989.

[40] Richards, W. F., "Microstrip Antennas," in *Antenna Handbook—Theory, Applications and Design*, Y. L. Lo and S. W. Lee, eds., New York: Van Nostrand Reinhold Company, 1988, pp. 10.1–10.74.

[41] Gupta, K. C., and A. Benalla, *Microstrip Antenna Design*, reprint volume, Norwood, MA: Artech House, 1988.

[42] James, J. R., P. S. Hall, and C. Wood, *Microstrip Antenna Theory and Design*, London: Peter Peregrinus, 1981.

[43] Bahl, I. J., and P. Bhartia, *Microstrip Antennas*, Norwood, MA: Artech House, 1980.

[44] Gupta, K. C., "Two-Port Transmission Characteristics of Rectangular Microstrip Patch Radiators," *1985 IEEE AP-S International Antennas Propagat. Symp. Digest*, pp. 71–74.

[45] Benalla, A., and K. C. Gupta, "Transmission Line Model for 2-Port Rectangular Microstrip Patches with Ports at the Non-Radiating Edges," *Electron. Lett.*, Vol. 23, 1987, pp. 882–884.

[46] Benalla, A., and K. C. Gupta, "Two-Dimensional Analysis of One-Port and Two-Port Microstrip Antennas," *Scientific Rept. 85*, Electromagnetics Laboratory, University of Colorado, May 1986, p. 48.

[47] Lo, Y. T., et al., "Theory and Experiment on Microstrip Antennas," *IEEE Trans. Antennas Propag.*, Vol. AP-27, 1979, pp. 137–145.

[48] Richards, W. F., et al., "An Improved Theory for Microstrip Antennas and Applications," *IEEE Trans. Antennas Propag.*, Vol. AP-29, 1981, pp. 38–46.

[49] Coffey, E. L., and T. H. Lehman, "A New Analysis Technique for Calculating the Self and Mutual Impedance of Microstrip Antennas," *Proc. Workshop Printed Circuit Antenna Technology*, New Mexico State University, 1979, pp. 31.1–31.21.

[50] Gupta, K. C., and P. S. Hall, "Analysis and Design of Integrated Circuit-Antenna Modules," New York: John Wiley and Sons, 2000.

[51] Itoh, T., ed., *Numerical Techniques for Microwave and Millimeter-Wave Passive Structures*, New York: John Wiley and Sons, 1989.

[52] Sadiku, M. N. O., *Numerical Techniques in Electromagnetics*, Boca Raton, FL: CRC Press, 1992.

[53] Taflove, A., *Computational Electrodynamics: The Finite-Difference Time-Domain Method*, Norwood MA: Artech House, 1995.

[54] Bandler, J. W., (Guest Editor), "Special Issue on Automated Circuit Design Using Electromagnetic Simulators," *IEEE Trans. Microwave Theory and Techniques*, Vol. 45, No. 5, Part II, May 1997.

[55] Bandler, J. W., et al. "Design Optimization of Interdigital Filters using Aggressive Space Mapping and Decomposition," *IEEE Trans. Microwave Theory and Techniques*, Vol. 45, No. 5, May 1997, pp. 761–769.

[56] Haeusler, M., "Radiation From Discontinuities of Coplanar Waveguides (CPW) and Design of a CPW Patch Antenna," Diplomaingenieur Thesis (directed by K. C. Gupta and Gerd Schaller), Friedrich Alexander Univ., Erlangen-Nuremberg (Germany), 1998.

[57] Balanis, C. A., *Advanced Engineering Electromagnetics*, New York: John Wiley and Sons, 1989.

[58] Wang, F., and Q. J. Zhang, "Knowledge-Based Neural Models for Microwave Design," *IEEE Trans. Microwave Theory and Techniques*, Vol. 45, Dec. 1997, pp. 1349–1358.

[59] Bandler, J. W., M. A. Ismail, J. E. Rayas-Sanchez and Q. J. Zhang, "Neuromodeling of Microwave Circuits Exploiting Space Mapping Technology," *IEEE MTT-S International Microwave Symp. Digest*, Anaheim, CA: June 1999, pp. 149–152.

3

Neural Network Structures

This chapter describes various types of neural network structures that are useful for RF and microwave applications. The most commonly used neural network configurations, known as multilayer perceptrons (MLP), are described first, together with the concept of basic backpropagation training, and the universal approximation theorem. Other structures discussed in this chapter include radial basis function (RBF) network, wavelet neural network, and self-organizing maps (SOM). Brief reviews of arbitrary structures for ANNs and recurrent neural networks are also included.

3.1 Introduction

A neural network has at least two physical components, namely, the processing elements and the connections between them. The processing elements are called neurons, and the connections between the neurons are known as links. Every link has a weight parameter associated with it. Each neuron receives stimulus from the neighboring neurons connected to it, processes the information, and produces an output. Neurons that receive stimuli from outside the network (i.e., not from neurons of the network) are called input neurons. Neurons whose outputs are used externally are called output neurons. Neurons that receive stimuli from other neurons and whose output is a stimulus for other neurons in the neural network are known as hidden neurons. There are different ways in which information can be processed by a neuron, and different ways of connecting the neurons to one another. Different neural network structures can be constructed by using different processing elements and by the specific manner in which they are connected.

A variety of neural network structures have been developed for signal processing, pattern recognition, control, and so on. In this chapter, we describe several neural network structures that are commonly used for microwave modeling and design [1, 2]. The neural network structures covered in this chapter include multilayer perceptrons (MLP), radial basis function networks (RBF), wavelet neural networks, arbitrary structures, self-organizing maps (SOM), and recurrent networks.

3.1.1 Generic Notation

Let n and m represent the number of input and output neurons of the neural network. Let x be an n-vector containing the external inputs (stimuli) to the neural network, y be an m-vector containing the outputs from the output neurons, and w be a vector containing all the weight parameters representing the connections in the neural network. The function $y = y(x, w)$ mathematically represents a neural network. The definition of w and the manner in which y is computed from x and w, determine the structure of the neural network.

To illustrate the notation, we consider the neural network model of an FET shown in Figure 3.1. The inputs and outputs of the FET neural model are given by,

$$x = [L\ W\ a\ N_d\ V_{GS}\ V_{DS}\ freq]^T \qquad (3.1)$$
$$y = [MS_{11}\ PS_{11}\ MS_{12}\ PS_{12}\ MS_{21}\ PS_{21}\ MS_{22}\ PS_{22}]^T \qquad (3.2)$$

where $freq$ is frequency, and MS_{ij} and PS_{ij} represent the magnitude and phase of the S-parameter S_{ij}. The input vector x contains physical/process/

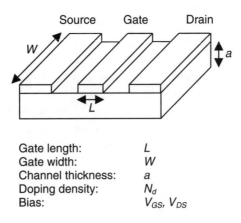

Gate length: L
Gate width: W
Channel thickness: a
Doping density: N_d
Bias: $V_{GS},\ V_{DS}$

Figure 3.1 A physics-based FET.

bias parameters of the FET. The original physics-based FET problem can be expressed as

$$y = f(x) \tag{3.3}$$

The neural network model for the problem is

$$y = y(x, w) \tag{3.4}$$

3.1.2 Highlights of the Neural Network Modeling Approach

In the FET example above, the neural network will represent the FET behavior only after learning the original $x - y$ relationship through a process called *training*. Samples of (x, y) data, called *training data*, should first be generated from original device physics simulators or from device measurements. Training is done to determine neural network internal weights w such that the neural model output best matches the training data. A trained neural network model can then be used during microwave design providing answers to the task it learned. In the FET example, the trained model can be used to provide S-parameters from device physical/geometrical and bias values during circuit design.

To further highlight features of the neural network modeling approach, we contrast it with two broad types of conventional microwave modeling approaches. The first type is the detailed modeling approach such as EM-based models for passive components and physics-based models for active components. The overall model, ideally, is defined by well-established theory and no experimental data is needed for model determination. However, such detailed models are usually computationally expensive. The second type of conventional modeling uses empirical or equivalent circuit-based models for passive and active components. The models are typically developed using a mixture of simplified component theory, heuristic interpretation and representations, and fitting of experimental data. The evaluation of these models is usually much faster than that of the detailed models. However, the empirical and equivalent circuit models are often developed under certain assumptions in theory, range of parameters, or type of components. The models have limited accuracy especially when used beyond original assumptions. The neural network approach is a new type of modeling approach where the model can be developed by learning from accurate data of the original component. After training, the neural network becomes a fast and accurate model of the original problem it learned. A summary of these aspects is given in Table 3.1.

An in-depth description of neural network training, its applications in modeling passive and active components and in circuit optimization will be

Table 3.1
A Comparison of Modeling Approaches for RF/Microwave Applications

Basis for Comparison	EM/Physics Models	Empirical and Equivalent Circuit Models	Pure Neural Network Models
Speed	Slow	Fast	Fast
Accuracy	High	Limited	Could be close to EM/physics models
Number of training data	0	A few	Sufficient training data is required, which could be large for high-dimensional problems
Circuit/EM theory of the problem	Maxwell, or semiconductor equations	Partially involved	Not involved

described in subsequent chapters. In the present chapter, we describe structures of neural networks, that is, the various ways of realizing $y = y(x, w)$. The structural issues have an impact on model accuracy and cost of model development.

3.2 Multilayer Perceptrons (MLP)

Multilayer perceptrons (MLP) are the most popular type of neural networks in use today. They belong to a general class of structures called feedforward neural networks, a basic type of neural network capable of approximating generic classes of functions, including continuous and integrable functions [3]. MLP neural networks have been used in a variety of microwave modeling and optimization problems.

3.2.1 MLP Structure

In the MLP structure, the neurons are grouped into layers. The first and last layers are called input and output layers respectively, because they represent inputs and outputs of the overall network. The remaining layers are called hidden layers. Typically, an MLP neural network consists of an input layer, one or more hidden layers, and an output layer, as shown in Figure 3.2.

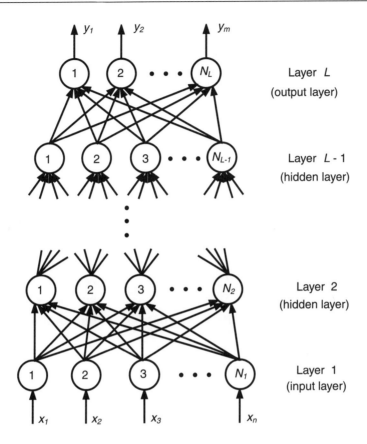

Figure 3.2 Multilayer perceptrons (MLP) structure.

Suppose the total number of layers is L. The 1st layer is the input layer, the Lth layer is the output layer, and layers 2 to $L - 1$ are hidden layers. Let the number of neurons in lth layer be N_l, $l = 1, 2, \ldots, L$.

Let w_{ij}^l represent the weight of the link between jth neuron of $l - 1$th layer and ith neuron of lth layer, $1 \leq j \leq N_{l-1}$, $1 \leq i \leq N_l$. Let x_i represent the ith external input to the MLP, and z_i^l be the output of ith neuron of lth layer. We introduce an extra weight parameter for each neuron, w_{i0}^l, representing the bias for ith neuron of lth layer. As such, $\boldsymbol{w}$ of MLP includes w_{ij}^l, $j = 0, 1, \ldots, N_{l-1}$, $i = 1, 2, \ldots, N_l$, $l = 2, 3, \ldots, L$, that is,

$$\boldsymbol{w} = [w_{10}^2 \; w_{11}^2 \; w_{12}^2 \ldots, w_{N_L N_{L-1}}^L]^T \tag{3.5}$$

3.2.2 Information Processing by a Neuron

In a neural network, each neuron—with the exception of neurons at the input layer—receives and processes stimuli (inputs) from other neurons. The processed information is available at the output end of the neuron. Figure 3.3 illustrates the way in which each neuron in an MLP processes the information. As an example, a neuron of the lth layer receives stimuli from the neurons of $l-1$th layer, that is, z_1^{l-1}, z_2^{l-1}, ..., $z_{N_{l-1}}^{l-1}$. Each input is first multiplied by the corresponding weight parameter, and the resulting products are added to produce a weighted sum γ. This weighted sum is passed through a neuron activation function $\sigma(\cdot)$ to produce the final output of the neuron. This output z_i^l can, in turn, become the stimulus for neurons in the next layer.

3.2.3 Activation Functions

The most commonly-used hidden neuron activation function is the sigmoid function given by

$$\sigma(\gamma) = \frac{1}{(1 + e^{-\gamma})} \tag{3.6}$$

As shown in Figure 3.4, the sigmoid function is a smooth switch function having the property of

$$\sigma(\gamma) \rightarrow \begin{cases} 1 & \text{as } \gamma \rightarrow +\infty \\ 0 & \text{as } \gamma \rightarrow -\infty \end{cases}$$

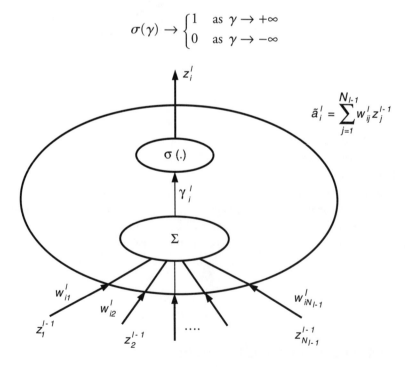

Figure 3.3 Information processing by ith neuron of lth layer.

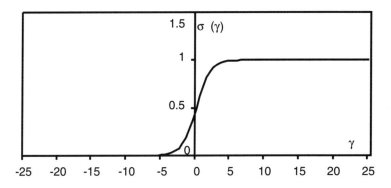

Figure 3.4 Sigmoid function.

Other possible hidden neuron activation functions are the arc-tangent function shown in Figure 3.5 and given by

$$\sigma(\gamma) = \left(\frac{2}{\pi}\right)\arctan(\gamma) \tag{3.7}$$

and the hyperbolic-tangent function shown in Figure 3.6 and given by

$$\sigma(\gamma) = \frac{(e^{\gamma} - e^{-\gamma})}{(e^{\gamma} + e^{-\gamma})} \tag{3.8}$$

All these logistic functions are bounded, continuous, monotonic, and continuously differentiable.

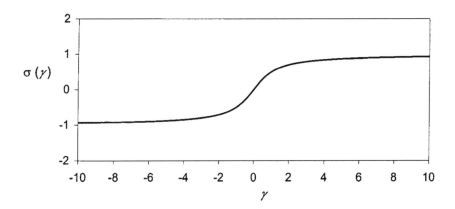

Figure 3.5 Arc-tangent function.

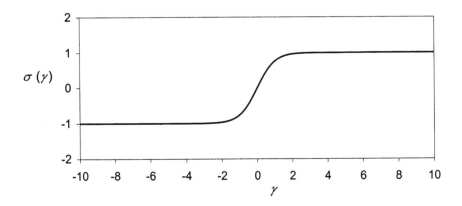

Figure 3.6 Hyperbolic-tangent function.

The input neurons simply relay the external stimuli to the hidden layer neurons; that is, the input neuron activation function is a relay function, $z_i^1 = x_i$, $i = 1, 2, \ldots, n$, and $n = N_1$. As such, some researchers only count the hidden and output layer neurons as part of the MLP. In this book, we follow a convention, wherein the input layer neurons are also considered as part of the overall structure. The activation functions for output neurons can either be logistic functions (e.g., sigmoid), or simple linear functions that compute the weighted sum of the stimuli. For RF and microwave modeling problems, where the purpose is to model continuous electrical parameters, linear activation functions are more suitable for output neurons. The linear activation function is defined as

$$\sigma(\gamma) = \gamma = \sum_{j=0}^{N_{L-1}} w_{ij}^L z_j^{L-1} \qquad (3.9)$$

The use of linear activation functions in the output neurons could help to improve the numerical conditioning of the neural network training process described in Chapter 4.

3.2.4 Effect of Bias

The weighted sum expressed as

$$\gamma_i^l = w_{i1}^l z_1^{l-1} + w_{i2}^l z_2^{l-1} + \ldots + w_{iN_{l-1}}^l z_{N_{l-1}}^{l-1} \qquad (3.10)$$

is zero, if all the previous hidden layer neuron responses (outputs) $z_1^{l-1}, z_2^{l-1}, \ldots, z_{N_{l-1}}^{l-1}$ are zero. In order to create a bias, we assume a fictitious neuron whose output is

$$z_0^{l-1} = 1 \qquad (3.11)$$

and add a weight parameter w_{i0}^{l-1} called bias. The weighted sum can then be written as

$$\gamma_i^l = \sum_{j=0}^{N_{l-1}} w_{ij}^l z_j^{l-1} \qquad (3.12)$$

The effect of adding the bias is that the weighted sum is equal to the bias when all the previous hidden layer neuron responses are zero, that is,

$$\gamma_i^l = w_{i0}^l, \text{ if } z_1^{l-1} = z_2^{l-1} = \ldots = z_{N_{l-1}}^{l-1} = 0 \qquad (3.13)$$

The parameter w_{i0}^l is the bias value for ith neuron in lth layer as shown in Figure 3.7.

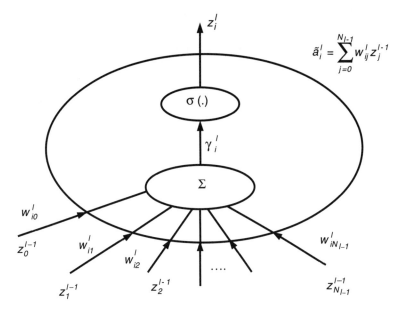

Figure 3.7 A typical ith hidden neuron of lth layer with an additional weight parameter called bias.

3.2.5 Neural Network Feedforward

Given the inputs $x = [x_1 \; x_2 \; \ldots \; x_n]^T$ and the weights w, neural network feedforward is used to compute the outputs $y = [y_1 \; y_2 \; \ldots \; y_m]^T$ from a MLP neural network. In the feedforward process, the external inputs are first fed to the input neurons (*1*st layer), the outputs from the input neurons are fed to the hidden neurons of the 2nd layer, and so on, and finally the outputs of $L - 1$th layer are fed to the output neurons (Lth layer). The computation is given by,

$$z_i^1 = x_i, \; i = 1, 2, \ldots, N_1, \; n = N_1 \qquad (3.14)$$

$$z_i^l = \sigma\left(\sum_{j=0}^{N_{l-1}} w_{ij}^l z_j^{l-1}\right), \; i = 1, 2, \ldots, N_l, \; l = 2, 3, \ldots, L \qquad (3.15)$$

The outputs of the neural network are extracted from the output neurons as

$$y_i = z_i^L, \; i = 1, 2, \ldots, N_L, \; m = N_L \qquad (3.16)$$

During feedforward computation, the neural network weights w are fixed. As an example, consider a circuit with four transmission lines shown in Figure 3.8. Given the circuit design parameters, $x = [l_1 \; l_2 \; l_3 \; l_4 \; R_1 \; R_2 \; R_3 \; R_4 \; C_1 \; C_2 \; C_3 \; C_4 \; V_{peak} \; \tau_{rise}]^T$, where V_{peak} and τ_{rise} are the peak amplitude and rise time of the source voltage, the signal delays at four output nodes *A, B, C, D*, represented by the output vector $y = [\tau_1 \; \tau_2 \; \tau_3 \; \tau_4]^T$ need to be computed. The original problem $y = f(x)$ is a nonlinear relationship between the circuit parameters and the delays. The conventional way to compute the delays is to solve the Kirchoff's current/voltage equations of the circuit, and this process is CPU-intensive, especially if the delay has to be evaluated repetitively for different x. In the neural network approach, a neural model can be developed such that each input neuron corresponds to a circuit parameter in x, and each output neuron represents a signal delay in y. The weights w in the model $y = y(x, w)$ are determined through a neural network training process. The model is used to compute the signal delays y for given values of x using neural network feedforward operation. The feedforward computation involves simple sum, product, and sigmoid evaluations, and not the explicit Kirchoff's current/voltage equations. A question arises: can such a simple feedforward computation represent the complicated Kirchoff's current/voltage equations or maybe even the Maxwell's 3-D EM

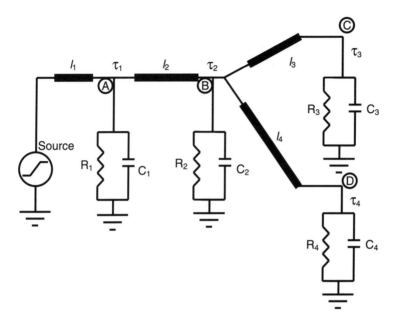

Figure 3.8 A circuit with four transmission lines.

equations? The universal approximation theorem presented in the following sub-section answers this exciting question.

3.2.6 Universal Approximation Theorem

The universal approximation theorem for MLP was proved by Cybenko [4] and Hornik et al. [5], both in 1989. Let I_n represent an n-dimensional unit cube containing all possible input samples x, that is, $x_i \in [0, 1]$, $i = 1, 2, \ldots, n$, and $C(I_n)$ be the space of continuous functions on I_n. If $\sigma(\cdot)$ is a continuous sigmoid function, the universal approximation theorem states that the finite sums of the form

$$y_k = y_k(x, w) = \sum_{i=1}^{N_2} w_{ki}^3 \sigma\left(\sum_{j=0}^{n} w_{ij}^2 x_j\right) \quad k = 1, 2, \ldots, m \quad (3.17)$$

are dense in $C(I_n)$. In other words, given any $f \in C(I_n)$ and $\epsilon > 0$, there is a sum $y(x, w)$ of the above form that satisfies $|y(x, w) - f(x)| < \epsilon$ for all $x \in I_n$. As such, there always exists a 3-layer perceptron that can approximate an arbitrary nonlinear, continuous, multi-dimensional function f with any desired accuracy.

However, the theorem does not state how many neurons are needed by the three-layer MLP to approximate the given function. As such, failure to develop an accurate neural model can be attributed to an inadequate number of hidden neurons, inadequate learning/training, or presence of a stochastic rather than a deterministic relation between inputs and outputs [5].

We illustrate how an MLP model matches an arbitrary one-dimensional function shown in Figure 3.9. A drop in the function y from 2.0 to 0.5 as x changes from 5 to 15, corresponds to a sigmoid $\sigma(-(x - 10))$ scaled by a factor of 1.5. On the other hand, a slower increase in the function from 0.5 to 4.5 as x changes from 20 to 60, corresponds to a sigmoid $\sigma(0.2(x - 40))$ scaled by a factor 4 ($= 4.5 - 0.5$). Finally, the overall function is shifted upwards by a bias of 0.5. The function can then be written as

$$y = y(x, w) = 0.5 + 1.5\sigma(-(x - 10)) + 4\sigma(0.2(x - 40)) \quad (3.18)$$
$$= 0.5 + 1.5\sigma(-x + 10) + 4\sigma(0.2x - 8)$$

and the structure of the neural network model is shown in Figure 3.10. In practice, the optimal values of the weight parameters w are obtained by a training process, which adjusts w such that the error between the neural model outputs and the original problem outputs is minimized.

Discussion

This example is an illustration of how a neural network approximates a simple function in a manner similar to polynomial curve-fitting. The real power of neural networks, however, lies in its modeling capacity when the nonlinearity and dimensionality of the original problem increases. In such cases, curve-fitting techniques using higher-order, higher-dimensional polynomial functions are very cumbersome and ineffective. Neural network models can handle such

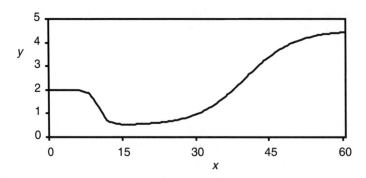

Figure 3.9 A one-dimensional function to be modeled by an MLP.

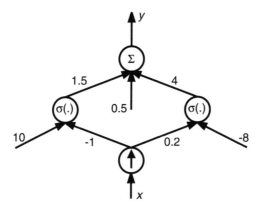

Figure 3.10 A neural network model for the function in Figure 3.9. In this figure, an arrow inside the input neuron means that the input neuron simply relays the value of the input (x) to the network. The hidden neurons use sigmoid activation functions, while the output neurons use linear functions.

problems more effectively, and can be accurate over a larger region of the input space. The most significant features of neural networks are:

- Neural networks are distributed models by nature. In other words, no single neuron can produce the overall x-y relationship. Each neuron is a simple processing element with switching activation function. Many neurons combined produce the overall x-y relationship. For a given value of external stimuli, some neurons are switched on, some are off, while others are in transition. It is the rich combination of the neuron switching states responding to different values of external stimuli that enables the network to represent a nonlinear input-output mapping.
- Neural networks have a powerful learning capability, that is, they can be trained to represent any given problem behavior. The weight parameters in the neural network represent the weighted connections between neurons. After training the neural network, the weighted connections capture/encode the problem information from the raw training data. Neural networks with different sets of weighted connections can represent a diverse range of input-output mapping problems.

3.2.7 Number of Neurons

The universal approximation theorem states that there exists a three-layer MLP that approximates virtually any nonlinear function. However, it did not specify what size the network should be (i.e., number of hidden neurons) for a given problem complexity. The precise number of hidden neurons required for a modeling task remains an open question. Although there is no clear-cut answer, the number of hidden neurons depends on the degree of nonlinearity and the dimensionality of the original problem. Highly nonlinear problems need more neurons and smoother problems need fewer neurons. Too many hidden neurons may lead to overlearning of the neural network, which is discussed in Chapter 4. On the other hand, fewer hidden neurons will not give sufficient freedom to the neural network to accurately learn the problem behavior. There are three possible solutions to address the question regarding network size. First, experience can help determine the number of hidden neurons, or the optimal size of the network can be obtained through a trial and error process. Second, the appropriate number of neurons can be determined by an adaptive process— or optimization process—that adds/deletes neurons as needed during training [6]. Finally, the ongoing research in this direction includes techniques such as constructive algorithms [7], network pruning [8], and regularization [9], to match the neural network model complexity with problem complexity.

3.2.8 Number of Layers

Neural networks with at least one hidden layer are necessary and sufficient for arbitrary nonlinear function approximation. In practice, neural networks with one or two hidden layers, that is, three-layer or four-layer perceptrons (including input and output layers) are commonly used for RF/microwave applications. Intuitively, four-layer perceptrons would perform better in modeling nonlinear problems where certain localized behavioral components exist repeatedly in different regions of the problem space. A three-layer perceptron neural network—although capable of modeling such problems—may require too many hidden neurons. Literature that favors both three-layer perceptrons and four-layer perceptrons does exist [10, 11]. The performance of a neural network can be evaluated in terms of generalization capability and mapping capability [11]. In the function approximation or regression area where generalization capability is a major concern, three-layer perceptrons are usually preferred [10], because the resulting network usually has fewer hidden neurons. On the other hand, four-layer perceptrons are favored in pattern classification tasks where decision boundaries need to be defined [11], because of their better mapping capability. Structural optimization algorithms that determine the optimal number of layers according to the training data have also been investigated [12, 13].

3.3 Back Propagation (BP)

The main objective in neural model development is to find an optimal set of weight parameters w, such that $y = y(x, w)$ closely represents (approximates) the original problem behavior. This is achieved through a process called training (that is, optimization in w-space). A set of training data is presented to the neural network. The training data are pairs of (x_k, d_k), $k = 1, 2, \ldots, P$, where d_k is the desired outputs of the neural model for inputs x_k, and P is the total number of training samples.

During training, the neural network performance is evaluated by computing the difference between actual neural network outputs and desired outputs for all the training samples. The difference, also known as the error, is quantified by

$$E = \frac{1}{2} \sum_{k \in T_r} \sum_{j=1}^{m} (y_j(x_k, w) - d_{jk})^2 \qquad (3.19)$$

where d_{jk} is the jth element of d_k, $y_j(x_k, w)$ is the jth neural network output for input x_k, and T_r is an index set of training data. The weight parameters w are adjusted during training, such that this error is minimized. In 1986, Rumelhart, Hinton, and Williams [14] proposed a systematic neural network training approach. One of the significant contributions of their work is the error back propagation (BP) algorithm.

3.3.1 Training Process

The first step in training is to initialize the weight parameters w, and small random values are usually suggested. During training, w is updated along the negative direction of the gradient of E, as $w = w - \eta \dfrac{\partial E}{\partial w}$, until E becomes small enough. Here, the parameter η is called the learning rate. If we use just one training sample at a time to update w, then a per-sample error function E_k given by

$$E_k = \frac{1}{2} \sum_{j=1}^{m} (y_j(x_k, w) - d_{jk})^2 \qquad (3.20)$$

is used and w is updated as $w = w - \eta \dfrac{\partial E_k}{\partial w}$. The following sub-section describes how the error back propagation process can be used to compute the gradient information $\dfrac{\partial E_k}{\partial w}$.

3.3.2 Error Back Propagation

Using the definition of E_k in (3.20), the derivative of E_k with respect to the weight parameters of the lth layer can be computed by simple differentiation as

$$\frac{\partial E_k}{\partial w_{ij}^l} = \frac{\partial E_k}{\partial z_i^l} \cdot \frac{\partial z_i^l}{\partial w_{ij}^l} \qquad (3.21)$$

and

$$\frac{\partial z_i^l}{\partial w_{ij}^l} = \frac{\partial \sigma}{\partial \gamma_i^l} \cdot z_j^{l-1} \qquad (3.22)$$

The gradient $\dfrac{\partial E_k}{\partial z_i^l}$ can be initialized at the output layer as

$$\frac{\partial E_k}{\partial z_i^L} = (y_i(\boldsymbol{x}_k, \boldsymbol{w}) - d_{ik}) \qquad (3.23)$$

using the error between neural network outputs and desired outputs (training data). Subsequent derivatives $\dfrac{\partial E_k}{\partial z_i^l}$ are computed by back-propagating this error from $l+1$th layer to lth layer (see Figure 3.11) as

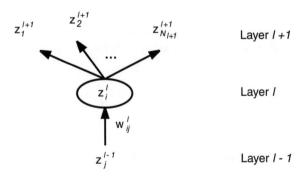

Figure 3.11 The relationship between ith neuron of lth layer, with neurons of layers $l-1$ and $l+1$.

$$\frac{\partial E_k}{\partial z_i^l} = \sum_{j=1}^{N_{l+1}} \frac{\partial E_k}{\partial z_j^{l+1}} \cdot \frac{\partial z_j^{l+1}}{\partial z_i^l} \quad (3.24)$$

For example, if the MLP uses sigmoid (3.6) as hidden neuron activation function,

$$\frac{\partial \sigma}{\partial \gamma} = \sigma(\gamma)(1 - \sigma(\gamma)) \quad (3.25)$$

$$\frac{\partial z_i^l}{\partial w_{ij}^l} = z_i^l(1 - z_i^l)z_j^{l-1} \quad (3.26)$$

and

$$\frac{\partial z_i^l}{\partial z_j^{l-1}} = z_i^l(1 - z_i^l)w_{ij}^l \quad (3.27)$$

For the same MLP network, let δ_i^l be defined as $\delta_i^l = \frac{\partial E_k}{\partial \gamma_i^l}$ representing local gradient at ith neuron of lth layer. The back propagation process is then given by,

$$\delta_i^L = (y_i(\boldsymbol{x}_k, \boldsymbol{w}) - d_{ik}) \quad (3.28)$$

$$\delta_i^l = \left(\sum_{j=1}^{N_{l+1}} \delta_j^{l+1} w_{ji}^{l+1} \right) z_i^l(1 - z_i^l), \, l = L - 1, L - 2, \ldots 2 \quad (3.29)$$

and the derivatives with respect to the weights are

$$\frac{\partial E_k}{\partial w_{ij}^l} = \delta_i^l z_j^{l-1} \; l = L, L - 1, \ldots, 2 \quad (3.30)$$

3.4 Radial Basis Function Networks (RBF)

Feedforward neural networks with a single hidden layer that use radial basis activation functions for hidden neurons are called radial basis function (RBF) networks. RBF networks have been applied to various microwave modeling purposes—for example, to model intermodulation distortion behavior of MESFETs and HEMTs [15].

3.4.1 RBF Network Structure

A typical radial-basis–function neural network is shown in Figure 3.12. The RBF neural network has an input layer, a radial basis hidden layer, and an output layer.

The parameters c_{ij}, λ_{ij}, are centers and standard deviations of radial basis activation functions. Commonly used radial basis activation functions are Gaussian and multiquadratic. The Gaussian function shown in Figure 3.13 is given by

$$\sigma(\gamma) = \exp(-\gamma^2) \qquad (3.31)$$

The multiquadratic function shown in Figure 3.14 is given by

$$\sigma(\gamma) = \frac{1}{(c^2 + \gamma^2)^\alpha}, \; \alpha > 0 \qquad (3.32)$$

where c is a constant.

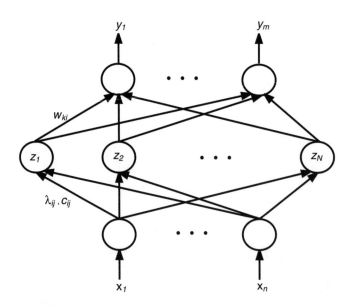

Figure 3.12 RBF neural network structure.

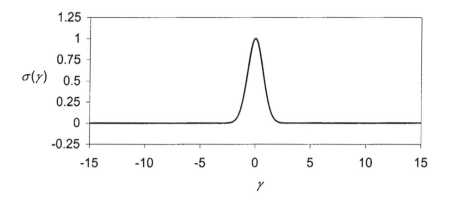

Figure 3.13 Gaussian function.

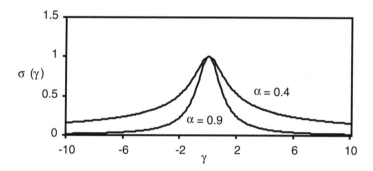

Figure 3.14 Multiquadratic function with $c = 1$.

3.4.2 Feedforward Computation

Given the inputs x, the total input to the ith hidden neuron γ_i is given by

$$\gamma_i = \sqrt{\sum_{j=1}^{n} \left(\frac{x_j - c_{ij}}{\lambda_{ij}}\right)^2}, \quad i = 1, 2, \ldots, N \quad (3.33)$$

where N is the number of hidden neurons. The output value of the ith hidden neuron is $z_i = \sigma(\gamma_i)$, where $\sigma(\gamma)$ is a radial basis function. Finally, the outputs of the RBF network are computed from hidden neurons as

$$y_k = \sum_{i=0}^{N} w_{ki} z_i, \quad k = 1, 2, \ldots, m \quad (3.34)$$

where w_{ki} is the weight of the link between ith neuron of the hidden layer and kth neuron of the output layer. Training parameters w of the RBF network include w_{k0}, w_{ki}, c_{ij}, λ_{ij}, $k = 1, 2, \ldots, m$, $i = 1, 2, \ldots, N$, $j = 1, 2, \ldots, n$.

For illustration, we use an RBF network to approximate the one-dimensional function shown in Figure 3.15. The function has a narrow peak at $x = 2$, that can be approximated by a Gaussian function $\sigma(x - 2)$. The wider valley at $x = 9$ is represented by a Gaussian $\sigma[(x - 9)/3]$ scaled by a factor of -2. Finally, a bias of value 1 is added. As such, an RBF network with two hidden neurons given by

$$y = \sigma(x - 2) - 2\sigma\left(\frac{x - 9}{3}\right) + 1 \qquad (3.35)$$

can approximate this function. The RBF network structure for this example is shown in Figure 3.16. In practice, the RBF weight parameters w_{ki}, c_{ij}, λ_{ij}, are determined through a training process.

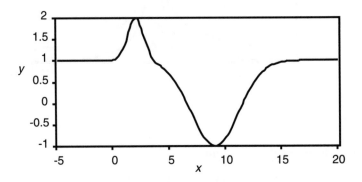

Figure 3.15 A one-dimensional function to be modeled by an RBF network.

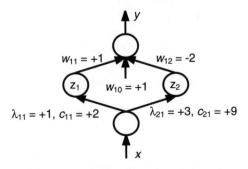

Figure 3.16 The RBF network structure for the function in Figure 3.15.

3.4.3 Universal Approximation Theorem

In 1991, Park and Sandberg proved the universal approximation theorem for RBF networks [16]. According to their work, an RBF neural network with a sufficient number of hidden neurons is capable of approximating any given nonlinear function to any degree of accuracy. In this section, we provide a simplified interpretation of the theorem. Let R^n be the space of n-dimensional real-valued vectors; $C(R^n)$ be the space of continuous functions on R^n; and μ be a finite measure on R^n. Given any function $f \in C(R^n)$ and a constant $\epsilon > 0$, there is a sum $y(x, w)$ of the form (3.34), for which $|f(x) - y(x, w)| < \epsilon$ for almost all $x \in R^n$. The size μ of the x region for which $|f(x) - y(x, w)| > \epsilon$ is very small, that is, $\mu < \epsilon$.

3.4.4 Two-Step Training of RBF Networks

Step 1: The centers c_{ij} of the hidden neuron activation functions can be initialized using a clustering algorithm. This step is called the unsupervised training and the purpose is to find the centers of clusters in the training sample distribution. A detailed explanation of the clustering algorithm is presented in Section 3.8. This provides better initial values for hidden neuron centers as compared to random initialization. While clustering for pattern-classification applications are well-defined [17], the question of effectively initializing RBF centers for microwave modeling still remains an open task.

Step 2: Update by gradient-based optimization techniques, such as $w = w - \eta \frac{\partial E}{\partial w}$, where w includes all the parameters in RBF networks (i.e. λ_{ij}, c_{ij}, and w_{ki}) until neural network learns the training data well. This step is similar to that of MLP training.

3.5 Comparison of MLP and RBF Neural Networks

Both MLP and RBF belong to a general class of neural networks called feedforward networks, where information processing in the network structure follows one direction—from input neurons to output neurons. However, the hidden neuron activation functions in MLP and RBF behave differently. First, the activation function of each hidden neuron in an MLP processes the inner product of the input vector and the synaptic weight vector of that neuron. On the other hand, the activation function of each hidden neuron in an RBF network processes the Euclidean norm between the input vector and the center of that neuron. Second, MLP networks construct global approximations to nonlinear input-output mapping. Consequently they are capable of generalizing in those regions of the input space where little or no training data is available. Conversely, RBF networks use exponentially decaying localized nonlinearities to construct local approximations to nonlinear input-output mapping. As a result, RBF neural networks learn at faster rates and exhibit reduced sensitivity to the order of presentation of training data [17]. In RBF, a hidden neuron influences the network outputs only for those inputs that are near to its center, thus requiring an exponential number of hidden neurons to cover the entire input space. From this perspective, it is suggested in [18] that RBF networks are suitable for problems with a smaller number of inputs.

For the purpose of comparison, both MLP and RBF networks were used to model a physics-based MESFET [2]. The physical, process, and bias parameters of the device (channel length L, channel width W, doping density N_d, channel thickness a, gate-source voltage V_{GS}, drain-source voltage V_{DS}) are neural network inputs. Drain-current (i_d) is the neural network output. Three sets of training data with 100, 300, and 500 samples were generated using OSA90 [19]. A separate set of data with 413 samples, called test data, was also generated by the same simulator in order to test the accuracy of the neural models after training is finished. The accuracy for the trained neural models is measured by the average percentage error of the neural model output versus test data, shown in Table 3.2. As can be seen from the table, RBF networks used more hidden neurons than MLP to achieve similar model accuracy. This can be attributed to the localized nature of radial basis activation functions. When training data is large and sufficient, RBF networks achieve better accuracy than MLP. As the amount of training data becomes less, the performance of the MLP degrades slowly as compared to the RBF networks. This shows that MLP networks have better generalization capability. However, the training process of RBF networks is usually easier to converge than the MLP.

Table 3.2
Model Accuracy Comparison Between MLP and RBF (from [2], Wang, F., et al., "Neural Network Structures and Training Algorithms for RF and Microwave Applications," Int. J. RF and Microwave CAE, pp. 216–240, © 1999, John Wiley and Sons. Reprinted with permission from John Wiley and Sons, Inc.).

Training Sample Size	No. of Hidden Neurons	MLP				RBF			
		7	10	14	18	25	20	30	40	50
100	Ave. Test Error (%)	1.65	2.24	2.60	2.12	2.91	6.32	5.78	6.15	8.07
300	Ave. Test Error (%)	0.69	0.69	0.75	0.69	0.86	1.37	0.88	0.77	0.88
500	Ave. Test Error (%)	0.57	0.54	0.53	0.53	0.60	0.47	0.43	0.46	0.46

3.6 Wavelet Neural Networks

The idea of combining wavelet theory with neural networks [20–22] resulted in a new type of neural network called wavelet networks. The wavelet networks use wavelet functions as hidden neuron activation functions. Using theoretical features of the wavelet transform, network construction methods can be developed. These methods help to determine the neural network parameters and the number of hidden neurons during training. The wavelet network has been applied to modeling passive and active components for microwave circuit design [23–25].

3.6.1 Wavelet Transform

In this section, we provide a brief summary of the wavelet transform [20]. Let R be a space of real variables and R^n be an n-dimensional space of real variables.

Radial Function

Let $\psi = \psi(\cdot)$ be a function of n variables. The function $\psi(x)$ is radial, if there exists a function $g = g(\cdot)$ of a single variable, such that for all $x \in R^n$, $\psi(x) = g(\|x\|)$. If $\psi(x)$ is radial, its Fourier transform $\hat{\psi}(\omega)$ is also radial.

Wavelet Function

Let $\hat{\psi}(\omega) = \eta(\|\omega\|)$, where η is a function of a single variable. A radial function $\psi(x)$ is a wavelet function, if

$$C_\psi = (2\pi)^n \int_0^\infty \frac{|\eta(\xi)|^2}{\xi} d\xi < \infty \tag{3.36}$$

Wavelet Transform

Let $\psi(x)$ be a wavelet function. The function can be shifted and scaled as $\psi\left(\dfrac{x-t}{a}\right)$, where t, called the translation (shift) parameter, is an n-dimensional real vector in the same space as x, and a is a scalar called the dilation parameter. Wavelet transform of a function $f(x)$ is given by

$$W(a, t) = \int_{R^n} f(x) a^{-n/2} \psi\left(\frac{x-t}{a}\right) dx \tag{3.37}$$

The wavelet transform transforms the function from original domain (x domain) into a wavelet domain (a, t domain). A function $f(x)$ having both smooth global variations and sharp local variations can be effectively represented in wavelet domain by a corresponding wavelet function $W(a, t)$. The original function $f(x)$ can be recovered from the wavelet function by using an inverse wavelet transform, defined as

$$f(x) = \frac{1}{C_\Psi} \int_{0^+}^{\infty} a^{-(n+1)} \int_{R^n} W(a, t) a^{-n/2} \psi\left(\frac{x-t}{a}\right) dt\, da \qquad (3.38)$$

The discretized version of the inverse wavelet transform is expressed as

$$f(x) = \sum_i W_i a_i^{-n/2} \psi\left(\frac{x-t_i}{a_i}\right) \qquad (3.39)$$

where (a_i, t_i) represent discrete points in the wavelet domain, and W_i is the coefficient representing the wavelet transform evaluated at (a_i, t_i). Based on (3.39), a three-layer neural network with wavelet hidden neurons can be constructed. A commonly used wavelet function is the inverse Mexican-hat function given by

$$\psi\left(\frac{x-t_i}{a}\right) = \sigma(\gamma_i) = (\gamma_i^2 - n) \exp\left(-\frac{\gamma_i^2}{2}\right) \qquad (3.40)$$

where

$$\gamma_i = \left\|\frac{x-t_i}{a_i}\right\| = \sqrt{\sum_{j=1}^{n} \left(\frac{x_j - t_{ij}}{a_i}\right)^2} \qquad (3.41)$$

A one-dimensional inverse Mexican-hat with $t_{11} = 0$ is shown in Figure 3.17 for two different values of a_1.

3.6.2 Wavelet Networks and Feedforward Computation

Wavelet networks are feedforward networks with one hidden layer, as shown in Figure 3.18. The hidden neuron activation functions are wavelet functions. The output of the ith hidden neuron is given by

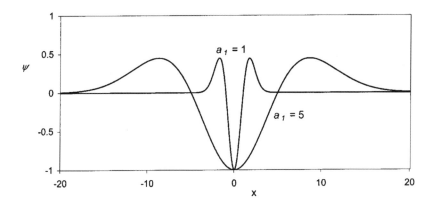

Figure 3.17 Inverse Mexican-hat function.

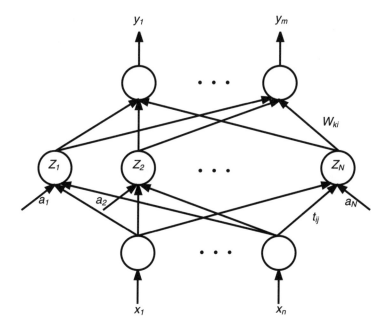

Figure 3.18 Wavelet neural network structure.

$$z_i = \sigma(\gamma_i) = \psi\left(\frac{x - t_i}{a_i}\right), \, i = 1, 2, \ldots, N \qquad (3.42)$$

where N is the number of hidden neurons, $x = [x_1 \, x_2 \, \ldots \, x_n]^T$ is the input vector, $t_i = [t_{i1} \, t_{i2} \, \ldots \, t_{in}]^T$ is the translation parameter, a_i is a dilation parameter, and $\psi(\cdot)$ is a wavelet function. The weight parameters of a wavelet network w include a_i, t_{ij}, w_{ki}, w_{k0}, $i = 1, 2, \ldots, N$, $j = 1, 2, \ldots, n$, $k = 1, 2, \ldots, m$.

In (3.42), γ_i is computed following (3.41). The outputs of the wavelet network are computed as

$$y_k = \sum_{i=0}^{N} w_{ki} z_i, \quad k = 1, 2, \ldots, m \qquad (3.43)$$

where w_{ki} is the weight parameter that controls the contribution of the ith wavelet function to the kth output. As an illustration, we use a wavelet network to model the one-dimensional function shown in Figure 3.19.

The narrow peak at $x = 4$ that looks like a hat can be approximated by a wavelet function with $t_{11} = 4$, $a_1 = 0.5$, and a scaling factor of -5. The wider valley at $x = 12$ that looks like an inverted hat can be represented by a wavelet with $t_{12} = 12$ and $a_2 = 1$ scaled by 3. As such, two hidden wavelet neurons can model this function. The resulting wavelet network structure is shown in Figure 3.20. In practice, all the parameters shown in the wavelet network are determined by the training process.

3.6.3 Wavelet Neural Network With Direct Feedforward From Input to Output

A variation of the wavelet neural network was used in [23], where additional connections are made directly from input neurons to the output neurons. The outputs of the model are given by

$$y_k = \sum_{i=0}^{N} w_{ki} z_i + \sum_{j=1}^{n} v_{kj} x_j, \quad k = 1, 2, \ldots, m \qquad (3.44)$$

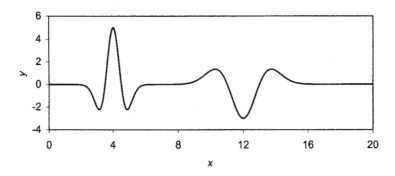

Figure 3.19 A one-dimensional function to be modeled by a wavelet network.

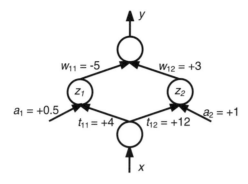

Figure 3.20 Wavelet neural network structure for the one-dimensional function of Figure 3.19.

where z_i is defined in (3.42) and v_{kj} is the weight linking jth input neuron to the kth output neuron.

3.6.4 Wavelet Network Training

The training process of wavelet networks is similar to that of RBF networks.

Step 1: Initialize translation and dilation parameters of all the hidden neurons, $t_i, a_i, i = 1, 2, \ldots, N$.

Step 2: Update the weights w of the wavelet network using a gradient-based training algorithm, such that the error between neural model and training data is minimized. This step is similar to MLP and RBF training.

3.6.5 Initialization of Wavelets

An initial set of wavelets can be generated in two ways. One way is to create a wavelet lattice by systematically using combinations of grid points along dilation and translation dimensions [20]. The grids of the translation parameters are sampled within the model input-space, or x-space. The grids for dilation parameters are sampled from the range $(0, a)$, where a is a measure of the "diameter" of the x-space that is bounded by the minimum and maximum values of each element in the x-vector in training data. Another way to create initial wavelets is to use clustering algorithms to identify centers and widths of clusters in the training data. The cluster centers and width can be used for initializing translation and dilation parameters of the wavelet functions. A detailed description of the clustering algorithm is given in Section 3.8.

The initial number of wavelets can be unnecessarily large, especially if the wavelet set is created using the wavelet lattice approach. This is because of the existence of many wavelets that are redundant with respect to the given problem. This leads to a large number of hidden neurons adversely affecting the training and the generalization performance of the wavelet network. A wavelet reduction algorithm was proposed in [20], which identifies and removes less important wavelets. From the initial set of wavelets, the algorithm first selects the wavelet that best fits the training data, and then repeatedly selects one of the remaining wavelets that best fits the data when combined with all previously selected wavelets. For computational efficiency, the wavelets selected later are ortho-normalized to ones that were selected earlier.

At the end of this process, a set of wavelet functions that best span the training data outputs are obtained. A wavelet network can then be constructed using these wavelets, and the network weight parameters can be further refined by supervised training.

3.7 Arbitrary Structures

All the neural network structures discussed so far have been layered. In this section, we describe a framework that accommodates arbitrary neural network structures. Suppose the total number of neurons in the network is N. The output value of the jth neuron in the network is denoted by $z_j, j = 1, 2, \ldots, N$. The weights of the links between neurons (including bias parameters) can be arranged into an $N \times (N + 1)$ matrix given by

$$\boldsymbol{w} = \begin{bmatrix} w_{10} & w_{11} & \cdots & w_{1N} \\ w_{20} & w_{21} & \cdots & w_{2N} \\ \vdots & \vdots & \cdots & \vdots \\ w_{N0} & w_{N1} & \cdots & w_{NN} \end{bmatrix} \quad (3.45)$$

where the 1st column represents the bias of each neuron. A weight parameter w_{ij} is nonzero if the output of the jth neuron is a stimulus (input) to the ith neuron. The features of an arbitrary neural network are:

- There is no layer-by-layer connection requirement;
- Any neuron can be linked to any other neuron;
- External inputs can be supplied to any predefined set of neurons;
- External outputs can be taken from any predefined set of neurons.

The inputs to the neural network are $x = [x_1 \ x_2 \ \ldots \ x_n]^T$, where the input neurons are identified by a set of indices—$i_1, i_2, \ldots, i_n$, $i_k \in \{1, 2, \ldots, N\}$, and $k = 1, 2, \ldots, n$. The outputs of the neural network are $y = [y_1 \ y_2 \ \ldots \ y_m]^T$, where the output neurons are identified by a set of indices $j_1, j_2, \ldots, j_m, j_k \in \{1, 2, \ldots, N\}$, and $k = 1, 2, \ldots, m$. All the neurons could be considered to be in one layer.

This formulation accommodates general feedforward neural network structures. Starting from the input neurons, follow the links to other neurons, and continue following their links and so on. If, in the end, we reach the output neurons without passing through any neuron more than once (i.e., no loop is formed), then the network is a feedforward network since the information processing is one-way, from the input neurons to the output neurons.

Figure 3.21 illustrates the method of mapping a given neural network structure into a weight matrix. In the weight matrix, the ticks represent non-zero value of weights, indicating a link between corresponding neurons. A blank entry in the (i, j)th location of the matrix means there is no link between the output of the jth neuron and the input of the ith neuron. If an entire row (column) is zero, then the corresponding neuron is an input (output)

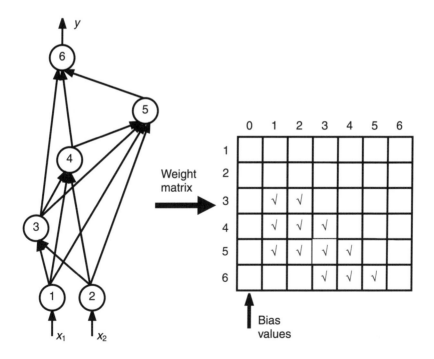

Figure 3.21 Mapping a given arbitrary neural network structure into a weight matrix.

neuron. A neuron is a hidden neuron if its corresponding row and column both have nonzero entries.

Feedforward for an Arbitrary Neural Network Structure

First, identify the input neurons by input neuron indices and feed the external input values to these neurons as

$$\begin{cases} z_{i_1} = x_1 \\ z_{i_2} = x_2 \\ \vdots \\ z_{i_n} = x_n \end{cases} \tag{3.46}$$

Then follow the nonzero weights in the weight matrix to process subsequent neurons, until the output neurons are reached. For example, if the kth neuron uses a sigmoid activation function, it can be processed as

$$\gamma_k = \sum_{j=1}^{N} w_{kj} z_j + w_{k0} \tag{3.47}$$

$$z_k = \sigma(\gamma_k) \tag{3.48}$$

where w_{k0} is the bias for kth neuron and $\sigma(\cdot)$ represents a sigmoid function. Once all the neurons in the network have been processed, the outputs of the network can be extracted by identifying the output neurons through output neuron indices as

$$\begin{cases} y_1 = z_{j1} \\ y_2 = z_{j2} \\ \vdots \\ y_m = z_{jm} \end{cases} \tag{3.49}$$

The arbitrary structures allow neural network structural optimization. The optimization process can add/delete neurons and connections between neurons during training [13, 26].

3.8 Clustering Algorithms and Self-Organizing Maps

In microwave modeling, we encounter situations where developing a single model for the entire input space becomes too difficult. A decomposition approach is often used to divide the input space, such that within each subspace, the problem is easier to model. A simple example is the large-signal transistor modeling where the model input space is first divided into saturation, breakdown, and linear regions; different equations are then used for each subregion. Such decomposition conventionally involves human effort and experience. In this section, we present a neural network approach that facilitates automatic decomposition through processing and learning of training data. This approach is very useful when the problem is complicated and the precise shape of subspace boundaries not easy to determine. A neural-network–based decomposition of E-plane waveguide filter responses with respect to the filter geometrical parameters was carried out in [27]. The neural network structure used for this purpose is the self-organizing map (SOM) proposed by Kohonen [28]. The purpose of SOM is to classify the parameter space into a set of clusters, and simultaneously organize the clusters into a map based upon the relative distances between clusters.

3.8.1 Basic Concept of the Clustering Problem

The basic clustering problem can be stated in the following way. Given a set of samples x_k, $k = 1, 2, \ldots, P$, find cluster centers c_p, $p = 1, 2, \ldots, N$, that represent the concentration of the distribution of the samples. The symbol x—defined as the input vector for the clustering algorithm—may contain contents different than those used for inputs to MLP or RBF. More specifically, x here represents the parameter space that we want classify into clusters. For example, if we want to classify various patterns of frequency response (e.g., S-parameters) of a device/circuit, then x contains S-parameters at a set of frequency points. Having said this, let us call the given samples of x the training data to the clustering algorithm. The clustering algorithm will classify (decompose) the overall training data into clusters—that is, training data similar to one another will be grouped into one cluster and data that are quite different from one another will be separated into different clusters. Let R_i be the index set for cluster center c_i that contains the samples close to c_i. A basic clustering algorithm is as follows:

Step 1: Assume total number of clusters to be N.

Set c_i = initial guess, $i = 1, 2, \ldots, N$.

Step 2: For each cluster, for example the ith cluster, build an index set R_i such that

$$R_i = \{k \mid \|x_k - c_i\| < \|x_k - c_j\|,$$
$$j = 1, 2, \ldots, N, \; i \neq j, \; k = 1, 2, \ldots, P\}$$

Update center c_i such that $c_i = \dfrac{1}{size(R_i)} \sum_{k \in R_i} x_k$ \qquad (3.50)

Step 3: If the changes in all c_i's are small, then stop. Otherwise go to Step 2.

After finishing this process, the solution (i.e., resulting cluster centers) can be used in the following way. Given an input vector x, find the cluster to which x belongs. This is done by simply comparing the distance of x from all the different cluster centers. The closest cluster center wins. This type of input-output relationship is shown in Figure 3.22.

An Example of Clustering Filter Responses

The idea of combining neural networks with SOM clustering for microwave applications was proposed in [27]. An E-plane waveguide filter with a metal insert was modeled using the combined approach. A fast model for the filter that represents the relationship between frequency domain response ($|S_{11}|$) and the input parameters (septa lengths and spacing, waveguide width, frequency) is needed for statistical design taking into account random variations and tolerances in the filter geometry [29].

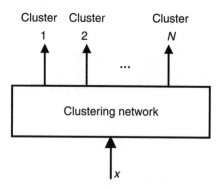

Figure 3.22 A clustering network where each output represents a cluster. The network divides the x-space into N clusters. When x is closest to the ith cluster, the ith output will be 1 and all other outputs will be 0's.

The training data was collected for 65 filters (each with a different geometry), at 300 frequency points per filter. However, the variation of $|S_{11}|$ with respect to filter geometry and frequency is highly nonlinear and exhibits sharp variations. In [27], the filter responses are divided into four groups (clusters), in order to simplify the modeling problem and improve neural model accuracy. Figure 3.23 illustrates the four types of filter responses extracted from the 65 filter responses using a neural-network–based automatic clustering algorithm.

In order to perform such clustering, the frequency responses of the filters are used as inputs to the clustering algorithm. Let Q be a feature vector of a filter containing output $|S_{11}|$ at 300 frequency points. For 65 filters, there are 65 corresponding feature vectors: Q_1, Q_2, Q_3, ..., Q_{65}. Using the feature vectors as training data (i.e., as x), employ the clustering algorithm to find

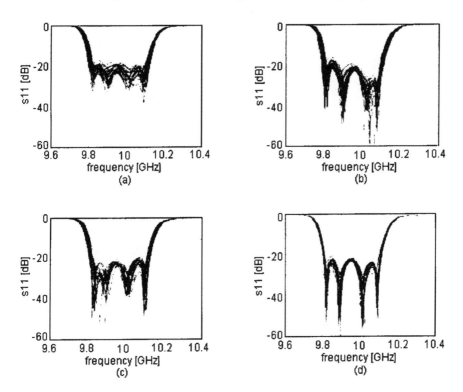

Figure 3.23 Four clusters of filter responses extracted using automatic clustering (from [27], Burrascano, P., S. Fiori, and M. Mongiardo, "A Review of Artificial Neural Network Applications in Microwave CAD," *Int. J. RF and Microwave CAE*, pp. 158–174, © 1999, John Wiley and Sons. Reprinted with permission from John Wiley and Sons, Inc.).

several clusters that correspond to several groups of filter responses. In this example, the filters were divided into 4 groups. The responses of the filters within each group are more similar to one another than those in other groups.

3.8.2 k-Means Algorithm

The k-means algorithm is a popular clustering method. In this algorithm, we start by viewing all the training samples as one cluster. The number of clusters are automatically increased in the following manner:

Step 1: Let $N = 1$ (one cluster)

$$c_1 = \frac{1}{P}\sum_{k=1}^{P} x_k$$

Step 2: Split cluster c_i, $i = 1, 2, \ldots, N$, into two new clusters,

$c_{i1} = \lambda c_i$
$c_{i2} = (1 - \lambda)c_i$, $0 < \lambda < 1$

Use them as initial values and run the basic clustering algorithm described earlier to obtain refined values of all the centers.
Step 3: If the solution is satisfactory, stop.
Otherwise go to Step 2.

This algorithm uses a basic binary splitting scheme. An improved k-means algorithm was proposed in [30], where only those clusters with large variance with respect to their member samples are split. In this way, the total number of clusters does not have to be 2^k. The variance of the ith cluster is defined as

$$v_i = \sum_{k \in R_i} (x_k - c_i)^T (x_k - c_i) \qquad (3.51)$$

3.8.3 Self-Organizing Map (SOM)

A self-organizing map (SOM) is a special type of neural network for clustering purposes, proposed by Kohonen [28]. The total number of cluster centers to be determined from training data is equal to the total number of neurons in SOM. As such, each neuron represents a cluster center in the sample distribution. The neurons of the SOM are arranged into arrays, such as a one-dimensional array or two-dimensional array. A two-dimensional array—also known as a map—is the most common arrangement. The clusters or neurons in the map have a double index (e.g., c_{ij}). There are two spaces in SOM: namely, the original problem space (i.e., x-space), and the map-space (i.e., (i, j)-space). Figure 3.24 shows a 5×5 map-space.

SOM has a very important feature called topology ordering, which is defined in the following manner. Let x_1 and x_2 be two points in x-space that belong to clusters c_{ij} and c_{pq}, respectively. If x_1 and x_2 are close to (or far away from) each other in x-space, then the corresponding cluster centers c_{ij} and c_{pq} are also close to (or far away from) each other in (i, j)-space. In other words, if $\|c_{ij} - c_{pq}\|$ is small (or large), then $|i - p| + |j - q|$ is also small (or large).

3.8.4 SOM Training

The self-organizing capability of SOM comes from its special training routine. The SOM training algorithm is quite similar to the clustering algorithms except that an additional piece of information called the topological closeness is learned though a neighborhood update. First, for each training sample x_k, $k = 1, 2, \ldots, P$, we find the nearest cluster c_{ij} such that

$$\|c_{ij} - x_k\| < \|c_{pq} - x_k\| \quad \forall p, q, \ i \neq p, j \neq q \qquad (3.52)$$

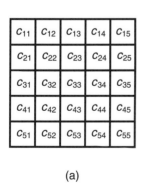
(a)

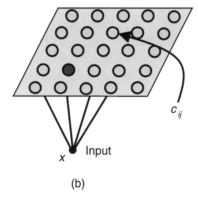
(b)

Figure 3.24 Illustration of a 5×5 map where each entry represents a cluster in the x-space.

In the next step, the center c_{ij} and its neighboring centers c_{pq} are all updated as

$$c_{pq} = c_{pq} + \alpha(t)(x_k - c_{pq}) \qquad (3.53)$$

The neighborhood of c_{ij} is defined as $|p - i| < N_c$ and $|q - j| < N_c$, where N_c is the size of the neighborhood, and $\alpha(t)$ is a positive constant similar to a learning rate that decays as training proceeds.

With such neighborhood updates, each neuron will be influenced by its neighborhood. The neighborhood size N_c is initialized to be a large value and gradually shrinks during the training process. The shrinking process needs to be slow and carefully controlled in order to find out the topological information of the sample distribution.

3.8.5 Using a Trained SOM

Suppose a SOM has been trained—that is, values of c_{ij} are determined. A trained SOM can be used in the following way. Given any x, find a map entry (i, j) such that $\|x - c_{ij}\| < \|x - c_{pq}\|$ for all p and q.

Waveguide Filter Example

Continue with the waveguide filter example [27] described in Section 3.8.1. The overall filter model consists of a general MLP for all filters, a SOM that classifies the filter responses into four clusters, and four specific MLP networks representing each of the four groups of filters, as shown in Figure 3.25. The feature vectors of the 65 filters (that is Q_1, Q_2, Q_3, ..., Q_{65}) are used as training vectors for SOM, and four clusters of filter responses were identified. The general MLP and the four specific MLP networks have the same definition of inputs and outputs. The inputs to the MLPs (denoted as x in Figure 3.25) include frequency as well as filter geometrical parameters such as septa lengths and spacing, and waveguide width. The output of the MLPs (denoted as y) is $|S_{11}|$. The general MLP was trained with all 19,500 training samples (S-parameter at 300 frequency points per-filter geometry for 65 filter geometries). Each of the four specific MLP networks was trained with a subset of the 19,500 samples, according to the SOM classification. These specific MLP networks can be trained to provide more accurate predictions of the S-parameter responses of the filters in the corresponding cluster. This is because within the cluster, the responses of different filters are similar to each other and the modeling problem is relatively easier than otherwise. After training is finished, the overall model can be used in this way. Given a new filter geometry (not

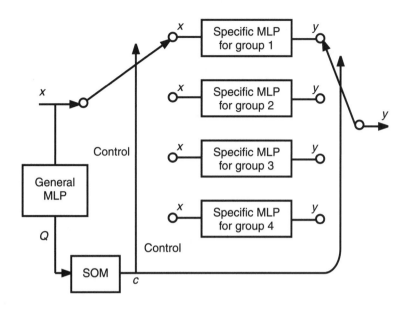

Figure 3.25 The overall neural network model using SOM to cluster the filter responses (after [27]).

seen during training), the general MLP first provides an estimation of the frequency response, which is fed to the SOM. The SOM identifies which cluster or group the filter belongs to. Finally, a specific MLP is selected to provide an accurate filter response. This overall process is illustrated in Figure 3.25.

3.9 Recurrent Neural Networks

In this section, we describe a new type of neural network structure that allows time-domain behaviors of a dynamic system to be modeled. The outputs of a dynamic system depend not only on the present inputs, but also on the history of the system states and inputs. A recurrent neural network structure is needed to model such behaviors [31–34].

A recurrent neural network structure with feedback of delayed neural network outputs is shown in Figure 3.26. To represent time-domain behavior, the time parameter t is introduced such that the inputs and the outputs of the neural network are functions of time as well. The history of the neural network outputs is represented by $y(t - \tau)$, $y(t - 2\tau)$, $y(t - 3\tau)$, and so forth, and the history of the external inputs is represented by $x(t - \tau)$, $x(t - 2\tau)$,

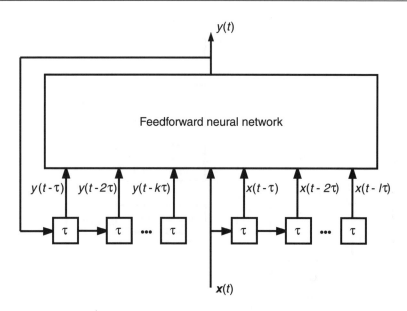

Figure 3.26 A recurrent neural network with feedback of delayed neural network output.

$x(t - 3\tau)$, and so forth. The dynamic system can be represented by the time-domain equation,

$$y(t) = f(y(t - \tau), y(t - 2\tau), \ldots, y(t - k\tau), \qquad (3.54)$$
$$x(t), x(t - \tau), \ldots, x(t - l\tau))$$

where k and l are the maximum number of delay steps for y and x, respectively.

The combined history of the inputs and outputs of the system forms an intermediate vector of inputs to be presented to the neural network module that could be any of the standard feedforward neural network structures (e.g., MLP, RBF, wavelet networks). An example of a three-layer MLP module for a recurrent neural network with two delays for both inputs and outputs is shown in Figure 3.27. The feedforward network together with the delay and feedback mechanisms results in a recurrent neural network structure. The recurrent neural network structure suits such time-domain modeling tasks as dynamic system control [31–34], and FDTD solutions in EM modeling [35].

A special type of recurrent neural network structure is the Hopfield network [36, 37]. The overall structure of the network starts from a single layer (as described in arbitrary structures). Let the total number of neurons be N. Each neuron, say the ith neuron, can accept stimulus from an external input x_i, outputs from other neurons, $y_j, j = 1, 2, \ldots, N, j \neq i$, and output of the neuron itself, that is, y_i. The output of each neuron is an external output

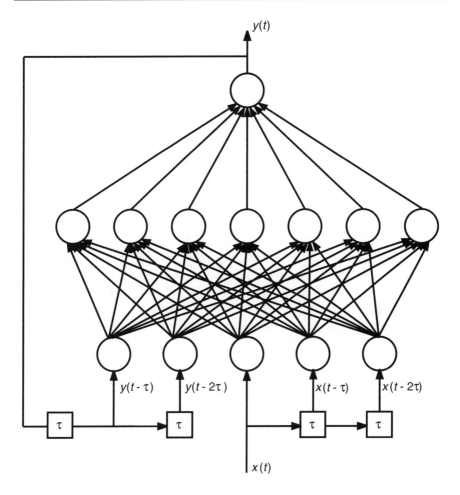

Figure 3.27 A recurrent neural network using MLP module and two delays for output feedback.

of the neural network. The input to the activation function in the ith neuron is given by

$$\gamma_i(t) = \sum_{j=1}^{N} w_{ij} y_j(t) + x_i(t) \qquad (3.55)$$

and the output of the ith neuron at time t is given by

$$y_i(t) = \sigma(\gamma_i(t-1)) \qquad (3.56)$$

for the discrete case, and

$$y_i(t) = \sigma(\lambda u_i) \tag{3.57}$$

for the continuous case. Here, λ is a constant and u_i is the state variable of the ith neuron governed by the simple internal dynamics of the neuron. The dynamics of the Hopfield network are governed by an energy function (i.e., Lyapunov function), such that as the neural network dynamically changes with respect to time t, the energy function will be minimized.

3.10 Summary

This chapter introduced a variety of neural network structures potentially important for RF and microwave applications. For the first time, neural network problems are described using a terminology and approach more oriented toward the RF and microwave engineer's perspective. The neural network structures covered in this chapter include MLP, RBF, wavelet networks, arbitrary structures, SOM, and recurrent networks. Each of these structures has found its way into RF and microwave applications, as reported recently by the microwave researchers. With the continuing activities in the microwave-oriented neural network area, we expect increased exploitation of various neural network structures for improving model accuracy and reducing model development cost.

In order to choose a neural network structure for a given application, one must first identify whether the problem involves modeling a continuous parametric relationship of device/circuits, or whether it involves classification/decomposition of the device/circuit behavior. In the latter case, SOM or clustering algorithms may be used. In the case of continuous behaviors, time-domain dynamic responses such as FDTD solutions require a recurrent neural model that is currently being researched. Nondynamic modeling problems, on the other hand, (or problems converted from dynamic to nondynamic using methods like harmonic balance) may be solved by feedforward neural networks such as MLP, RFB, wavelet, and arbitrary structures.

The most popular choice is the MLP, since the structure and its training are well-established, and the model has good generalization capability. RBF and wavelet networks can be used when the problem behavior contains highly nonlinear phenomena or sharp variations. In such cases, the localized nature of the RBF and wavelet neurons makes it easier to train and achieve respectable model accuracy. However, sufficient training data is needed. For CAD tool developers, an arbitrary neural network framework could be very attractive since it represents all types of feedforward neural network structures and facilitates structural optimization during training. One of the most important research directions in the area of microwave-oriented neural network structures is the combining of RF/microwave empirical information and existing equivalent circuit models with neural networks. This has led to an advanced class of structures known as knowledge-based artificial neural networks, described in Chapter 9.

References

[1] Zhang, Q. J., F. Wang, and V. K. Devabhaktuni, "Neural Network Structures for RF and Microwave Applications," *IEEE AP-S Antennas and Propagation Int. Symp.*, Orlando, FL, July 1999, pp. 2576–2579.

[2] Wang, F., et al., "Neural Network Structures and Training Algorithms for RF and Microwave Applications," *Int. Journal of RF and Microwave CAE*, Special Issue on Applications of ANN to RF and Microwave Design, Vol. 9, 1999, pp. 216–240.

[3] Scarselli, F., and A. C. Tsoi, "Universal Approximation using Feedforward Neural Networks: A Survey of Some Existing Methods, and Some New Results," *Neural Networks*, Vol. 11, 1998, pp. 15–37.

[4] Cybenko, G., "Approximation by Superpositions of a Sigmoidal Function," *Math. Control Signals Systems*, Vol. 2, 1989, pp. 303–314.

[5] Hornik, K., M. Stinchcombe, and H. White, "Multilayer Feedforward Networks are Universal Approximators," *Neural Networks*, Vol. 2, 1989, pp. 359–366.

[6] Devabhaktuni, V. K., et al., "Robust Training of Microwave Neural Models," *IEEE MTT-S Int. Microwave Symposium Digest*, Anaheim, CA, June 1999, pp. 145–148.

[7] Kwok, T. Y., and D. Y. Yeung, "Constructive Algorithms for Structure Learning in Feedforward Neural Networks for Regression Problems," *IEEE Trans. Neural Networks*, Vol. 8, 1997, pp. 630–645.

[8] Reed, R., "Pruning Algorithms—A Survey," *IEEE Trans. Neural Networks*, Vol. 4, Sept. 1993, pp. 740–747.

[9] Krzyzak, A., and T. Linder, "Radial Basis Function Networks and Complexity Regularization in Function Learning," *IEEE Trans. Neural Networks*, Vol. 9, 1998, pp. 247–256.

[10] de Villiers, J., and E. Barnard, "Backpropagation Neural Nets with One and Two Hidden Layers," *IEEE Trans. Neural Networks*, Vol. 4, 1992, pp. 136–141.

[11] Tamura, S., and M. Tateishi, "Capabilities of a Four-Layered Feedforward Neural Network: Four Layer Versus Three," *IEEE Trans. Neural Networks*, Vol. 8, 1997, pp. 251–255.

[12] Doering, A., M. Galicki, and H. Witte, "Structure Optimization of Neural Networks with the A-Algorithm," *IEEE Trans. Neural Networks*, Vol. 8, 1997, pp. 1434–1445.

[13] Yao, X., and Y. Liu, "A New Evolution System for Evolving Artificial Neural Networks," *IEEE Trans. Neural Networks*, Vol. 8, 1997, pp. 694–713.

[14] Rumelhart, D. E., G. E. Hinton, and R. J. Williams, "Learning Internal Representations by Error Propagation," in *Parallel Distributed Processing*, Vol. 1, D.E. Rumelhart and J. L. McClelland, Editors, Cambridge, MA: MIT Press, 1986, pp. 318–362.

[15] Garcia, J. A., et al., "Modeling MESFET's and HEMT's Intermodulation Distortion Behavior using a Generalized Radial Basis Function Network," *Int. Journal of RF and*

Microwave CAE, Special Issue on Applications of ANN to RF and Microwave Design, Vol. 9, 1999, pp. 261–276.

[16] Park, J., and I. W. Sandberg, "Universal Approximation Using Radial-Basis–Function Networks," *Neural Computation*, Vol. 3, 1991, pp. 246–257.

[17] Haykin, S., *Neural Networks: A Comprehensive Foundation*, NY: IEEE Press, 1994.

[18] Moody, J., and C. J. Darken, "Fast Learning in Networks of Locally-Tuned Processing Units," *Neural Computation*, Vol. 1, 1989, pp. 281–294.

[19] *OSA90 Version 3.0*, Optimization Systems Associates Inc., P.O. Box 8083, Dundas, Ontario, Canada L9H 5E7, now HP EESof (Agilent Technologies), 1400 Fountaingrove Parkway, Santa Rosa, CA 95403.

[20] Zhang, Q. H., "Using Wavelet Network in Nonparametric Estimation," *IEEE Trans. Neural Networks*, Vol. 8, 1997, pp. 227–236.

[21] Zhang, Q. H., and A. Benvensite, "Wavelet Networks," *IEEE Trans. Neural Networks*, Vol. 3, 1992, pp. 889–898.

[22] Pati, Y. C., and P. S. Krishnaprasad, "Analysis and Synthesis of Fastforward Neural Networks using Discrete Affine Wavelet Transformations," *IEEE Trans. Neural Networks*, Vol. 4, 1993, pp. 73–85.

[23] Harkouss, Y., et al., "Modeling Microwave Devices and Circuits for Telecommunications System Design," *Proc. IEEE Intl. Conf. Neural Networks*, Anchorage, Alaska, May 1998, pp. 128–133.

[24] Bila, S., et al., "Accurate Wavelet Neural Network Based Model for Electromagnetic Optimization of Microwaves Circuits," *Int. Journal of RF and Microwave CAE*, Special Issue on Applications of ANN to RF and Microwave Design, Vol. 9, 1999, pp. 297–306.

[25] Harkouss, Y., et al., "Use of Artificial Neural Networks in the Nonlinear Microwave Devices and Circuits Modeling: An Application to Telecommunications System Design," *Int. Journal of RF and Microwave CAE*, Special Issue on Applications of ANN to RF and Microwave Design, Guest Editors: Q. J. Zhang and G. L. Creech, Vol. 9, 1999, pp. 198–215.

[26] Fahlman, S. E., and C. Lebiere, "The Cascade-Correlation Learning Architecture," in *Advances in Neural Information Processing Systems 2*, D. S. Touretzky, ed., San Mateo, CA: Morgan Kaufmann, 1990, pp. 524–532.

[27] Burrascano, P., S. Fiori, and M. Mongiardo, "A Review of Artificial Neural Networks Applications in Microwave CAD," *Int. Journal of RF and Microwave CAE*, Special Issue on Applications of ANN to RF and Microwave Design, Vol. 9, April 1999, pp. 158–174.

[28] Kohonen, T., "Self Organized Formulation of Topologically Correct Feature Maps," *Biological Cybermatics*, Vol. 43, 1982, pp. 59–69.

[29] Burrascano, P., et al., "A Neural Network Model for CAD and Optimization of Microwave Filters," *IEEE MTT-S Int. Microwave Symposium Digest*, Baltimore, MD, June 1998, pp. 13–16.

[30] Wang, F., and Q. J. Zhang, "An Improved K-Means Clustering Algorithm and Application to Combined Multi-Codebook/MLP Neural Network Speech Recognization," *Proc. Canadian Conf. Electrical and Computer Engineering*, Montreal, Canada, September 1995, pp. 999–1002.

[31] Aweya, J., Q. J. Zhang, and D. Montuno, "A Direct Adaptive Neural Controller for Flow Control in Computer Networks," *IEEE Int. Conf. Neural Networks*, Anchorage, Alaska, May 1998, pp. 140–145.

[32] Aweya, J., Q. J. Zhang, and D. Montuno, "Modelling and Control of Dynamic Queues in Computer Networks using Neural Networks," *IASTED Int. Conf. Intelligent Syst. Control*, Halifax, Canada, June 1998, pp. 144–151.

[33] Aweya, J., Q. J. Zhang, and D. Montuno, "Neural Sensitivity Methods for the Optimization of Queueing Systems," *1998 World MultiConference on Systemics, Cybernetics and Infomatics*, Orlando, Florida, July 1998 (invited), pp. 638–645.

[34] Tsoukalas, L. H., and R. E. Uhrig, *Fuzzy and Neural Approaches in Engineering*, NY: Wiley-Interscience, 1997.

[35] Wu, C., M. Nguyen, and J. Litva, "On Incorporating Finite Impulse Response Neural Network with Finite Difference Time Domain Method for Simulating Electromagnetic Problems," *IEEE AP-S Antennas and Progpagation International Symp.*, 1996, pp. 1678–1681.

[36] Hopfield, J., and D. Tank, "Neural Computation of Decisions in Optimization Problems," *Biological Cybernetics*, Vol. 52, 1985, pp. 141-152.

[37] Freeman, J. A., and D. M. Skapura, *Neural Networks: Algorithms, Applications and Programming Techniques*, Reading, MA: Addison-Wesley, 1992.

4

Training of Neural Networks

A neural network in the beginning stages will not represent any device/circuit behavior unless it is trained with corresponding device/circuit data. In this chapter, we present a systematic description of important issues in microwave-oriented neural model development, including data generation, data scaling, training, validation, and testing. Training algorithms that are relevant to RF/microwave neural modeling are described, including backpropagation, conjugate gradient, Quasi-Newton, Levenberg-Marquardt, Genetic Algorithms, and Simulated Annealing. Comparisons between these training methods are discussed.

4.1 Microwave Neural Modeling: Problem Statement

Let x represent an n-vector containing physical/process/bias parameters of a microwave device/circuit (e.g., gate length and gate width of an FET, width and spacing of a transmission line). Let y represent an m-vector containing the electrical behaviors of the device/circuit under consideration (e.g., drain current of an FET, or mutual inductance of a transmission line). The theoretical physics/EM relationship between x and y can be represented as

$$y = f(x) \qquad (4.1)$$

The relation f can be highly nonlinear and multidimensional. In practice, the theoretical model for this relationship may not yet be available, (for example, for a new type of semiconductor device). The existing problem theory may be too complicated to implement, or the theoretical model may be computationally intensive for online microwave design and repetitive optimization (for example, a 3-D full-wave EM analysis inside a Monte Carlo statistical design loop).

To overcome these problems, we aim to develop a fast and accurate neural model for the relation f by teaching/training a neural network with a set of measured/simulated data called the training data. The training data is denoted by input-output sample pairs, $\{(x_k, d_k), k \in T_r\}$, where d_k represents the measured/simulated output y for the input x_k, and T_r represents the index set of training data. The measured/simulated outputs and inputs are related by

$$d_k = f(x_k) \qquad (4.2)$$

Let the neural network model be defined as

$$y = y(x, w) \qquad (4.3)$$

where w represents N_w trainable (adjustable) parameters inside the neural network, also known as the weight vector. For training purposes, we define an error function $E(w)$ as

$$E(w) = \sum_{k \in T_r} E_k(w) \qquad (4.4)$$

where $E_k(w)$ is the error between neural network prediction and the kth training sample given by

$$E_k(w) = \left[\frac{1}{p}\sum_{j=1}^{m} |y_j(x_k, w) - d_{jk}|^p\right] \qquad (4.5)$$

where p represents least pth optimization, d_{jk} is the jth element of d_k, and $y_j(x_k, w)$ is the jth output of the neural network for input sample x_k. The primary objective of neural network training is to find w such that $E(w)$ is minimized. Training of RF/microwave-oriented neural networks involves typical judgements with regard to data generation/processing, choice of error criteria, and selection of training algorithms [1, 2]. The resulting models are expected to capture continuous, nonlinear, multidimensional relationships, as opposed to the neural network models for discrete (binary) pattern-classification and signal-processing applications.

4.2 Key Issues in Neural Model Development

The preliminary step toward developing a neural model is the identification of the inputs and outputs of the problem to be modeled, so that the corresponding training data can be generated. In general, the model outputs are determined based on the purpose of the model—for example, real and imaginary parts of S-parameters for linear circuit components, currents and charges for large-signal component models (harmonic balance), and cross-sectional RLCG parameters for high-speed VLSI interconnect analysis. Other factors influencing the choice of outputs are ease of data generation, ease of incorporation of the neural model into circuit simulators, and so forth. The neural model inputs are the device/circuit parameters that affect the outputs—for example, circuit design variables, physical/process parameters of the components, and independent parameters such as frequency in passive components and bias for active components. As an illustration, to develop a small-signal FET model, we choose $x = [V_{GS}\ V_{DS}\ freq]^T$ and $y = [RS_{11}\ IS_{11}\ RS_{12}\ \ldots\ IS_{22}]^T$. Here, V_{GS} and D_{DS}, are gate-to-source and drain-to-source voltages, and RS_{ij} and IS_{ij} are real and imaginary parts of S-parameters.

4.2.1 Data Generation

The first step in neural model development is the generation and collection of data for training and testing the neural models. Data generation generally involves using a data generator to obtain the output d_k for each input sample x_k. The total number of samples, P, to be generated is chosen so that the developed neural model best represents the original problem. There are two types of data generators for microwave applications, namely measurement and simulation. The choice of data generator depends on both the application and the availability of the data generator.

Data Generation by Measurement

For a given input x_k, measure the output, d_k, $k = 1, 2, \ldots, P$, (e.g., given bias and frequency, measure S-parameters of an FET using an HP-network analyzer).

Data generation by measurement has a number of advantages. Measurement can be used even if the problem theory-equation does not exist, or if the theory is too complicated to implement or is CPU intensive. Measured data represents the entire problem behavior including secondary effects (e.g., parasitic effects, fringe effects). And the process of measurement does not involve any theoretical assumptions. But data generation by measurement has its disadvantages as well. Some inputs may be difficult to change (e.g., gate length of an FET), and some outputs may be difficult to measure (e.g., drain charge in FET). Measurement equipment may have errors/tolerances.

Data Generation by Simulation

For a given input x_k, simulate the output, d_k, $k = 1, 2, \ldots, P$, (e.g., given bias and frequency, simulate S-parameters of an FET using HP-ADS/HP-Libra).

Data generation by simulation also has a number of advantages. Any input parameter can be changed easily, because it is only a numerical change and does not involve any physical/manual change. Any response can be computed as long as its evaluation is supported by the simulator. Errors introduced into simulation data due to floating-point truncation/round-off operations are much smaller compared to the errors that can be present in measurement data due to equipment tolerances. But data generation by simulation has disadvantages as well. Theory must exist for the implementation of a simulator. For larger problems, longer computation time may be needed. Simulators are limited by the assumptions of the existing problem theory (or the assumptions made during implementation of the simulator), and as such d_k may not represent the entire problem behavior. A comparison between data generation by measurement and data generation by simulation is shown in Table 4.1.

4.2.2 Range and Distribution of Samples in Model Input Parameter Space

Because a set of samples x_k, $k = 1, 2, \ldots, P$, are needed in order to generate data, we need to define the range and distribution of these samples. Suppose the range of input parameter space over which the neural model would be used during microwave design is $[x_{min}, x_{max}]$. Validation data and test data should be generated in the range $[x_{min}, x_{max}]$. Training data could be generated in the same range as well. We suggest that, wherever feasible, the training data be sampled slightly beyond the model utilization range, for example $[x_{min} - \Delta, x_{max} + \Delta]$. This is to ensure good performance of the neural model at the boundaries of input parameter space.

Once the range of input parameters is fixed, the next step is the selection of a sampling strategy. Suggested sample distributions include: uniform grid distribution, nonuniform grid distribution, star distribution, central-composite distribution, and random distribution.

Uniform Grid Distribution

In uniform grid distribution, each input parameter is sampled at equal intervals. An example of this kind of sampling for a three-input parameter case is shown in Figure 4.1. Uniform grid distribution is the simplest sampling strategy. Suppose the number of grids along each input dimension x_i is n_i. The total

Table 4.1
A Comparison of Neural-Network–Based Microwave Model Development, Using Data from Two Types of Data Generators, Namely, the Measurement and the Simulation

Basis of Comparison	Neural Model Development Using Measurement Data	Neural Model Development Using Simulation Data
Availability of Problem Theory-Equations	Model can be developed even if the theory-equations are not known, or if they're difficult to implement in CAD.	Model can be developed only for the problems that have theory that is implemented in a simulator.
Assumptions	No assumptions are involved and the model could include all the effects, e.g., 3D-fullwave effects, fringing effects etc.	Often involves assumptions and the model will be limited by the assumptions made by the simulator, e.g., 2.5D EM.
Input Parameter Sweep	Data generation could be either expensive or infeasible, if a geometrical parameter needs to be sampled or changed.	Relatively easier to sweep any parameter in the simulator, because the changes are numerical and not physical/manual.
Sources of Small and Large/Gross Errors	Equipment limitations and tolerances.	Accuracy limitations and nonconvergence of simulations.
Feasibility of Getting Desired Output	Development of models is possible for measurable responses only. For example, drain charge of an FET may not be easy to measure.	Any response can be modeled as long as it can be computed by the simulator.

number of samples is $P = \prod_{i=1}^{n} n_i$. Let D be the index set of all samples given by $D = \{1, 2, \ldots, P\}$. The kth sample is obtained as

$$x_k = x_{min} + \begin{bmatrix} (k_1 - 1)\Delta x_1 \\ \vdots \\ (k_n - 1)\Delta x_n \end{bmatrix}, \quad 1 \leq k_i \leq n_i \qquad (4.6)$$

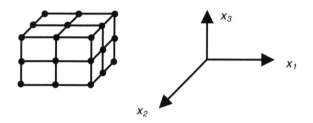

Figure 4.1 Uniform grid distribution for a three-input parameter case.

where $\Delta x_i = \dfrac{x_{\max,i} - x_{\min,i}}{n_i - 1}$. Here $x_{\max,i}$ and $x_{\min,i}$ are the maximum and minimum values of the ith element of x. Each combination of $k_1, k_2, \ldots, k_n$ defines one sample.

For an FET modeling example, suppose the neural model input and output parameters are $x = [V_{GS}\ V_{DS}\ freq]^T$, $y = [RS_{11}\ IS_{11}\ \ldots\ IS_{22}]^T$, and the range in which the model is expected to be used is:

$$\begin{bmatrix} -5 \\ 0 \\ 1 \end{bmatrix} \leq x \leq \begin{bmatrix} 0 \\ 10 \\ 20 \end{bmatrix} \quad (4.7)$$

The range of x for training data can be chosen as

$$\begin{bmatrix} -5 - 0.5 \\ 0 \\ 1 - 0.5 \end{bmatrix} \leq x \leq \begin{bmatrix} 0 \\ 10 + 1 \\ 20 + 2 \end{bmatrix} \quad (4.8)$$

where the addition/subtraction of small values along each dimension reflects the sampling of training data slightly beyond the model utilization range. If we assume $n_1 = 6$, $n_2 = 12$, and $n_3 = 20$, then the input samples are $x_1 = [-5.5\ 0\ 0.5]^T$, $x_2 = [-4.4\ 0\ 0.5]^T$, and so forth. At each of these input values x_k, measurements/simulations are performed to obtain the S-parameters α_k. In this particular example, the total number of samples is $P = 6*12*20 = 1440$. Uniform distribution generally leads to a large number of samples.

Nonuniform Grid Distribution

In nonuniform grid distribution, each input parameter can be sampled at unequal intervals, as shown in Figure 4.2. Suppose the problem behavior to be modeled is highly nonlinear in certain subregions of the input space. In such situations, nonuniform grid distribution is used and dense

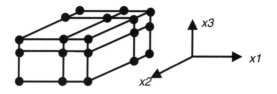

Figure 4.2 Nonuniform grid distribution for a three-input parameter case.

samples are chosen from the corresponding subregions. Suppose $x_i^{(j)}$ is the jth grid point along the ith dimension—then the samples are given by $x_k = [x_1^{(k_1)}, x_2^{(k_2)}, \ldots, x_n^{(k_n)}]^T$ for different combinations of $[k_1 \, k_2 \ldots k_n]^T$, $1 \leq k_i \leq n_i$. The total number of samples is $P = \prod_{i=1}^{n} n_i$.

As an example, nonuniform grid distribution could be used for modeling the I-V curves of a FET, as shown in Figure 4.3. The I-V curves exhibit larger variations when V_{DS} is small (e.g., between 0 and 1 V), and exhibit linear behavior when V_{DS} is large. A nonuniform grid distribution with more samples in the small V_{DS} region and sparse samples in the large V_{DS} region, for example, $V_{DS} = \{0.1, 0.2, 0.5, 1, 2, 4, 6\}$, is effective. The total number of data samples can be reduced in this way and sufficient nonlinear information of the problem is still retained.

Design of Experiments (DOE) and Central Composite Distribution

Central composite distribution is a sampling strategy developed from the concept of DOE [3]. The DOE methodology can be used to systematically study

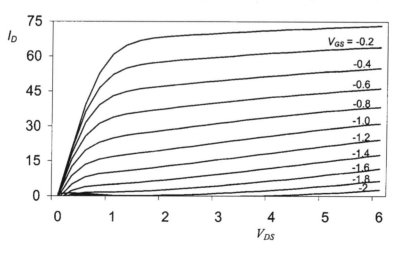

Figure 4.3 Typical IV curves of an FET where nonuniform sampling is effective.

the complex input-output relationships of a component/process. In building a model, one would like to perform as few EM simulations or measurements as possible in order to achieve the desired accuracy. This implies that one could start with a low-order experimental design and sequentially build up higher-order designs by adding simulation or measurement points. The DOE methodology also provides a reasonable distribution of simulation points by spreading the information throughout the region of the input variables.

There are two major design patterns or sample distributions, namely, the 2^n factorial experimental design and the central composite experimental design. The 2^n factorial experimental design requires 2^n corner points, where n is the number of model input parameters. In addition to these corner points, the central composite design requires the center point and $2n$ axial points. If, with these samples, the input-output relationship has not been sufficiently captured by the model, additional samples are added to include the higher-order nonlinearities. Figure 4.4 shows the design patterns in the case of two design variables. In a composite distribution, a sample may be obtained by perturbing any number of the elements of x at a time.

Star Distribution

The star distribution [4] shown in Figure 4.5 has very few sample points. A middle point of the input space x_{mid} is determined first. Samples are then obtained by perturbing one element of x_{mid} at a time, either toward its maximum value or toward its minimum value. The middle point can be calculated as

$$x_{mid} = \frac{1}{2}(x_{max} + x_{min}) \quad (4.9)$$

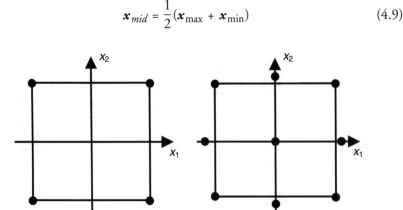

Figure 4.4 A two-input parameter example: (a) 2^n factorial experimental design and (b) central composite experimental design.

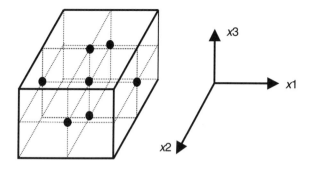

Figure 4.5 Star distribution for a three-input parameter case.

In general, the kth sample is obtained by

$$x_k = x_{mid} + \Delta x \quad (4.10)$$

$$= x_{mid} + \begin{bmatrix} 0 \\ \vdots \\ \pm\frac{1}{2}(x_{max,k} - x_{min,k}) \\ \vdots \\ 0 \end{bmatrix}$$

where the kth element of Δx has a nonzero value.

The total number of samples for star distribution including the middle point is $P = 1 + 2n$. Star distribution is used in situations where data generation is very expensive, and the model behavior is assumed to vary smoothly within $[x_{min}, x_{max}]$. Data generation could turn out to be expensive, due to long computation times in the case of simulations, or due to difficulties in changing device geometry in the case of measurements. Star distribution can also be used for the training of knowledge-based neural networks that embed problem knowledge, and can thus afford to be trained with fewer samples.

Random Distribution

In random distribution, each input sample x_k is a vector of random variables between x_{min} and x_{max}. Let x_i be a random number in the range $[x_{min,i}, x_{max,i}]$. A random number generator (e.g., x = rand() in C language) is used to create different combinations of x_i, $i = 1, 2, \ldots, n$. Such combinations can, in turn, produce n-dimensional random vectors $x_k \in [x_{min}, x_{max}]$. A simple form of the distribution function used in random distribution is uniform and independent, where two random variables, x_i and x_j, $i \neq j$, are independent. The total number of samples P is dictated by user applications. Random distribution is used when input parameter space is of high dimension (i.e., n is very large).

Discussion

The sample distribution and the number of samples should be selected in such a way so that the original problem $y = f(x)$ in the input space $[x_{\min}, x_{\max}]$ is sufficiently represented by the data samples. The total number of required data samples depends on the nonlinearity of the problem (that is, highly nonlinear problems require more data). The exact relationship between the size of the training data and the modeling problem is an interesting area for research [5]. Conceptually, in the case of grid distribution, the theoretical factor behind the number of samples is Shannon's sampling theorem, which determines the number of samples per input dimension. For random distribution, the total number of samples P is chosen such that the statistical confidence for the neural network to accurately represent the original data matches the user-dictated confidence level (e.g., 90%).

Suppose that the number of samples per dimension in a grid distribution is $n_1 = n_2 = \ldots = n_n = k$. The total number of data samples is $P = k^n$, that is, P increases with n as $exp(n)$, resulting in a huge amount of data. In a random distribution, P increases with n but not as fast as $exp(n)$. As such, random distribution is preferred when n is too large, in order to reduce the cost of data generation.

4.2.3 Data Splitting

Neural model development typically requires three sets of data, namely the training data, validation data, and test data. We define T_r, V, and T_e as the index sets of training data, validation data, and test data, respectively. Training data is used to guide the training process—that is, to update the neural network weight parameters during training. Validation data is used to monitor the quality of the neural model during training, and the training can be terminated when the desired quality is reached. Test data is used to examine the final quality of the developed model, including the generalization capability. We define the following notation.

P: Total number of data samples generated
D: Index set for the entire data, $D = \{1, 2, \ldots, P\}$
T_r: Index set of training data

V: Index set of validation data
T_e: Index set of test data

Ideally, each data set (T_r, V, and T_e) should be an adequate representation of the original problem $y = f(x)$ in the input parameter range of interest, and these three sets should not overlap. In practice, we split D into T_r, V, and T_e, depending on the quantity of the data available. There are three possible cases, as described below.

Extreme Case (Sufficient Data): When data is quite sufficient, split D into three nonoverlapping sets T_r, V, and T_e—that is, $D = T_r \cup V \cup T_e$, $T_r \cap V = \phi$, $T_e \cap V = \phi$, $T_r \cap T_e = \phi$. Standard neural network theory emphasizes that T_r, V, and T_e, must be different and uncorrelated.

Extreme Case (Limited Data): When data is very limited, duplicate D such that $T_r = V = T_e = D$. This is not considered to be a good practice for neural network training. It is an acceptable way of using the data if we know in advance that the problem behavior is smooth between the sample points and if the minimal number of hidden neurons is used (in order to avoid overlearning). Overlearning is explained at length in Section 4.2.6.

Intermediate Case: When data is neither sufficient nor unlimited, split D into two sets and use them in the following ways:

- Option 1: Use one set for training and validation ($T_r = V$) and the other for testing.
- Option 2: Use one set for training (T_r) and the other for validation and testing ($V = T_e$).
  It may be noted that in both cases the training set and the test set are different. Splitting of D, for example, can be performed as:
  - 50%–50% split;
  - 80%–20% split between T_r and T_e. Here, the neural network test error level may be low because of training with more data. However, since the neural model is tested with fewer data, the error level cannot be trusted;
  - 20%–80% split between T_r and T_e. Here, the neural network test error level may be high because of training with less data. However, since the neural model is tested with a sufficient amount of data, the error level can be trusted.

Once the data is divided into three sets, we can proceed to define the corresponding neural network errors. Using the pth norm, neural network training error is defined as

$$E_{T_r}(\boldsymbol{w}) = \frac{1}{p} \sum_{k \in T_r} \sum_{j=1}^{m} |y_j(\boldsymbol{x}_k, \boldsymbol{w}) - d_{jk}|^p \qquad (4.11)$$

The validation error $\boldsymbol{E}_V(\boldsymbol{w})$ and the test error $E_{T_e}(\boldsymbol{w})$ can be similarly defined. The objective of the neural network training can now be restated: The purpose of training is to find $\boldsymbol{w}$ to minimize $\boldsymbol{E}_V(\boldsymbol{w})$, but the update of $\boldsymbol{w}$ is based on the information of $E_{T_r}(\boldsymbol{w})$ and $\dfrac{\partial E_{T_r}}{\partial \boldsymbol{w}}$. For example, $\boldsymbol{w}$ is updated as $\boldsymbol{w} = \boldsymbol{w} - \eta \dfrac{\partial E_{T_r}}{\partial \boldsymbol{w}}$. After training, the final quality of the neural model is measured by the test error.

4.2.4 Data Scaling

Contrary to the typical '0' and '1' (binary) situations in pattern recognition applications, the orders of magnitude of input/output parameter values in microwave applications can be very different from one another. As such, scaling of training data is desirable for efficient neural network training. Scaling of data samples can be performed on the input parameter values, output parameter values, or both. Input/output scaling and descaling for neural model training and in the usage of developed models is shown in Figure 4.6.

Various scaling schemes described here include linear scaling, log arithmetic scaling, and two-sided log arithmetic scaling. Let x, x_{min}, and x_{max} represent a generic element in the vectors $\boldsymbol{x}$, $\boldsymbol{x}_{min}$, and $\boldsymbol{x}_{max}$ of original data, respectively. Let $\bar{x}$, $\bar{x}_{min}$, and $\bar{x}_{max}$ represent a generic element in the vectors $\bar{\boldsymbol{x}}$, $\bar{\boldsymbol{x}}_{min}$, and $\bar{\boldsymbol{x}}_{max}$ of scaled data, respectively, where $[\bar{x}_{min}, \bar{x}_{max}]$ represent the input parameter range after scaling.

Linear Scaling

The linear scaling shown in Figure 4.7 is given by

$$\bar{x} = \bar{x}_{min} + \frac{x - x_{min}}{x_{max} - x_{min}} (\bar{x}_{max} - \bar{x}_{min}) \qquad (4.12)$$

The corresponding descaling is given by

$$x = x_{min} + \frac{\bar{x} - \bar{x}_{min}}{\bar{x}_{max} - \bar{x}_{min}} (x_{max} - x_{min}) \qquad (4.13)$$

Training of Neural Networks

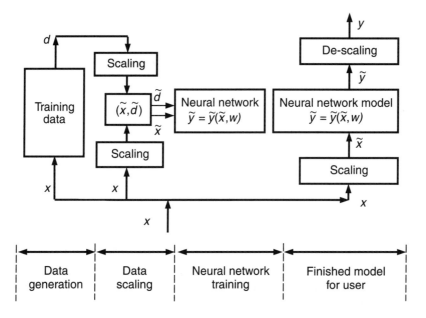

Figure 4.6 Input/output scaling and descaling for neural model training and in the usage of developed models.

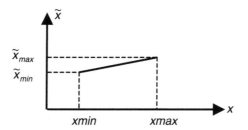

Figure 4.7 Linear scaling.

Linear scaling of inputs improves the condition of the trainable parameters (weights), and the balance between different x's. Linear scaling of outputs can balance different y's whose magnitudes are very different.

Log Arithmetic Scaling

A simple form of log arithmetic scaling shown in Figure 4.8 is given by

$$\tilde{x} = ln(x - x_{min}) \tag{4.14}$$

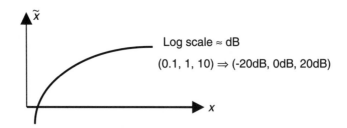

Figure 4.8 Log arithmetic scaling.

The corresponding descaling is given by

$$x = x_{min} + e^{\tilde{x}} \qquad (4.15)$$

Applying log arithmetic scaling to outputs with large variations provides a balance between large and small magnitudes of the same output in different regions of the model.

Two-Sided Log Arithmetic Scaling

The two-sided log arithmetic scaling shown in Figure 4.9 is given by

$$\tilde{x} = \pm \frac{\ln\left(1 + \frac{\pm(x - x_{mid})}{x_{max}}\alpha\right)}{\ln(1 + \alpha)} \tilde{x}_{max}, \quad \begin{cases} + \text{ if } x > x_{mid} \\ - \text{ if } x < x_{mid} \end{cases} \qquad (4.16)$$

where α is a user-defined constant representing the usable range of log function, $x_{max} = \max|x - x_{mid}|$, and x_{mid} is the reference level of x such that

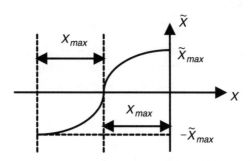

Figure 4.9 An illustration of the relationship of x and $\tilde{x}$ in a two-sided log arithmetic scaling scheme.

$$\tilde{x} = \begin{cases} > 0 & \text{if } x > x_{mid} \\ 0 & \text{if } x = x_{mid} \\ < 0 & \text{if } x < x_{mid} \end{cases} \tag{4.17}$$

Using this scaling we have,

$$\tilde{x} = \begin{cases} \tilde{x}_{max} & \text{if } x = x_{mid} + x_{max} \\ 0 & \text{if } x = x_{mid} \\ -\tilde{x}_{max} & \text{if } x = x_{mid} - x_{max} \end{cases} \tag{4.18}$$

The descaling is given by

$$x = x_{mid} \pm \frac{\left\{\exp\left[\ln(1+\alpha)\frac{\pm\tilde{x}}{\tilde{x}_{max}}\right] - 1\right\}}{\alpha} x_{max}, \quad \begin{cases} + \text{ if } \tilde{x} > 0 \\ - \text{ if } \tilde{x} < 0 \end{cases} \tag{4.19}$$

The two-sided log arithmetic scaling scheme is similar to the log arithmetic scaling scheme except that the former is used to avoid the overshadowing of mid-range values of the response by greatly increasing and greatly decreasing trends.

No Scaling

In this case, the original data and the scaled data are the same, that is, $x = \tilde{x}$. Such a scaling scheme is necessary for training knowledge-based neural networks, where the knowledge function is required to preserve the physical meaning of the input variables (e.g., Ohm's law and Faraday's law). Knowledge-based neural networks will be described in Chapter 9.

The input/output scaling can make the problem better conditioned for training, and it is more effective than blind optimization scaling in neural network weight (w) space. Assuming—in order to simplify the notation—that scaling has already been done, we use notations x and d in place of $\tilde{x}$ and $\tilde{d}$ to denote scaled model inputs and scaled model outputs in the following sections.

4.2.5 Initialization of Neural Model Weight Parameters

The neural network weight parameters need to be initialized in order to provide a good starting point for training (optimization) procedure. The random-weight method is the most widely used strategy for MLP weight initialization, in which the weights are initialized with small random values (e.g., in the range [−0.5, 0.5]). To improve convergence of training, one can use different distributions (uniform or Gaussian), different ranges, and different variances for the random number generators while randomly initializing the neural network weights [6]. Another scheme suggests that the range of the random values be inversely proportional to the square root of number of stimuli a neuron receives on average [7]. RBF and wavelet networks can be initialized by estimating the parameters of hidden neuron activation functions (i.e., centers and radii of RBF, translation and dilation parameters for wavelets) with the help of an unsupervised learning process based on training data. A good starting point for knowledge-based neural networks can be provided using physical/electrical knowledge and experience of the problem.

4.2.6 Overlearning and Underlearning

Once the neural network is initialized, the training process can start. The ability of a neural network to estimate output y accurately when presented with input x never seen during training is called generalization ability. To understand the phenomena of overlearning and underlearning, we define normalized training error as

$$\hat{E}_{T_r}(\boldsymbol{w}) = \left[\frac{1}{size(T_r) \cdot m} \sum_{k \in T_r} \sum_{j=1}^{m} \left| \frac{y_j(\boldsymbol{x}_k, \boldsymbol{w}) - d_{jk}}{d_{max,j} - d_{min,j}} \right|^p \right]^{1/p} \quad (4.20)$$

where $d_{max,j}$ and $d_{min,j}$ are the maximum and minimum values of the jth element from all vectors of $\boldsymbol{d}_k$, $k \in T_r$.

The normalized validation error $\hat{E}_V$ can be similarly defined.

Overlearning is a phenomenon in which the neural network memorizes the training data but cannot generalize well. In other words, the training error $\hat{E}_{T_r}$ is small, but the validation error $\hat{E}_V \gg \hat{E}_{T_r}$. Figure 4.10 shows a neural model that matches the training data very well but does not match the validation data. Possible reasons for overlearning include the presence of too many hidden neurons, or insufficient training data. Too many hidden neurons lead to too much freedom in the *x-y* relationship represented by a neural network. Suggested remedies for overlearning include deleting a certain number of hidden neurons, or adding more samples to the training data.

Underlearning, on the other hand, is a situation in which the neural network has difficulties even learning the training data itself (i.e., $\hat{E}_{T_r} \gg 0$). Figure 4.11 shows a neural model that cannot even match the training data. Possible reasons for underlearning include the presence of insufficient hidden

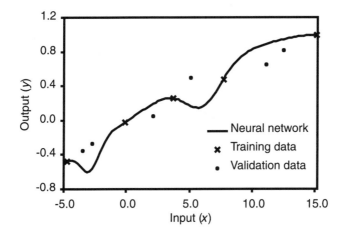

Figure 4.10 A neural network model illustrating the overlearning phenomenon.

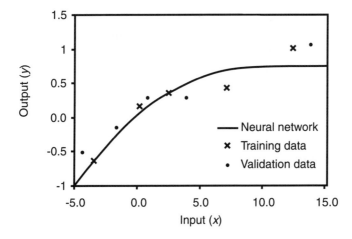

Figure 4.11 A neural network model illustrating the underlearning phenomenon.

neurons, insufficient training, or training procedure stuck in a local minimum. Suggested remedies include adding more hidden neurons, continuing training, or perturbing the current solution w, thereby escaping from the local minimum and a continuation of training.

Good learning of a neural network is observed when both $\hat{E}_{T_r}$ and $\hat{E}_V$ have small values and are close to each other. Figure 4.12 shows that with good learning, a neural model can match both training data and validation data very well.

It is not always possible to view the neural model behavior curve to determine whether overlearning or underlearning occurred, especially for high-

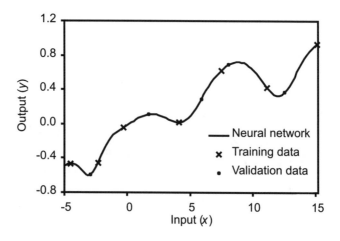

Figure 4.12 A neural network model illustrating good-learning.

dimensional problems. An easy way to observe these phenomena is to have a look at the training and validation error curves during the training process, as shown in Figure 4.13. In this figure, the term "epoch" represents the iteration count during training or a cycle where all the training samples are presented once to the neural network for updating w. $\hat{E}_V^*$ is the minimum value of $\hat{E}_V$, that is, $\hat{E}_V^* \triangleq \min_{epoch} \hat{E}_V(epoch)$. Overlearning begins at the point where the validation error starts increasing after reaching $\hat{E}_V^*$, while the training error continues to decrease. Figure 4.14 shows the effect of the number of training samples on the overlearning phenomenon. When fewer training samples are available, the overlearning occurs during the earlier stages of the training process. On the other hand, when there are enough training samples, overlearning is less likely to occur before an acceptable model accuracy is reached.

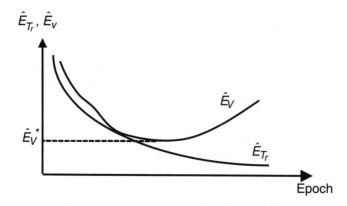

Figure 4.13 A typical learning curve.

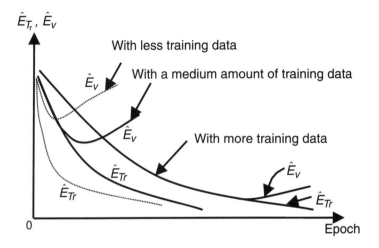

Figure 4.14 Effect of number of training samples on the training and validation errors.

Relaxed Definitions

If the entire set of data contains errors (for example, measurement errors), or if the higher-order effects in the original problem are neglected in the model, then $\hat{E}_{T_r} = 0$ is not achievable or is not needed. In this case, there is no question of perfect learning, and we aim for the best possible learning.

Best possible learning with respect to a given validation data set is achieved when $\hat{E}_{T_r} \approx \hat{E}_V \approx \hat{E}_V^*$, even though $\hat{E}_V^*$ is not small enough. Figure 4.15 illustrates such a situation.

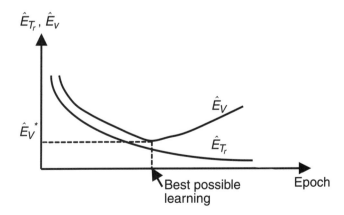

Figure 4.15 Training and validation error curves showing the point of best possible learning.

Overlearning can be detected when $\hat{E}_V \gg \hat{E}_{T_r}$ is observed throughout the training process, regardless of whether $\hat{E}_{T_r}$ is small enough or not. Figure 4.16 illustrates such a situation.

Quality of Learning Versus Size of Training Data

Here we examine the effect of the size of training data on the neural network training process. Let $\hat{E}_{T_r}^*$ be the minimum training error after training for a sufficient number of epochs and $\hat{E}_V^*$ be the minimum validation error in the history of the training process. Figure 4.17 illustrates the variation of $\hat{E}_{T_r}^*$ and $\hat{E}_V^*$ with the size of training data. It can be seen from the figure that $\hat{E}_{T_r}^* \ll \hat{E}_V^*$, when the size of training data is small. What this means is that

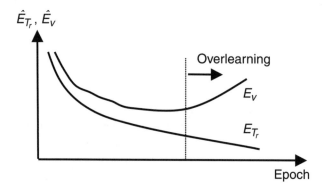

Figure 4.16 Training and validation error curves showing the phenomenon of overlearning.

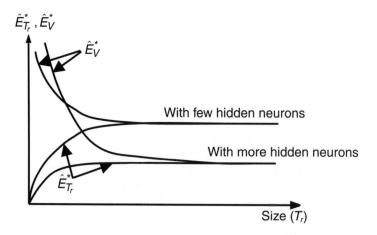

Figure 4.17 Variation of training and validation errors versus the amount of training data.

with less training data, the network is more likely to overlearn. The figure also shows that the training and validation error curves come closer to each other as the amount of training data increases. It means that with large amounts of training data, good learning is more likely to be achieved (i.e., $\hat{E}^*_{T_r} = \hat{E}^*_V$). However, beyond a certain point, a further increase in the amount of training data will not substantially improve the validation error. This shows that for a given problem and a user-desired validation error, a certain amount of training data would be sufficient, and additional data would not provide any extra information to the network [8]. However, as far as the quality of the trained neural model is concerned, the use of extra training data is always harmless (when the amount of data is already large), beneficial (when the amount of data is medium), and extremely essential (when the amount of training data is small). Reasons for avoiding large training data include the cost of data generation and/or the prolonged neural network training time.

Quality of Learning Versus Number of Hidden Neurons

Another factor affecting the neural network training is the number of hidden neurons. When the number of hidden neurons is small, the best possible validation error will be high because the neural network does not have enough freedom to follow the nonlinearities in the problem behavior of all training samples. More hidden neurons mean more freedom, and so the neural network will be able to capture all the tendencies in training data and achieve a smaller validation error. However, too many hidden neurons will lead to overlearning. The effect of the number of hidden neurons on training and validation errors is further illustrated in Figure 4.18.

Types of Training Processes

Neural network training processes can be categorized into sample-by-sample training and batch-mode training. In sample-by-sample training—also called online training—the neural network weights (w) are updated each time a training sample is presented to the network. In batch-mode training—also known as offline training—the weights are updated after all the training samples are presented. For RF/microwave applications, batch-mode training is usually more effective unless the amount of training data is too large.

The training process can also be categorized into supervised training and unsupervised training. In supervised training, both x and y data are used to train the neural network, as in MLP training. In unsupervised training, either x data or y data is used, as in SOM training.

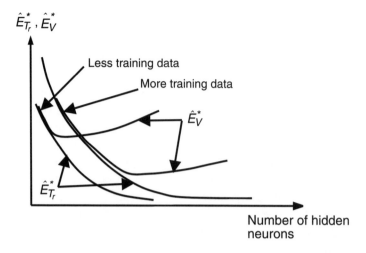

Figure 4.18 Variation of training and validation errors versus the number of hidden neurons.

4.2.7 Quality Measures for a Neural Model

The quality of a trained neural model is tested with an independent set of data, that is, the test data. We define a relative error δ_{kj} for jth output of neural model on kth data sample as

$$\delta_{kj} = \frac{y_j(x_k, w) - d_{jk}}{d_{\max,j} - d_{\min,j}}, j = 1, \ldots, m, k \in T_e \qquad (4.21)$$

A quality measure based on pth-norm measure is then defined as

$$M_p = \left[\sum_{k \in T_e} \sum_{j=1}^{m} |\delta_{kj}|^p \right]^{1/p} \qquad (4.22)$$

When $p = 1$, the average test error can be directly calculated from M_1 as

$$Average\ Test\ Error = \frac{M_1}{Size(T_e)m} = |\overline{\delta}| \qquad (4.23)$$

When $p = 2$, the pth-norm measure is the Euclidean distance between the neural model prediction and the test data. When $p = \infty$, the pth-norm measure is the maximum test error, also known as worst-case error among the entire test data and all model outputs, that is

$$\text{Worst-case Error} = M_\infty = \max_{k \in T_e} \max_{j=1}^{m} |\delta_{kj}| \qquad (4.24)$$

In (4.23), $|\overline{\delta}|$ is the mean value of absolute error $|\delta_{kj}|$ for all k and j. The standard deviation of the absolute error can be another quality criterion indicating how close the error populations are clustered around the average error. Standard deviation of the test error is given by

$$\sigma^2 = \frac{\sum_{j=1}^{m} \sum_{k \in T_e} (|\delta_{kj}| - |\overline{\delta}|)^2}{Size(T_e) \cdot m} \qquad (4.25)$$

Statistical measures such as error mean and correlation coefficient of the error δ_{kj} can also be used to indicate the quality of a neural model. The error mean $\overline{\delta}$ can be defined as

$$\text{mean}(\delta) = \frac{1}{Size(T_e) m} \sum_{j=1}^{m} \sum_{k \in T_e} \delta_{kj} = \overline{\delta} \qquad (4.26)$$

Note that $\overline{\delta}$ is not same as the average test error of (4.23). If $\overline{\delta} = 0$, the neural model is unbiased with respect to the test data, and if $\overline{\delta} \neq 0$ the neural network model is biased. Another quality measure that indicates the correlation between neural model and test data is the correlation coefficient defined as

$$\rho = \frac{\sum_{j=1}^{m} \sum_{k \in T_e} (y_j(x_k, w) - \overline{y(x, w)})(d_{jk} - \overline{d_{jk}})}{\left(\left[\sum_{j=1}^{m} \sum_{k \in T_e} (y_j(x_k, w) - \overline{y(x, w)})^2 \right] \left[\sum_{j=1}^{m} \sum_{k \in T_e} (d_{jk} - \overline{d_{jk}})^2 \right] \right)^{1/2}} \qquad (4.27)$$

where $\overline{y(x, w)}$ represents the mean of the neural network outputs given by

$$\overline{y(x, w)} = \frac{1}{Size(T_e) \cdot m} \sum_{j=1}^{m} \sum_{k \in T_e} y_j(x_k, w) \qquad (4.28)$$

and $\overline{d_{jk}}$ represents the mean of the measured/simulated outputs in the test data given by

$$\overline{d_{jk}} = \frac{1}{Size(T_e) \cdot m} \sum_{j=1}^{m} \sum_{k \in T_e} d_{jk} \qquad (4.29)$$

Several graphs can be used to show the statistical distribution of errors. A graph showing the correlation of neural model and test data can be seen in Figure 4.19. A histogram of δ (see Figure 4.20) can be used to display the quality of neural model including the mean values shown therein. The histogram of absolute error $|\delta|$ shown in Figure 4.21 can also be a useful graphical representation of the quality of a neural model.

4.3 Neural Network Training

The most important and time-consuming step in model development is neural network training. A neural network learns the microwave behavior through

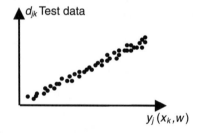

Figure 4.19 A typical scattering plot showing the correlation of neural model and test data. Different points in the figure are obtained by varying k, and j, $k = 1, 2, \ldots, Size(T_e)$, $j = 1, 2, \ldots, m$. Each point in the figure corresponds to an index pair (k, j).

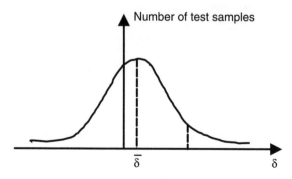

Figure 4.20 A typical histogram of δ.

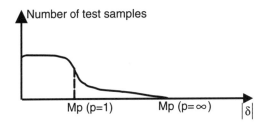

Figure 4.21 A typical histogram of the absolute error $|\delta|$.

this process. The neural network would be taught with measured/simulated samples from the training set. Conventional training of neural networks is an optimization process in the weight space using optimization-based training algorithms. The most popular training algorithm in neural network literature is back propagation (BP). However, for most of the microwave modeling problems, second-order training algorithms such as conjugate-gradient, and Quasi-Newton are more efficient. From the optimization point of view, the training process adjusts w such that the error between neural model predictions and desired outputs is minimized, that is, $\min E(w)$, where $E(w) = E_{T_r}(w)$.

The objective function $E(w)$ (see (4.4) and (4.5)) is a nonlinear function of the adjustable weight parameters w. Due to the complexity of $E(w)$, iterative algorithms are used to explore the weight space. In iterative methods, we start with an initial guess of w and then iteratively update w as

$$w_{next} = w_{now} + \eta h \qquad (4.30)$$

where w_{now} and w_{next} are the current and new values of w, h is the update direction, and η is a positive step size regulating the extent to which w can be updated in that direction. There are different training algorithms and each algorithm has its own scheme for updating the neural network weights.

4.3.1 Categorization of Training Techniques

The training techniques can be categorized in two ways. In one way, the training methods are categorized based upon whether or not the gradient information $\dfrac{\partial E_{T_r}}{\partial w}$ is used, leading to gradient-based and nongradient-based techniques. Another way is to classify the training methods based upon the ability of w to escape from traps of local minimum, leading to local minimization and global minimization methods. Categorization of various training techniques described in this chapter is shown in Table 4.2.

4.3.2 Gradient-Based Methods

Gradient-based methods are the most important type of training techniques. In these methods, the update direction h is determined from $E_{T_r}(w)$ and $\dfrac{\partial E_{T_r}(w)}{\partial w}$. Training techniques that belong to this category are back propagation, conjugate-gradient, Quasi-Newton, and Levenberg-Marquardt. Let ϵ be the error threshold (accuracy criterion) desired by the user, and *max_epoch* be the maximum allowable number of epochs before terminating the training process. Major steps in gradient-based training techniques are:

Step 1: Set w = initial guess and *epoch_number* = 0.
Step 2: Evaluate validation error $E_V(epoch)$. If $E_V(epoch) < \epsilon$ or if *epoch* > *max_epoch*, stop.
Step 3: Compute $E_{T_r}(w)$ and $\dfrac{\partial E_{T_r}(w)}{\partial w}$ using all or part of training samples.
Step 4: Use an optimization technique to determine Δw and update the weights as $w = w + \Delta w$.
Step 5: If all training samples are used, increment *epoch* = *epoch* + 1 and go to Step 2. Otherwise, go to Step 3.

As mentioned earlier, different algorithms have different schemes for finding Δw. Once the update direction h is determined, the step size η decides the change Δw. In the basic backpropagation algorithm, the step size can be

Table 4.2
Categorization of Various Training Techniques

	Local Methods	Global Methods
Gradient-based methods	Back propagation Conjugate-gradient Quasi-Newton Levenberg-Marquardt	
Nongradient-based methods	Simplex	Genetic algorithm Simulated annealing

a small fixed constant set by the user—for example, the user sets $\eta = 0.1$ and the step size remains 0.1 throughout training. The step size can also be adaptive during training—for example, the user initially sets $\eta = 0.1$. Later, η can be changed during training as:

$\eta = \eta * 125\%$ if $E_{T_r}(w)$ decreases steadily during the recent epochs

$\eta = \eta * 80\%$ otherwise

Another method—the line minimization method—finds the best value of η along h. This is very important for gradient-based techniques like the Conjugate-gradient and Quasi-Newton methods.

4.3.3 Line Minimization

Line minimization is used in conjunction with gradient-based training techniques in order to determine the best step size for a fixed update direction. We define a function of a single variable as

$$\Phi(\eta) \stackrel{\Delta}{=} E_{T_r}(w + \eta h) \qquad (4.31)$$

Given the current values of w and h, line minimization uses one-dimensional optimization methods in order to find an η that minimizes $\Phi(\eta)$. Two such methods are discussed in the following subsections—the golden section method and the quadratic interpolation method.

4.3.3.1 Golden Section Method (0.618 method)

The golden section method (see Figure 4.22) has three steps.

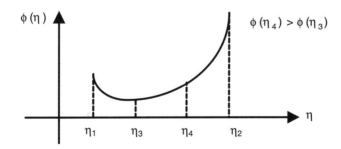

Figure 4.22 Distribution of sections in golden section method.

Step 1: Initialization
$\eta_1 = \eta_{\min}$, $\eta_2 = \eta_{\max}$
$\eta_3 = \eta_2 - 0.618(\eta_2 - \eta_1)$, $\eta_4 = \eta_1 + 0.618(\eta_2 - \eta_1)$
Compute $\Phi(\eta)$ at $\eta = \eta_1, \eta_2, \eta_3, \eta_4$

Step 2: If $\Phi(\eta_3) > \Phi(\eta_4)$, then let
$\eta_1 = \eta_3$, $\eta_2 = \eta_2$
$\eta_3 = \eta_4$, $\eta_4 = \eta_1 + 0.618(\eta_2 - \eta_1)$
Recompute $\Phi(\eta_4)$
Otherwise
$\eta_1 = \eta_1$, $\eta_2 = \eta_4$
$\eta_4 = \eta_3$, $\eta_3 = \eta_2 - 0.618(\eta_2 - \eta_1)$
Recompute $\Phi(\eta_3)$

Step 3: If $|\eta_2 - \eta_1|$ is very small, then stop. Solution is any one of $\eta_1, \eta_2, \eta_3, \eta_4$.
Otherwise, go to Step 2.

4.3.3.2 Quadratic Interpolation Method

The quadratic interpolation method uses the information $(\eta_1, \Phi(\eta_1))$, $(\eta_1 + \Delta\eta, \Phi(\eta_1 + \Delta\eta))$, and $(\eta_1 + 2\Delta\eta, \Phi(\eta_1 + 2\Delta\eta))$ to construct a quadratic function $\hat{\Phi}(\eta) = a + b\eta + c\eta^2$. Distribution of these points is shown in Figure 4.23. It then finds the minimum of $\hat{\Phi}(\eta)$ by letting $\frac{\partial \hat{\Phi}}{\partial \eta} = 2c\eta + b = 0$, that is, $\eta^* = -\frac{b}{2c}$. We then set $\eta_1 = \eta^*$ and find three new points, $\eta_1, \eta_1 + \Delta\eta, \eta_1 + 2\Delta\eta$. These are completely new points and $\Delta\eta$ could be different as well. This process is continued until the difference between η_1's of two consecutive iterations is small enough.

4.3.4 Local Minimum and Global Minimum

A neural network training solution w^* is a local minimum if $E_{T_r}(w^*) < E_{T_r}(w)$ for all w in the small neighborhood of w^*; that is, if

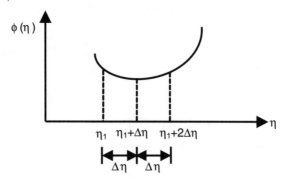

Figure 4.23 Distribution of points in quadratic interpolation method.

$\|w - w^*\| < \epsilon$. The solution w^* is a global minimum if $E_{T_r}(w^*) < E_{T_r}(w)$, for all the points in w space.

Global minimum is actually one of the many local minimum values. In neural network training, if the training error remains high and flat for a number of epochs (see Figure 4.24), it indicates the underlearning phenomenon. If the underlearning is not due to an insufficient number of hidden neurons, it is because the training process is trapped in a local minimum. We can either perturb w, try a new initial guess and restart the training process; or switch to a global optimization method like a genetic algorithm.

4.4 Back Propagation Algorithm and Its Variants

One of the most popular algorithms for neural network training is back propagation (BP), proposed by Rumelhart, Hinton, and Williams in 1986 [9]. The basic concept of back propagation training was discussed in Chapter 3, when we introduced MLP. In this section, variants of BP are presented. Back propagation is a stochastic algorithm based upon the steepest descent principle [10], in which the weights of the neural network are updated along the negative gradient direction in the weight space. The update formulae are given by

$$\Delta w_{now} = w_{next} - w_{now} = -\eta \frac{\partial E_k(w)}{\partial w}\bigg|_{w=w_{now}} \quad (4.32)$$

or

$$\Delta w_{now} = w_{next} - w_{now} = -\eta \frac{\partial E_{T_r}(w)}{\partial w}\bigg|_{w=w_{now}} \quad (4.33)$$

Figure 4.24 Training error E_{T_r} versus number of epochs when w is trapped in a local minimum.

wherein η (called the *learning rate*) controls the step size of weight update. Update formula (4.32) is called *update sample-by-sample*, in which the weights are updated after each training sample is presented to the network. Formula (4.33) is called *batch mode update*, in which the weights are updated after all training samples are presented to the network.

The basic back propagation algorithm suffers from slower convergence and possible weight oscillation. Sample-by-sample training—in which E_{T_r} and $\dfrac{\partial E_{T_r}}{\partial w}$ are computed with one data sample at a time—leads to a stochastic process. Since $\dfrac{\partial E_{T_r}}{\partial w}$ may change between samples, w can also oscillate, since the present and past updates of w may partially cancel each other out. To remedy such a situation, we can keep η small and add a momentum term. By keeping η small, the training process tends to use more epochs and the updating of w becomes a more stable stochastic process. The addition of a momentum term to (4.32) and (4.33), as proposed by [9], results in a significant improvement to the basic back propagation, reducing the weight oscillation. The update formulae with the momentum term include

$$\Delta w_{now} = -\eta \left.\frac{\partial E_k(w)}{\partial w}\right|_{w=w_{now}} + \alpha \Delta w_{old} = -\eta \left.\frac{\partial E_k(w)}{\partial w}\right|_{w=w_{now}} + \alpha(w_{now} - w_{old}) \quad (4.34)$$

$$\Delta w_{now} = -\eta \left.\frac{\partial E_{T_r}(w)}{\partial w}\right|_{w=w_{now}} + \alpha \Delta w_{old} = -\eta \left.\frac{\partial E_{T_r}(w)}{\partial w}\right|_{w=w_{now}} + \alpha(w_{now} - w_{old}) \quad (4.35)$$

Here α is the momentum factor controlling the influence of the previous weight update direction on the current weight update, and w_{old} denotes the previous value of w. This technique is also known as the generalized delta-rule [11]. Other approaches to weight oscillation reduction include: invoking a correction term that employs the difference between gradients [12], and imposing constraints on weights in order to achieve better consistency between weight updates in different epochs [13].

As neural network research moved from the state-of-the-art paradigm to real-world applications, the training time and computational requirements associated with training became significant considerations [14–16]. Some real-world applications of neural networks demand fast and efficient training algorithms. An important means of improving efficiency in BP is the use of adaptation schemes that allow the learning rate and momentum factor to be adaptive during training. There are several schemes for such adaptation.

- η can be adapted following stochastic theory, for example [17], as

$$\eta = \frac{c}{epoch} \qquad (4.36)$$

where c is a constant.

- The Search Then Converge (STC) scheme in [18] updates η as

$$\eta = \eta_0 \frac{1 + \left(\frac{c}{\eta_0}\right) \cdot \left(\frac{epoch}{\tau}\right)}{1 + \left(\frac{c}{\eta_0}\right) \cdot \left(\frac{epoch}{\tau}\right) + \tau\left(\frac{epoch}{\tau}\right)^2} \qquad (4.37)$$

where η_0, τ, c are user-defined constants.

- η adapts itself based on training errors [19], for example, $\eta = \eta * 125\%$ if $E_{T_r}(w)$ decreases steadily during the recent epochs; $\eta = \eta * 80\%$ otherwise. Momentum factor α can be adapted similarly.

One of the most interesting works in this area is the delta-bar-delta rule proposed by Jacobs [20]. Jacobs developed an algorithm based on a set of heuristics in which the learning rate for different weights was different and adapted separately during the learning process (i.e., a different η_i for each weight w_i in w). Each weight w_i is then updated as

$$w_i = w_i - \eta_i \frac{\partial E_{T_r}(w)}{\partial w_i} \qquad (4.38)$$

The adaptation of η_i is determined from the current and past derivatives of the training error with respect to the corresponding weight parameter, for example

$$\begin{cases} \Delta \eta_i = \gamma & \text{if } \left.\frac{\partial E_{T_r}}{\partial w_i}\right|_{epoch} \cdot \left.\frac{\partial E_{T_r}}{\partial w_i}\right|_{epoch-1} > 0 \\ \Delta \eta_i = -\phi \eta_i & \text{otherwise} \end{cases} \qquad (4.39)$$

where γ and ϕ are user-supplied constants.

As an example, let us compute $\dfrac{\partial E_{T_r}}{\partial w} = \sum_{k \in T_r} \dfrac{\partial E_k}{\partial w}$ for the MLP neural network. Let δ_i^l be the local gradient representing a derivative of E_k up to the ith neuron of lth layer. First, set the local gradient at the output layer as

$$\delta_i^L = (y_i(\boldsymbol{x}_k, \boldsymbol{w}) - d_{ik}) \qquad (4.40)$$

Then proceed to back-propagate the error through the hidden layers as

$$\delta_i^l = \left[\sum_{j=1}^{N_{l+1}} \delta_j^{l+1} \cdot w_{ji}^{l+1} \right] \cdot z_i^l (1 - z_i^l), \; l = L-1, L-2, \ldots, 2 \qquad (4.41)$$

The derivative of the error function with respect to a weight parameter w_{ij} is then given by,

$$\dfrac{\partial E_k}{\partial w_{ij}^l} = \delta_i^l \cdot z_j^{l-1} \qquad (4.42)$$

As seen in the above example, the need for computer memory (RAM) is reduced because the local gradient for only one layer (i.e., $l + 1$th layer) needs to be stored. BP is thus suitable for training large neural networks. With the sample-by-sample training where weight update occurs following presentation of each training sample, the stochastic nature of BP makes it suitable for applications with large amounts of training data.

There has been a tremendous amount of research focusing on the improvement of the back propagation algorithm. The sparsity of the hidden neuron activation pattern (some neurons on, some off, others in transition) has been utilized in [16], [21], and [22] to reduce the computations involved during training. Various learning rate adaptation techniques have been proposed. For example, a scheme in which the learning rate is adapted so as to reduce the energy value of the gradient direction in a close-to-optimal way [23], an enhanced back propagation algorithm [24] with a scheme to adapt the learning rate according to values of the weight parameters, and a learning algorithm inspired by the principle of "forced dynamics" for the total error function [25]. [26] presents an interesting adaptation scheme based on the concept of dynamic learning rate optimization, in which the first- and second-order derivatives of the objective function with respect to the learning rate are computed from the information gathered during the forward and backward propagation. Another work [12], considered to be an extension of Jacob's heuristics, corrects the values of weights near the bottom of the error surface ravine with a new acceleration algorithm. And yet another way to improve the training efficiency of BP would be the gradient reuse algorithm [27]. In this method, gradients that are computed during training get reused until the resulting weight updates no longer lead to a reduction in the training error.

4.5 Training Algorithms Using Gradient-Based Optimization Techniques

The BP algorithm based upon the steepest descent principle is relatively easy to implement. The error surface of the training objective function, however, contains planes with gentle slopes, as a result of commonly used logistic activation functions. The values of the error gradient are too small for the weights to move rapidly on these planes, and thus the rate of convergence is slowed down. The rate of convergence also gets slow when the steepest descent method encounters a "narrow valley" in the error surface, in which the direction of gradient moves close to the perpendicular direction of the valley.

Because supervised learning of neural networks can be viewed as a function optimization problem, higher-order optimization methods using gradient information can be used for neural network training in order to improve the rate of convergence. Compared to the heuristic BP algorithm, these methods have a sound theoretical basis and guaranteed convergence. Early work in this area was demonstrated in [28] and [29], with the development of second-order training algorithms for neural networks.

4.5.1 Conjugate Gradient Training Method

The conjugate gradient methods originally derived from quadratic minimization, in which the minimum of the objective function E_{T_r} can be efficiently found within N_w iterations. With initial gradient $g_{initial} = \left.\frac{\partial E_{T_r}}{\partial w}\right|_{w=w_{initial}}$, and direction vector $h_{initial} = -g_{initial}$, the conjugate gradient method recursively constructs two vector sequences [30]

$$g_{next} = h_{now} + \lambda_{now} H h_{now} \quad (4.43)$$

$$h_{next} = -g_{next} + \gamma_{now} h_{now} \quad (4.44)$$

$$\lambda_{now} = \frac{g_{now}^T g_{now}}{b_{now}^T H b_{now}} \tag{4.45}$$

$$\gamma_{now} = \frac{g_{next}^T g_{next}}{g_{now}^T g_{now}} \tag{4.46}$$

or,

$$\gamma_{now} = \frac{(g_{next} - g_{now})^T g_{next}}{g_{now}^T g_{now}} \tag{4.47}$$

where b is the conjugate direction and H is the Hessian matrix of the objective function E_{T_r}. Here, (4.46) is called the Fletcher-Reeves formula [31] and (4.47) is called the Polak-Ribiere formula [32]. To avoid the need for intensive Hessian matrix computation in determining the conjugate direction, we proceed from w_{now}, along the direction b_{now} to the local minimum of E_{T_r} at w_{next} through line minimization, and then set $g_{next} = \frac{\partial E_{T_r}}{\partial w}\bigg|_{w=w_{next}}$. This g_{next} can be used as the vector of (4.43). (4.45) is no longer needed. In this method, the descent direction runs along the conjugate direction, which can be accumulated without matrix computations. Thus, conjugate gradient methods are very efficient and scale well with the neural network size. The conjugate direction in a two-dimensional weight space is shown in Figure 4.25.

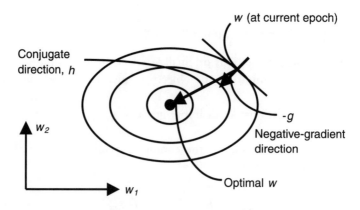

Figure 4.25 Illustration of conjugate direction in a two-dimensional weight space assuming an ideal quadratic error surface.

Two critical issues must be considered in applying conjugate gradient methods to neural network training. First, the computation required for exact one-dimensional optimization is expensive, because every function evaluation involves a complete cycle of sample presentations, as well as an error back propagation whenever gradient information is required. Therefore, an efficient approximation in one-dimensional optimization must be used. Second, because the neural network error function is not quadratic with respect to weight parameters, the convergence of the method is not assured *a priori*—it depends upon the degree to which a local quadratic approximation can be applied to the training error surface. In [23], an inexact line search was proposed using a modified definition of the conjugate search direction. To further reduce computational complexities, a scaled conjugate gradient (SCG) algorithm was introduced in [33].

If the contour of the training error can be represented as a quadratic function of w—for example, as $E_{T_r} = a + bw_1 + cw_2 + dw_1^2 + ew_2^2 + fw_1w_2$ in two-dimensional weight space ($N_w = 2$)—the optimal value of w will be reached in no more than two epochs. Conjugate gradient methods are generally faster than BP. As far as the memory requirement is concerned, only a few vectors of length N_w need to be stored. Therefore, conjugate-gradient methods are effective for training large neural networks.

4.5.2 Quasi-Newton Training Method

The Quasi-Newton method was also derived from the quadratic objective function. The inverse of the Hessian matrix $B = H^{-1}$ is used to bias the gradient direction, following Newton's method. In the Quasi-Newton training method, the neural network weights are updated as

$$w_{next} = w_{now} - \eta B_{now} g_{now} \qquad (4.48)$$

$$B_{now} = B_{old} + \Delta B_{now} \qquad (4.49)$$

The matrix B is not computed. It is successively estimated from the history of gradient directions, employing rank 1 or rank 2 updates, following each line search in a sequence of search directions [34]. There are two commonly used rank 2 formulas for computing ΔB_{now},

$$\Delta B_{now} = \frac{\Delta w \Delta w^T}{\Delta w^T \Delta g} - \frac{B_{old} \Delta g \Delta g^T B_{old}}{\Delta g^T B_{old} \Delta g} \qquad (4.50)$$

or

$$\Delta B_{now} = \left(1 + \frac{\Delta g^T B_{old} \Delta g}{\Delta w^T \Delta g}\right) \frac{\Delta w \Delta w^T}{\Delta w^T \Delta g} - \frac{\Delta w \Delta g^T B_{old} + B_{old} \Delta g \Delta w^T}{\Delta w^T \Delta g} \quad (4.51)$$

where

$$\Delta w = w_{now} - w_{old} \quad (4.52)$$

$$\Delta g = g_{now} - g_{old} \quad (4.53)$$

(4.50) is called the Davidon-Fletcher-Powell (DFP) formula [35] and (4.51) is called the Broyden-Fletcher-Goldfarb-Shanno (BFGS) formula [36].

Standard Quasi-Newton methods require N_w^2 units of space to store the approximation of the inverse Hessian matrix. Therefore the method will not be efficient for large neural networks. A line search is needed to compute a reasonably accurate step length. Limited-memory (LM) or one-step BFGS is a simplification in which the inverse Hessian approximation is reset to the identity matrix after every iteration, thus avoiding the need for storage [37]. Parallel implementation of second-order, gradient-based MLP training algorithms featuring full- and limited-memory BFGS algorithms have also been investigated [38]. Through the estimation of an inverse Hessian matrix, the Quasi-Newton has a faster convergence rate than the conjugate-gradient method.

4.5.3 Levenberg-Marquardt and Gauss-Newton Training Methods

Neural network training is usually formulated as a nonlinear least-squares problem. Methods dedicated to least-squares, such as Gauss-Newton, can be employed to train the neural network weight parameters. Let e be a vector containing the individual error terms in (4.5). For example

$$e = [e_{11} \ e_{12} \ \ldots \ e_{mP}]^T \quad (4.54)$$

where

$$e_{jk} = y_j(\pmb{x}_k, \pmb{w}) - d_{jk}, j \in \{1, 2, \ldots, m\}, k \in T_r \quad (4.55)$$

Let J be the Jacobian matrix containing the derivatives of error e with respect to w. The Gauss-Newton update formula can be expressed as [15]

$$w_{next} = w_{now} - (J_{now}^T J_{now})^{-1} J_{now}^T e_{now} \quad (4.56)$$

In the preceding formula, $J_{now}^T J_{now}$ is positive definite unless J_{now} is rank deficient. The Levenberg-Marquardt [39] method can be applied when J_{now} is rank deficient and the weight update is given by

$$w_{next} = w_{now} - (J_{now}^T J_{now} + \mu I)^{-1} J_{now}^T e_{now} \quad (4.57)$$

where μ is a non-negative number. A modified Levenberg-Marquardt training algorithm using a diagonal matrix instead of the identity matrix I in (4.57) was proposed [40] for efficient training of multilayer feedforward neural networks. In [41], the training samples are divided into several groups called localbatches, in order to reduce the size of J. Training is performed successively using these local batches. The computational requirements and memory complexity of Levenberg-Marquardt methods could also be reduced by utilizing the deficient Jacobian matrix [42].

4.6 Nongradient-Based Training: Simplex Method

The simplex method [43] is a representative method of the nongradient-based training techniques, which only require function evaluations. It uses the error function $E_{T_r}(w)$, without the derivative information $\dfrac{\partial E_{T_r}(w)}{\partial w}$. We define a simplex formed by $N_w + 1$ points in w-space, where N_w is the total number of neural network weights. The $N_w + 1$ points, $w^{(i)}$, $i = 1, 2, \ldots, N_w + 1$, are the vertices of the simplex.

The training error $E_{T_r}(w^{(i)})$ is computed at all vertices of the simplex. Comparing these values, the worst point $w^{(h)}$ (i.e., $E_{T_r}(w^{(h)}) > E_{T_r}(w^{(i)})$ for all i, $i \neq h$), can be selected. The centroid of all the points excluding $w^{(h)}$ (see Figure 4.26) can be computed as

$$w^{(o)} = \frac{1}{N_w} \sum_{\substack{i=1 \\ i \neq h}}^{N_w+1} w^{(i)} \quad (4.58)$$

In the simplex method, optimization means searching for new points where the training error is minimal. There are three operations used to create a new point—namely the reflection, the expansion, and the contraction. The reflection operation is given by

$$w^{(r)} = w^{(o)} + \alpha(w^{(o)} - w^{(h)}), \; \alpha > 0 \quad (4.59)$$

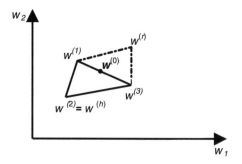

Figure 4.26 A simplex with three vertices in a two-dimensional w-space. Corresponding centroid and reflection points are also shown.

If $w^{(r)}$ turns out to be the best point of the present simplex, then use the expansion operation

$$w^{(e)} = w^{(o)} + \beta(w^{(r)} - w^{(o)}), \quad \beta > 1 \qquad (4.60)$$

Otherwise, if $w^{(r)}$ is still a worst point with little improvement over the previous worst point, then use the contraction operation

$$w^{(c)} = w^{(o)} + \gamma(w^{(o)} - w^{(h)}), \quad -1 < \gamma < 0 \qquad (4.61)$$

Otherwise, if $w^{(r)}$ is the worst, contract all $w^{(i)}$ around $w^{(l)}$, where $w^{(l)}$ is the best point in the simplex. Figure 4.27 illustrates the new points generated by different operations.

A new simplex is then formed by replacing $w^{(h)}$ with the new point, and this concludes one epoch. Repeat this process for subsequent epochs until one of the following stop criteria is met.

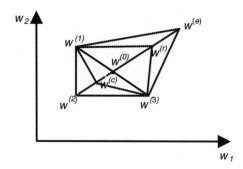

Figure 4.27 New points generated by the three different operations in the simplex method.

- Maximum number of epochs is reached;
- Best point in the current simplex satisfies the conditions $E_{T_r}(w) < \epsilon$ and $E_V(epoch) < $ user criteria;
- Simplex size is small enough;
- The standard deviation of E_{T_r} among all vertices of the current simplex is small enough.

Standard deviation of E_{T_r} is given by

$$\sigma^2 = \frac{1}{N_w + 1} \sum_{i=1}^{N_w+1} (E_{T_r}(w^{(i)}) - \overline{E_{T_r}(w)})^2 \qquad (4.62)$$

where $\overline{E_{T_r}(w)}$ is the mean value of E_{T_r} for all points in the simplex. The simplex training method is generally much slower than gradient-based methods. The simplex algorithm can be particularly useful for training neural networks with user-supplied neuron activation functions (e.g., empirical functions), where the gradient information may not be supplied to training.

4.7 Training With Global Optimization Methods

4.7.1 Genetic Algorithms

A genetic algorithm [45] is a stochastic optimization algorithm, derived from the concepts of the biological theory of evolution. Genetic algorithms use the information from $E_{T_r}(w)$ only (without gradient information). They are capable of escaping from the traps of the local minimum and finding the global minimum. Genetic algorithms have the following steps.

Step 1: Set up an initial population of w points, $w^{(i)}$, $i = 1, 2, \ldots, K$, where K is the size of the population.
Step 2: Evaluate the fitness of each point. The fitness function is defined in such a way that points with lower (higher) values of $E_{T_r}(w)$ will have higher (lower) fitness. For example,

$$fitness(w^{(i)}) = -aE_{T_r}(w^{(i)}) + \beta \qquad (4.63)$$

where a and β are constants.

Step 3: Choose lucky parents. A random selection process is adapted such that two w points with higher fitness values in the population are more likely to be selected. Let the selected points also called parents be w^A and w^B.

Step 4: Generate an offspring from the parents as $w^{(new)} = function(w^A, w^B)$. This step is usually achieved by a crossover operation between w^A and w^B.

The generation of offspring is well-defined for discrete optimization problems where w is a binary vector. For continuous optimization problems, however, where w is a real-valued vector, w must be converted into a binary vector with a preset resolution. For illustration, in the following example we convert real-valued parent vectors into binary vectors using 4-bit resolution, and then perform a crossover operation. Conversion to binary vectors:

$$w^A = \begin{bmatrix} 1.5 \\ 0.2 \end{bmatrix} \Rightarrow [1\ 1\ 1\ 1\ 0\ 0\ 1\ 0]^T$$

$$w^B = \begin{bmatrix} 1.0 \\ 1.0 \end{bmatrix} \Rightarrow [1\ 0\ 1\ 0\ 1\ 0\ 1\ 0]^T$$

Crossover operation: choose a random bit position among the 8-bits of the binary vectors—for example 3—as the division point. Take the first 3 bits of parent A and the second 5 bits of parent B to generate the new vector or offspring. Convert the binary offspring vector back into a real-valued vector.

$$[1\ 1\ 1\ 0\ 1\ 0\ 1\ 0]^T \Rightarrow \begin{bmatrix} 1.4 \\ 1.0 \end{bmatrix} = w^{(new)}$$

Step 5: Evaluate the fitness of $w^{(new)}$. Replace the point with least fitness value in the current generation with $w^{(new)}$ in order to form a new population. This concludes one epoch. If one of the following conditions is met, then stop. Otherwise, go to Step 3.
- The allocated CPU time is used up for neural network training;
- The training error E_{T_r} or the validation error E_V among all points in the current population is small enough.

It is possible that after a certain number of epochs, many points in the population become similar and the new offspring may be similar to an already existing point in the population. In such cases, a mutation should be performed by inverting the binary value of a randomly selected bit in the parent.

As optimization continues through the above steps, the overall fitness of the population improves and the training error E_{T_r} decreases. If the parent selection and crossover are done randomly, mutations are performed when necessary, and sufficient CPU is allocated, then chances of reaching a global minimum approach 100%. Nonbinary integer-valued vectors can generally also be used during crossover and mutation to represent parents and offspring. Genetic algorithms are much slower in practice than gradient-based methods like the conjugate-gradient method. Genetic algorithms would be needed mainly in cases where it would be too easy to get trapped in local minimum during neural network training.

4.7.2 Simulated Annealing (SA) Algorithms

A simulated annealing algorithm was proposed in [46], inspired by the idea of the physical annealing process in crystal growth. The SA algorithm uses the information from $E_{T_r}(w)$ only and is capable of finding the global minimum. In simulated annealing, the value of the objective function that we want to minimize is analogous to the energy of a thermodynamic system. At high temperatures, SA allows function evaluations from far away points and is likely to accept a new point with higher energy. This corresponds to the situation in which the molecules of a liquid have high mobility and can move freely at high temperatures. At low temperatures, SA evaluates the objective function only at local points and its likelihood of accepting a new point with higher energy is much lower. This is analogous to the situation in which low-mobility atoms orient themselves with regard to neighboring atoms at low temperatures. In the SA algorithm, a fictitious temperature parameter is used. The following basic steps are involved in a general SA method.

Step 1: Choose a starting point w and set a high starting temperature T. Set the iteration count $k = 1$.

Step 2: Evaluate the objective function, that is, the training error function $E = E_{T_r}(w)$.

Step 3: Generate a new point w_{new} by perturbing the current w with a small value Δw—that is, $w_{new} = w + \Delta w$. Methods to generate Δw are well-defined for discrete optimization cases. In our case, however, w is continuous and creating a Δw for continuous cases is still an open subject. A modified version of the simplex method is used in [30] to compute Δw.

Step 4: Evaluate the new value of the objective function $E_{new} = E_{T_r}(w_{new})$.

Step 5: Set w to w_{new} and E to E_{new} with a probability determined by an acceptance function, given by $h(\Delta E, T) = \exp\left(-\frac{\Delta E}{cT}\right)$, where $\Delta E = E_{new} - E$, and c is a system-dependent constant. It may be noted that w_{new} can be accepted even if $\Delta E > 0$—that is, even if w_{new} leads to a higher value of the training error. This is more likely the case when the temperature T is high. SA algorithm can thus escape the traps of local minimum.

Step 6: Reduce the temperature T according to the annealing schedule—for example, $T = \eta T$, where η is a constant between 0 and 1.

Step 7: Set $k = k + 1$. If k reaches maximum allowable iterations, or if E_V is small enough, stop. Otherwise go to Step 3.

If sufficient computation time is given, the generation of Δw is suitable and random, and the temperature reduction process is appropriate, then the chances of finding the global minimum approach 100%. In practice, however, the simulated annealing technique requires a much longer computational time than gradient-based local optimization methods. Simulated annealing is used in cases where it is too easy for the neural network training to be trapped in a local minimum.

4.8 Training Algorithms Utilizing Decomposed Optimization

As seen in earlier discussions, the implementation of powerful second-order optimization techniques for neural network training result in significant advantages in training. The second-order methods are typically much faster than BP, but could require the storage of an inverse Hessian matrix and its computation or approximation. For large neural networks, training turns out to be a very large-scale optimization. Decomposition is an important way to solve large-scale optimization problems. Several training algorithms that decompose the training process by training the neural network layer-by-layer have been proposed [47–49]. In [48], the weights w^L of the output layer and the output vector z^{L-1} of the previous layer are treated as two sets of variables. First, an optimal solution pair (w^L, z^{L-1}) is determined to minimize the sum-squared-error between the neural network outputs and the desired outputs. The current solution z^{L-1} is then set as the desired output of the previous hidden layer, and optimal weight vectors of the remaining hidden layers are recursively obtained.

Linear programming can be used to solve large-scale linearized optimization problems. Neural network training was linearized and formulated as a kind of constrained linear programming in [50]. In this work, weights are updated with small local changes, with a requirement that none of the individual sample errors increase, and an objective of maximizing the overall reduction in error. In [47], a layer-by-layer optimization of a neural network was presented that does not rely on the evaluation of local gradients, with linearization of the nonlinear hidden neurons. A special penalty term was added to the training objective function in order to limit the unavoidable linearization error, and the layers are optimized alternately through an iterative process.

A combination of linear and nonlinear programming techniques could reduce the degree of nonlinearity of the error function with respect to the hidden layer weights, and could also reduce the chances of being trapped in a local minimum. A hybrid-training algorithm has thus been attempted for large-scale problems [51]. For a feedforward neural network with either linear or linearized output layer weights, the hybrid algorithms adapt nonlinear hidden layer weights using BP [52] or BFGS [51], while employing a linear least-mean square error (LMS) algorithm to compute the optimum linear output layer weights.

RBF networks are usually trained by a decomposed process. The nonlinear hidden layer weights of the RBF network (i.e., centers and widths of radial basis funcions) can be fixed through unsupervised training, for example as a k-means clustering algorithm [53]. The output layer weights can then be obtained through a linear LMS algorithm. But the development of heuristics that initially assign the hidden layer weights of an MLP is very difficult due to MLP's black-box characteristics. Consequently, it is not possible to train MLP with such a decomposed strategy. The hybrid linear/nonlinear training algorithm [51] integrates the best features of the linear LMS algorithm of RBF and nonlinear optimization techniques of MLP into one routine. According to [51], the advantages of this technique are a reduced number of independent parameters, and a guaranteed global minimum with respect to the output layer weights.

4.9 Comparisons of Different Training Techniques

In this section, different training techniques are compared. First, based on the ability to find local and global minimums, we make the following observations. Conjugate-gradient, Quasi-Newton, and Levenberg-Marquardt algorithms converge to a local minimum that is nearest to the initial guess. The stochastic BP and simplex methods may escape the nearest local minimum, but they'll ultimately end up in some other local minimum. Genetic algorithms and simulated annealing both have the ability to find the global minimum.

Second, training techniques are compared on the basis of computation speed and memory requirements, and are ranked as shown in Figure 4.28. A detailed comparison of the training algorithms is also presented in Table 4.3.

4.10 Feedforward Neural Network Training: Examples

In this section, three conductor microstripline and physics-based MESFET modeling examples are used to compare the performance of various training algorithms for feedforward neural networks—namely, the adaptive back propagation, conjugate-gradient, Quasi-Newton, and Levenberg-Marquardt algorithms [2].

For the microstripline example, there are five input neurons corresponding to conductor width (w), spacing between conductors (s_1, s_2), substrate height (h), and relative permittivity (ϵ_r), as shown in Figure 4.29. There are six output neurons corresponding to the self-inductance of each conductor (L_{11}, L_{22}, L_{33}), and the mutual inductance between any two conductors (L_{12}, L_{23}, L_{13}). A total of 600 training samples and 640 test samples were generated using LINPAR [54]. A three-layer MLP structure with 28 hidden neurons was used, and the training results are shown in Table 4.4. The CPU time was given for a 200 MHz Pentium.

The total CPU time used by the Levenberg-Marquardt method is around 20 minutes. The adaptive back propagation algorithm used a large number of epochs and achieved a respectable accuracy of 0.252% in 4 hours. The Quasi-Newton method achieved a similar accuracy within 35 minutes. This confirms the faster convergence of the second-order methods. The Quasi-Newton method has a very fast convergence rate when closer to the optimal solution.

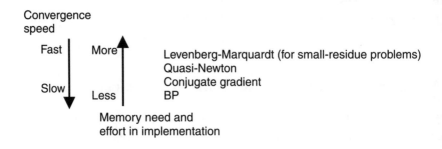

Figure 4.28 Ranking of different training techniques based on speed and memory requirements.

Table 4.3
Qualitative Comparison of the Performance of Various Neural Network Training Algorithms

Part A

Training Algorithm	Convergence Speed (Number of Epochs)	CPU/Epoch	Total CPU (Small Neural Network)	Total CPU (Large Neural Network)
Levenberg-Marquardt	Few	LU of a large matrix	Small	Huge
Quasi-Newton	Few	Matrix/Vector product	Small	Huge
Conjugate-gradient	Medium	Vector/Vector product	Small	Large
Back propagation	Large	Scalar operations	Large	Large
Simplex method	Large	Scalar operations	Large	Large
Genetic algorithm/ Simulated Annealing	Very Large	Scalar operations	Large	Huge

Part B

Training Algorithm	Ease of Implementation	Memory Requirement	Likelihood of Reaching Global Minimum
Levenberg-Marquardt Quasi-Newton Conjugate-gradient	Needs effort	Matrix Matrix Vectors	Depends on initial guess of the neural network weight parameters
Back propagation	Easiest	Small vectors	Possible
Simplex method	Easy	Vectors	Possible
Genetic algorithm/ simulated annealing	Needs effort	Vectors	Global solution

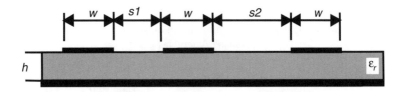

Figure 4.29 A three-conductor microstripline.

Table 4.4
Comparison of Various Training Algorithms for Microstripline Example
(from [2], Wang, F., et al., "Neural Network Structures and Training Algorithms for RF and Microwave Applications," *Int. J. RF and Microwave CAE*, pp. 216–240, © 1999, John Wiley and Sons. Reprinted with permission from John Wiley and Sons, Inc.).

Training Algorithm	No. of Epochs	Training Error (%)	Avg. Test Error (%)	CPU (in Sec)
Adaptive back propagation	10,755	0.224	0.252	13,724
Conjugate-gradient	2,169	0.415	0.473	5,511
Quasi-Newton	1,007	0.227	0.242	2,034
Levenberg-Marquardt	20	0.276	0.294	1,453

It is usually slow at the beginning of training process, however. A good strategy would be to use the conjugate-gradient method in the first stage of training, followed by Quasi-Newton training. When the MLP network already trained by the conjugate-gradient was further trained by the Quasi-Newton method, the model test error was reduced to 0.167%, and the total training time ended up being 2 hours.

For the MESFET example, the inputs to the neural model include frequency (f), channel thickness (a), gate-bias voltage (V_{GS}), and drain-bias voltage (V_{DS}). This particular MESFET has a fixed gate-length of 0.55 micron and a gate-width of 1 mm. The outputs include real and imaginary parts of the two-port S-parameters S_{11}, S_{12}, S_{21}, and S_{22}. Training and test samples are obtained using the simulator OSA90 [55] with the Katibzadeh and Trew model [56]. The MESFET behavior is more nonlinear in comparison to the microstripline example. A three-layer MLP structure with 60 hidden neurons was used, and the training results are shown in Table 4.5. This example demonstrates that as the size of the neural network increases, the speed of the Levenberg-Marquardt decreases (as compared to Quasi-Newton). This is due to the repeated inversion or LU decomposition of the large matrix in (4.57).

Table 4.5
Comparison of Various Training Algorithms for MESFET Example
(from [2], Wang, F., et al., "Neural Network Structures and Training Algorithms for RF and Microwave Applications," *Int. J. RF and Microwave CAE*, pp. 216–240, © 1999, John Wiley and Sons. Reprinted with permission from John Wiley and Sons, Inc.).

Training Algorithm	No. of Epochs	Training Error (%)	Ave. Test Error (%)	CPU (sec)
Adaptive back propagation	15,319	0.98	1.04	11,245
Conjugate-gradient	1,605	0.99	1.04	4,391
Quasi-Newton	570	0.88	0.89	1,574
Levenberg-Marquardt	12	0.97	1.03	4,322

References

[1] Devabhaktuni, V., et al., "Robust Training of Microwave Neural Models," *IEEE MTT-S Intl. Microwave Symp. Digest*, Anaheim, CA, June 1999, pp. 145–148.

[2] Wang, F., et al., "Neural Network Structures and Training Algorithms for RF and Microwave Applications," *Int. Journal of RF and Microwave CAE*, Special Issue on Applications of ANN to RF and Microwave Design, Vol. 9, 1999, pp. 216–240.

[3] Watson, P. M., and K. C. Gupta, "EM_ANN Models for Microstrip Vias and Interconnects in Multilayer Circuits," *IEEE Trans. Microwave Theory and Techniques*, Vol. 44, Dec. 1996, pp. 2495–2503.

[4] Bandler, J. W., M. A. Ismail, J. E. Rayas-Sanchez, and Q. J. Zhang, "Neuromodeling of Microwave Circuits Exploiting Space Mapping Technology," *IEEE Intl. Microwave Symp Digest*, Anaheim, CA, June 1999, pp. 149–152.

[5] Rao, N. S. V., et al. "Learning Algorithms for Feedforward Networks Based on Finite Samples, "*IEEE Trans. Neural Networks*, Vol. 7, 1996, pp. 926–940.

[6] Thimm, G., and E. Fiesler, "High-Order and Multilayer Perceptron Initialization," *IEEE Trans. Neural Networks*, Vol. 8, 1997, pp. 349–359.

[7] Kim, Y. K., and J. B. Ra, "Weight Value Initialization for Improving Training Speed in the Backpropagation Network, " *Proc. IEEE Int. Conf. Neural Networks*, Vol. 3, 1991, pp. 2396–1401.

[8] Burrascano, P., S. Fiori, and M. Mongiardo, "A Review of Artificial Neural Networks Applications in Microwave CAD," *Int. Journal of RF and Microwave CAE*, Special Issue on Applications of ANN to RF and Microwave Design, Guest Editors, Q. J. Zhang, G. L. Creech, Vol. 9, April 1999, pp. 158–174.

[9] Rumelhart, D. E., G. E. Hinton, and R. J. Williams, "Learning Internal Representations by Error Propagation," in *Parallel Distributed Processing*, Vol. 1, D. E. Rumelhart and J. L. McClelland, Editors, Cambridge, MA: MIT Press, 1986, pp. 318–362.

[10] Luenberger, D. G., *Linear and Nonlinear Programming*, Reading, MA: Addison-Wesley, 1989.

[11] Haykin, S., *Neural Networks: A Comprehensive Foundation*, New York, NY: IEEE Press, 1994.

[12] Ochiai, K., N. Toda, and S. Usui, "Kick-out Learning Algorithm to Reduce the Oscillation of Weights," *Neural Networks*, Vol. 7, 1994, pp. 797–807.

[13] Perantonis, S. J., and D. A. Karras, "An Efficient Constrained Learning Algorithm with Momentum Acceleration," *Neural Networks*, Vol. 8, 1995, pp. 237–249.

[14] Karayiannis, N. B., and A. N. Venetsanopoulos, *Artificial Neural Networks: Learning Algorithms, Performance Evaluation, and Applications*, Boston, MA: Kluwer Academic Publishers, 1993.

[15] Shepherd, A. J., *Second-Order Methods for Neural Networks*, London: Springer, 1997.

[16] Wang, F., and Q. J. Zhang, "A Sparse Matrix Approach to Neural Network Training," In *Proc. IEEE Intl. Conf. Neural Networks*, Perth, Australia, November 1995, pp. 2743–2747.

[17] Robbins, H., and S. Monro, "A Stochastic Approximation Method," *Annals of Mathematical Statistics*, Vol. 22, 1951, pp. 400–407.

[18] Darken, C., J. Chang, and J. Moody, "Learning Rate Schedules for Faster Stochastic Gradient Search," In *Neural Network for Signal Processing*, Vol. 2, S. Y. Kung, F. Fallside, J. A. Sorensen, and C. A. Kamm, Eds., IEEE Workshop, IEEE Press, 1992, pp. 3–13.

[19] *Neural Network Toolbox: For Use with Matlab*, The MathWorks Inc., Natick, Massachusetts, 1993.

[20] Jacobs, R. A., "Increased Rate of Convergence through Learning Rate Adaptation," *Neural Networks*, Vol. 1, 1988, pp. 295–307.

[21] Zaabab, A. H., Q. J. Zhang, and M. S. Nakhla, "Device and Circuit-Level Modeling using Neural Networks with Faster Training Based on Network Sparsity," *IEEE Trans. Microwave Theory and Techniques*, Vol. 45, 1997, pp. 1696–1704.

[22] Wang, F., and Q. J. Zhang, "An Adaptive and Fully Sparse Training Approach for Multilayer Perceptrons," *Proc. IEEE Intl. Conf. Neural Networks*, Washington, DC, June 1996, pp. 102–107.

[23] Battiti, R., "Accelerated Backpropagation Learning: Two Optimization Methods," *Complex Systems*, Vol. 3, 1989, pp. 331–342.

[24] Arisawa, M., and J. Watada, "Enhanced Backpropagation Learning and its Application to Business Evaluation," In *Proc. IEEE Intl. Conf. Neural Networks*, Vol. I, Orlando, Florida, July 1994, pp. 155–160.

[25] Parlos, A. G., et al., "An Accelerated Learning Algorithm for Multilayer Perceptron Networks," *IEEE Trans. Neural Networks*, Vol. 5, 1994, pp. 493–497.

[26] Yu, X. H., G. A. Chen, and S. X. Cheng, "Dynamic Learning Rate Optimization of the Backpropagation Algorithm," *IEEE Trans. Neural Networks*, Vol. 6, May 1995, pp. 669–677.

[27] Hush, D. R., and J. M. Salas, "Improving the Learning Rate of Backpropagation with the Gradient Reuse Algorithm," In *Proc. IEEE Int. Conf. Neural Networks*, Vol. I, San Diego, California, 1988, pp. 441–447.

[28] Parker, D. B., "Optimal Algorithms for Adaptive Networks: Second Order Backpropagation, Second Order Direct Propagation and Second Order Hebbian Learning," In *Proc. IEEE First Intl. Conf. Neural Networks*, Vol. II, San Diego, California, 1987, pp. 593–600.

[29] Watrous, R. L., "Learning Algorithms for Connectionist Networks: Applied Gradient Methods of Nonlinear Optimization," In *Proc. IEEE First Intl. Conf. Neural Networks*, Vol. II, San Diego, California, 1987, pp. 619–627.

[30] Press, W. H., et al., *Numerical Recipes: The Art of Scientific Computing*, Cambridge, UK: Cambridge University Press, 1992.

[31] Fletcher, R., and C. M. Reeves, "Function Minimization by Conjugate Gradients," *Computer Journal*, Vol. 6, 1964, pp. 149–154.

[32] Polak, E., and G. Ribiere, "Note sur la Convergence de Méthode de Directions Conjuguées," *Revue Francaise Informat. Rechercher Operationnelle*, Vol. 16, 1969, pp. 35–43.

[33] Moller, M. F., "A Scaled Conjugate Gradient Algorithm for Fast Supervised Learning," *Neural Networks*, Vol. 6, 1993, pp. 525–533.

[34] Cuthbert, T. R., Jr., "Quasi-Newton Methods and Constraints," In *Optimization using Personal Computers*, NY: John Wiley & Sons, 1987, pp. 233–314.

[35] Davidon, W. C., "Variable Metric Method for Minimization," Research and Development Report ANL-5990, U.S. Atomic Energy Commission, Argonne National Laboratories, 1959.

[36] Broyden, C. G., "Quasi-Newton Methods and their Application to Function Minimization," *Math. Comp.* Vol. 21, 1967, pp. 368–381.

[37] Nakano, K. R., "Partial BFGS Update and Efficient Step-Length Calculation for Three-Layer Neural Networks," *Neural Computation*, Vol. 9, 1997, pp. 123–141.

[38] McLoone, S., and G. W. Irwin, "Fast Parallel Off-Line Training of Multilayer Perceptrons," *IEEE Trans. Neural Networks*, Vol. 8, May 1997, pp. 646–653.

[39] Shepherd, A. J., "Second-Order Optimization Methods," In *Second-Order Methods for Neural Networks*, London: Springer-Verlag, 1997, pp. 43–72.

[40] Kollias, S., and D. Anastassiou, "An Adaptive Least Squares Algorithm for the Efficient Training of Artificial Neural Networks," *IEEE Trans. Circuits and Systems*, Vol. 36, 1989, pp. 1092–1101.

[41] Bello, M. G., "Enhanced Training Algorithms, and Integrated Training/Architecture Selection for Multilayer Perceptron Networks," *IEEE Trans. Neural Networks*, Vol. 3, Nov. 1992, pp. 864–875.

[42] Zhou G., and J. Si, "Advanced Neural Network Training Algorithm with Reduced Complexity based on Jacobian Deficiency," *IEEE Trans. Neural Networks*, Vol. 9, May 1998, pp. 448–453.

[43] Nelder, J. A., and R. Mead, "A Simplex Method for Function Minimization," *Computer Journal*, Vol. 7, 1965, pp. 308–313.

[44] Rao, S. S., *Engineering Optimization, Theory, and Practice*, New York: John Wiley and Sons, 1996.

[45] Goldberg, D. E., *Genetic Algorithms in Search, Optimization and Machine Learning*, Reading, MA: Addison-Wesley, 1989.

[46] Kirkpatrick, S., C. D. Gelatt, and M.P. Vecchi, "Optimization by Simulated Annealing," *Science*, Vol. 220, 1983, pp. 671–680.

[47] Ergezinger, S., and E. Thomsen, "An Accelerated Learning Algorithm for Multilayer Perceptrons: Optimization Layer by Layer," *IEEE Trans. Neural Networks*, Vol. 6, 1995, pp. 31–42.

[48] Wang, G. J., and C. C. Chen, "A Fast Multilayer Neural Network Training Algorithm Based on the Layer-by-Layer Optimizing Procedures," *IEEE Trans. Neural Networks*, Vol. 7, 1996, pp. 768–775.

[49] Lengelle, R., and T. Denceux, "Training MLPs Layer-by-Layer using an Objective Function for Internal Representations," *Neural Networks*, Vol. 9, 1996, pp. 83–97.

[50] Shawe-Taylor, J. S., and D. A. Cohen, "Linear Programming Algorithm for Neural Networks," *Neural Networks*, Vol. 3, 1990, pp. 575–582.

[51] Mcloone, S., et al., "A Hybrid Linear/Nonlinear Training Algorithm for Feedforward Neural Networks," *IEEE Trans. Neural Networks*, Vol 9, July 1998, pp. 669–684.

[52] Verma, B., "Fast Training of Multilayer Perceptrons," *IEEE Trans. Neural Networks*, Vol. 8, Nov. 1997, pp. 1314–1320.

[53] Wang, F., and Q. J. Zhang, "An Improved k-means Clustering Algorithm and Application to Combined Muticodebook/MLP Neural Network Speech Recognition," In *Proc. Canadian Conf. Elec. Comp. Engg.*, Vol. II, Montreal, Canada, 1995, pp. 999–1002.

[54] Djordjevic, A., R. F. Harrington, T. Sarkar, and M. Bazdar, *Matrix Parameters for Multiconductor Transmission Lines: Software and User's Manual*, Boston, MA: Artech House, 1989.

[55] *OSA90 Version 3.0*, Optimization Systems Associates Inc., P.O. Box 8083, Dundas, Ontario, Canada L9H 5E7, now HP EEsof (Agilent Technologies), 1400 Fountaingrove Parkway, Santa Rosa, CA 95403.

[56] Khatibzadeh, M. A., and R. J. Trew, "A Large Signal, Analytic Model for the GaAs MESFET," *IEEE Trans. Microwave Theory and Techniques*, Vol. 36, 1988, pp. 231–238.

5

Models for RF and Microwave Components

This chapter[1] includes several examples of the techniques described in earlier chapters for the development of CAD models for RF and microwave passive components. Examples of components include: one-port and two-port microstrip vias, vertical interconnects used in multiplayer circuits, CPW lines, CPW bends, other discontinuities occurring in CPW circuits, planar spiral inductors, multiconductor-multilayered transmission structures, microstrip patch antennas, and waveguide filter structures. Modeling of active RF and microwave devices is discussed separately in Chapter 7.

5.1 Modeling Procedure

This section discusses some general issues related to the development of ANN models for passive RF and microwave components. Examples of various models are given in the following sections of this chapter. Selection of model inputs and outputs as well as the generation of training data are discussed in this section. Error measures for model validation are also given. Once the ANN models are developed, they need to be linked to a commercial microwave simulator so that they can be used for circuit design and optimization.

---

1. The first three sections of this chapter are based on P. M. Watson's Ph.D. thesis [9] at University of Colorado, Boulder.

5.1.1 Selection of Model Inputs and Outputs

Selection of input parameters for models of RF and microwave passive components is relatively straightforward. Inputs are generally important physical (geometrical) parameters of the component, which should vary in frequency. Ranges of input variables are determined by desired model usage requirements. In the design of RF and microwave circuits, it is often desirable to connect active and passive elements together. This can be accomplished by representing model outputs in terms of S-parameters (scattering parameters), which are related to the powers reflected and transmitted by a given component. S-parameters are also a convenient way to link the developed models with commercial microwave simulators for design and optimization. For this reason, most ANN model outputs are given as S-parameters. The magnitude and phase of each S-parameter are separate outputs. Therefore, if the component to be modeled is a one-port device, the output parameters would be $|S_{11}|$ and $\angle S_{11}$. For a more detailed discussion of S-parameter theory and applications, one may refer to any text on microwave circuits (e.g., [1]).

For low-loss transmission line models, the outputs of the ANN model are the characteristic impedance, Z_o, and the phase constant, β. Models may be developed to yield attenuation constant α as well. Knowing these two parameters, and the length of the line, the S-parameters of a section of the line can be found. In this way, the length of the line is not included as a variable input parameter for the ANN model.

5.1.2 Training Data Generation

Training data for passive RF and microwave component models generally comes from either measurements of actual components or from electromagnetic (EM) simulation. Using actual measurements can be costly as it requires design and fabrication of many components for characterization. However, by using actual measurement data, one may obtain a model that is valid for a specific fabrication process, which may be desirable.

EM simulation is an attractive alternative to using measured data and is widely used in model development. One of the EM simulators used for this purpose is HP-Momentum [2]. EM simulators yield S-parameters for given component geometries, which are then used to train ANN models. These models trained by EM simulations are therefore termed electromagnetically trained artificial neural network (EM-ANN) models.

One drawback of using EM simulation for the generation of training data is that it can be very time consuming. Design of experiments (DOE) techniques [3] and knowledge-based methods (discussed in Chapter 9) are therefore used for reducing the number of EM simulations necessary for model development.

5.1.3 Error Measures

Error measures for model validation can be either absolute or relative in nature. The choice of error measure to be used is determined by the type of accuracy desired and the model usage requirements.

Many EM-ANN model errors are reported as the absolute average and the standard deviation of error for each output. This is useful for S-parameters, as it allows a designer to determine the amount of signal reflection and/or transmission of the model on average and to estimate some useful error bounds.

In some cases, relative errors are important, as with the characteristic impedance, Z_o, of a transmission line. When Z_o is small, a small absolute error can produce large mismatch errors when impedance matching is carried out. However, a larger absolute error is allowable for higher Z_o values. Therefore relative errors are considered in this case.

5.1.4 Integration of EM-ANN Models with Circuit Simulators

Once the EM-ANN models have been trained and verified, they provide accuracy approaching that of the EM simulator used for generating the training data. The models can then either be used in stand-alone mode or integrated into a commercial microwave circuit simulator, depending on the type and desired usage of the model.

Several EM-ANN models [4–7] have been integrated into a commercially available microwave circuit simulator, HP-MDS [8], using the configuration shown in Figure 5.1 [9]. Models can be inserted into other circuit simulators in a similar manner. During simulation, the circuit simulator passes input variables such as frequency and the physical parameters of a component to the user-defined linear model subroutine used for linking models to the circuit simulator. This subroutine is responsible for returning S-parameters of the component back to HP-MDS for further circuit simulation. The input variables are then passed on to a feedforward ANN subroutine by the user-defined linear model subroutine so that model S-parameters can be computed. The feedforward ANN subroutine implements the algorithm for finding the output of a 2-layer ANN. EM-ANN models are stored as files, which are read by the feedforward ANN subroutine. Each model file contains the number of input and output parameters, the number of neurons in the hidden layer, the maximum and minimum values (used for normalization) for each of the input and output parameters, and the weight connection matrices. Models can also be

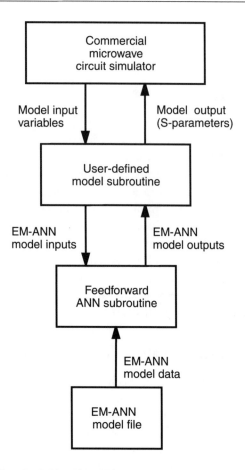

Figure 5.1 Flow of data for linking EM-ANN models to commercial microwave circuit simulators (from reference [9]).

grouped into libraries. A library is a collection of models that can be linked to the circuit simulator as a group. This offers a convenient way to create and distribute models for use by others.

5.2 Models for Vias and Multilayer Interconnects

Efforts to lower the cost and reduce the weight and volume of microwave circuits have resulted in high-density, multilayer circuits where a large number of vertical via interconnects are used. Within this increased complexity, accurate and efficient characterizations of via interconnect discontinuities and single-layer ground vias must be carried out in order to achieve accurate simulation results [10]. Several recent efforts have focused on the analytical and numerical evaluation of via discontinuities using quasi-static and full-wave techniques [10–26]. Quasi-static models are valid only at lower frequencies. Full-wave EM simulation and characterization can lead to accurate results, but it comes at much higher computational expense that prevents their use in practical interactive CAD.

EM-ANN methodology has been used successfully [4–7] for modeling via elements in microstrip circuits and multilayer via interconnects. Modeling examples include, one- and two-port microstrip vias, a stripline-to-stripline multilayer interconnect, a microstrip-to-microstrip multilayer interconnect, and a microstrip-to-CPW multilayer interconnect. An ANN model for microstrip transmission lines has been used for developing these interconnect models. Prior knowledge, as discussed later in Chapter 9, has been used to develop models for a 2-port microstrip via and to extend the frequency range of the stripline-to-stripline multilayer interconnect model. The developed EM-ANN models have been linked to a commercial microwave circuit simulator (HP-MDS [8]) where they can be used for circuit simulation and optimization.

5.2.1 Microstrip Transmission Line Model

An ANN model has been created to provide the characteristic impedance (Z_0) and the effective dielectric constant (ϵ_{eff}) for microstrip transmission lines. This modeling example demonstrates a situation where it is better to train for a specified maximum relative error rather than for an absolute error value, as mentioned earlier. For example, when creating a model for the characteristic impedance of a microstrip line, the output values may range from 10 Ω to 100 Ω. A 1 Ω absolute error corresponds to a 10% error at 10 Ω but only a 1% error at 100 Ω. Therefore the goal for this model was to have the sum of the average and standard deviation of the residual errors for Z_0 be less than 1%.

Figure 5.2 shows the geometry of the microstrip line to be modeled. Variable input parameters and corresponding ranges are given in Table 5.1.

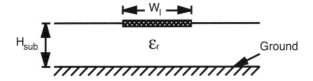

Figure 5.2 Cross-section of microstrip transmission line geometry where W_l is the width of the microstrip line, H_{sub} is the substrate height, and ϵ_r is the relative dielectric constant of the substrate.

Table 5.1
Variable Input Parameters for Microstrip Transmission Line Model

Input Parameter	Minimum Value	Maximum Value
Frequency	1 GHz	18 GHz
$\log_{10}(W_l/H_{sub})$	−1	1
ϵ_r	2	13

Output parameters are Z_o and ϵ_{eff}. Training data was provided using *linecalc* [27]. Since the training data was not time-consuming to obtain, DOE techniques were not used. Instead, a uniform grid of points was simulated to provide the training data. The training/test dataset consisted of 155 examples while the verification dataset contained 100 examples. Models were trained using both absolute and relative error criteria for comparison. The final models were trained using all 155 training/test examples and both used 10 hidden layer neurons.

The residual error results for both absolute error training and relative error training are given in Tables 5.2 and 5.3, respectively. Looking at these Z_o results, we note that the training for absolute error criterion provides lower absolute error, but higher relative (percentage) error. The highest relative errors are for low Z_o values as expected and do not meet the goal of having the average and standard deviation results sum to less than 1%. When using relative error measures during training, lower relative errors are obtained and the accuracy goal is achieved.

5.2.2 Broadband GaAs One-Port Microstrip Via

Figure 5.3 shows the structure and some parameters of the one-port via under consideration. The height of the substrate (H_{sub}), the dielectric constant (ϵ_r),

Table 5.2
Error Results (Average and Standard Deviation) Between the EM-ANN Model and *linecalc* for the Microstrip Transmission Line for Absolute Error Training

	Z_o (Ω)	Z_o (%)	ϵ_{eff}	ϵ_{eff} (%)
Train/test dataset				
Average error	0.206	1.161	0.012	0.377
Standard dev.	0.205	1.157	0.012	0.376
Verification dataset				
Average error	0.338	0.774	0.015	0.293
Standard dev.	0.520	0.875	0.013	0.223

Table 5.3
Error Results (Average and Standard Deviation) Between the EM-ANN Model and
linecalc for the Microstrip Transmission Line for Relative Error Training

	Z_o (Ω)	Z_o (%)	ϵ_{eff}	ϵ_{eff} (%)
Train/test dataset				
Average error	0.300	0.456	0.014	0.244
Standard dev.	0.344	0.334	0.015	0.186
Verification dataset				
Average error	0.399	0.518	0.011	0.186
Standard dev.	0.679	0.464	0.011	0.150

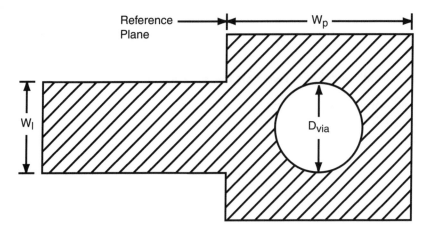

Figure 5.3 GaAs microstrip ground via geometry. Substrate thickness = 4 mil, ϵ_r = 12.9, tan δ = 0.002, σ_{metal} = 4.1 × 10^7 S/m, and t_{metal} = 0.1 mil.

and all loss parameters are considered constant for this example. The width of the incoming microstrip line, W_l, the side of the square shaped via pad, W_p, and the diameter of the via hole, D_{via}, are the three variable design parameters. Input variables for the EM-ANN model and their ranges are given in Table 5.4. Output variables are the magnitude and phase of S_{11} referenced to 50 Ω port termination.

EM simulations were performed from 5 to 55 GHz in 10 GHz steps using a commercially available full-wave electromagnetic simulator (HP-Momentum [2]). Via structures for 15 DOE central composite points—as well as for 14 additional training/testing points spaced midway between the previous points—were simulated. In addition, 16 structures were simulated for independent verification of the model upon completion of the training.

Table 5.4
Variable Input Parameters for GaAs Microstrip Ground Via Modeling

Input Parameter	Minimum Value	Maximum Value
Frequency	5 GHz	55 GHz
W_l/W_p	0.3	1.0
D_{via}/W_p	0.2	0.8
W_l/H_{sub}	0.1	2.0

Best results were obtained using 10 neurons in the hidden layer and 15 central composite points (DOE method), in addition to the 14 interior points for training the network. Residual error results for the EM-ANN model are given in Table 5.5.

5.2.3 Broadband GaAs Two-Port Microstrip Via

In addition to the 1-port via described above, a 2-port via has also been modeled. The same training points were used. The structure of the via is shown in Figure 5.4 and the input variables and corresponding ranges are given in Table 5.4. Output variables are the magnitudes and phases of S_{11} and S_{21}, again referenced to 50 Ω. As with the 1-port via, best results were obtained using 15 central composite points and the 14 interior points for training. Ten neurons were required for the hidden layer. Residual error results are given in Table 5.6.

It may be noted that the EM-ANN via models are able to achieve accuracy comparable to EM simulation over the entire 5–55 GHz range. Because a full-wave analysis is used, all the dielectric, conductor, and radiation losses, as well

Table 5.5
Error Results (Average and Standard Deviation) Between the EM-ANN Model and EM Simulation for the 1-Port Microstrip Via

	$\|S_{11}\|$	$\angle S_{11}$ (°)
Train/test dataset		
Average error	0.00105	0.456
Standard dev.	0.00102	0.522
Verification dataset		
Average error	0.00162	0.537
Standard dev.	0.00225	0.536

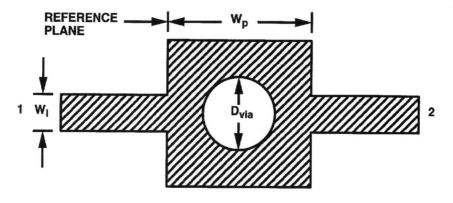

Figure 5.4 Two-port GaAs microstrip grounding via. Substrate thickness = 4 mil, ϵ_r = 12.9, $\tan \delta$ = 0.002, σ_{metal} = 4.1 × 10^7 S/m, and t_{metal} = 0.1 mil.

Table 5.6
Error Results (Average and Standard Deviation) Between the EM-ANN Model and EM Simulation for the 2-Port Microstrip Via

	$|S_{11}|$	$\angle S_{11}$ (°)	$|S_{21}|$	$\angle S_{21}$ (°)
Train/test dataset				
Average error	0.00361	0.413	0.00680	0.442
Standard dev.	0.00334	0.402	0.00492	0.474
Verification dataset				
Average error	0.00420	0.434	0.00881	0.771
Standard dev.	0.00385	0.356	0.00920	0.774

as parasitic effects, are included. The developed models may now be used in linear analysis, and also in nonlinear analysis where harmonic frequency components are generated.

5.2.4 Stripline-to-Stripline Multilayer Interconnect

Figure 5.5 shows the structure of a 50 Ω stripline-to-stripline multilayer interconnect for which an EM-ANN model has been developed. Reference planes are set at $W_{line}/2$ from the center of the via. The variable design parameters include the diameter of the via, D_{via}, and the diameter of the ground access opening, D_{gnd}. All other parameters are fixed. Model input variables and their ranges are given in Table 5.7. Output variables are the magnitudes and phases of S_{11} and S_{21}.

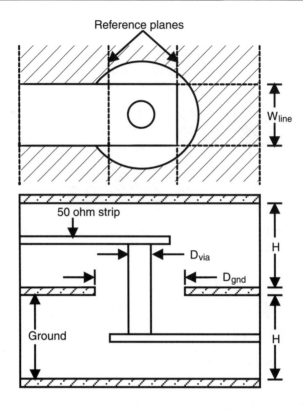

Figure 5.5 Stripline-to-stripline interconnect structure with W_{line} = 13.675 mil, Z_0 = 50 ohms, ϵ_r = 2.94, $\tan\delta$ = 0.0012, t_{metal} = 1.4 mil, σ_{metal} = 5.7 × 10^7 S/m, and H = 20 mil.

Table 5.7
Variable Parameters for Stripline-to-Stripline Multilayer Interconnect Model

Input Parameter	Minimum Value	Maximum Value
Frequency	1 GHz	26 GHz
D_{via}/W_{line}	0.365	0.8
D_{gnd}/D_{via}	1.25	6

EM simulations were performed from 1 GHz to 26 GHz in 5 GHz steps. Interconnect structures for 9 central composite points and 8 additional training/testing points (selected by the DOE approach) were simulated. In addition, 12 structures were simulated for model verification purposes.

Using the 9 central composite points plus the 8 additional points for training the model yielded the best results. Nine neurons were used in the hidden layer. Residual error results are given in Table 5.8.

Table 5.8
Error Results (Average and Standard Deviation) Between the EM-ANN Model and EM Simulation for the Stripline-to-Stripline Multilayer Interconnect

	$|S_{11}|$	$\angle S_{11}$ (°)	$|S_{21}|$	$\angle S_{21}$ (°)
Train/test dataset				
Average error	0.00172	1.043	0.00069	0.238
Standard dev.	0.00164	0.855	0.00087	0.230
Verification dataset				
Average error	0.00151	1.540	0.00057	0.220
Standard dev.	0.00128	1.306	0.00054	0.160

5.2.5 Microstrip-to-Microstrip Multilayer Interconnect

Figure 5.6 shows the structure of a 50 Ω microstrip-to-microstrip multilayer interconnect for which an EM-ANN model has been developed. Reference planes are set at $W_{bot}/2$ and $W_{top}/2$ from the center of the via for the bottom and top microstrip lines, respectively. The variable design parameters include the diameter of the via, D_{via}, and the length of the overhang for the top microstrip line, L_{OH}. All other parameters are fixed. Model input variables

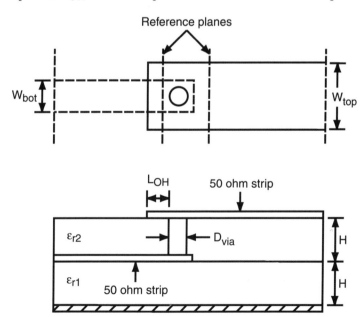

Figure 5.6 Microstrip-to-microstrip interconnect with Z_0 = 50 ohms, W_{bot} = 23 mil, W_{top} = 125 mil, ϵ_{r1} = 10.2, ϵ_{r2} = 2.2, $\tan\delta$ = 0.0012, t_{metal} = 1.4 mil, σ_{metal} = 5.7 × 10^7 S/m, and H = 25 mil.

and their ranges are given in Table 5.9. Output parameters for this ANN model are the magnitudes and phases of S_{11}, S_{21}, and S_{22}.

EM simulations were performed from 2 GHz to 12 GHz in 2 GHz steps. Interconnect structures for 9 central composite points and 8 additional training/testing points were simulated. In addition, 16 structures were simulated for model verification purposes.

Using the 9 central composite points (DOE method) plus the 8 additional interior points for training the model yielded the best results. 8 neurons were used in the hidden layer. Residual error results are given in Table 5.10.

5.2.6 Integration of EM-ANN Models with a Network Simulator

After training, the EM-ANN models are integrated into a microwave network simulator (HP-MDS [8]). Figure 5.7 compares the EM-ANN 1-port via model (NET1) with HP-Momentum results and the microstrip-via (*MSVIA*) element available in HP-MDS. Note that the *MSVIA* reference plane is at the center of the hole, while the ANN model reference plane is at the edge of the pad. Therefore, a more accurate model (whose performance is also shown in Figure 5.7) is constructed by adding additional HP-MDS elements such as *MSSTEP*,

Table 5.9
Variable Parameters for Microstrip-to-Microstrip Interconnect Model

Input Parameter	Minimum Value	Maximum Value
Frequency	2 GHz	12 GHz
D_{via}/W_{bot}	0.2	0.9
L_{OH}	1 mil	15 mil

Table 5.10
Error Results (Average and Standard Deviation) Between the EM-ANN Model and EM Simulation for the Microstrip-to-Microstrip Multilayer Interconnect

	$\|S_{11}\|$	$\angle S_{11}$ (°)	$\|S_{21}\|$	$\angle S_{21}$ (°)	$\|S_{22}\|$	$\angle S_{22}$ (°)
Train/test dataset						
Average error	0.0009	0.745	0.0003	0.083	0.0011	0.798
Standard dev.	0.0008	0.493	0.0002	0.082	0.0009	0.645
Verification dataset						
Average error	0.0010	0.752	0.0003	0.083	0.0013	0.813
Standard dev.	0.0009	0.659	0.0003	0.082	0.0010	0.768

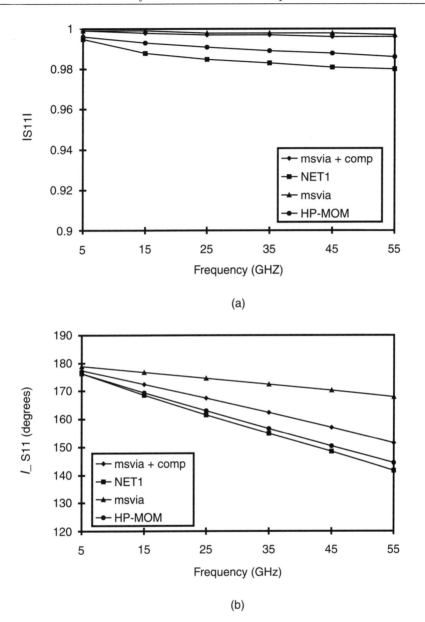

Figure 5.7 Comparison of EM-ANN model (NET1), HP-Momentum (HP-MOM), HP-MDS via element (msvia), and *MSVIA* with added components (msvia + comp). GaAs via with ϵ_r = 12.9, H_{sub} = 4 mil, t_{metal} = 0.1 mil, σ_{metal} = 4.1 × 10^7 S/m, tan δ = 0.002, W_l/W_p = 0.3875, D_{via}/W_p = 0.4, and W_l/H_{sub} = 0.3375 (from [5], © 1996 IEEE, reprinted with permission).

MSTL, and *MSOC* to account for pad length and step-in-width. However, this modified HP-MDS model does not accurately characterize the via hole over the entire range of input variables, whereas the EM-ANN via model does. As expected, the EM-ANN model results agree with the HP-Momentum [2] simulations. Simulation times for NET1, *MSVIA*, and HP-Momentum on an HP 700 workstation are shown in Table 5.11. Note that the EM-ANN model does not require a significant increase in simulation time over the conventional HP-MDS model.

5.3 EM-ANN Models for CPW Components

The use of coplanar waveguides (CPW) in RF and microwave integrated circuits offers several advantages due to the physical configuration of the CPW line. These include the ease of mounting shunt and series lumped components, low radiation losses, low dispersion, and the avoidance of the need for fragile substrates. Designers of CPW circuits currently face difficulties due to the unavailability of accurate and efficient models for CPW discontinuities such as bends, T-junctions, steps-in-width, short and open stubs, and so forth. Accurate characterization and modeling of these components are vital for accurate circuit simulation and increased first-pass design success.

Much effort has been expended in recent years toward the development of accurate and efficient methods for electromagnetic (EM) simulation of CPW discontinuities [28–60]. The time-consuming nature of EM simulation, however, limits the use of these tools for interactive CAD and circuit optimization. Equivalent circuit models available for CPW discontinuities require certain assumptions to be made which may or may not be valid over the desired range of operation. CPW discontinuities have also been characterized by measurements [61, 62], generating a library of data valid only for the structures measured. This method of characterization is also very time-consuming.

Table 5.11
Comparison of Simulation Times for the GaAs Via Described in Figure 5.6. The Times for *MSVIA* and NET1 Are Averaged over 100 Frequency Points. HP-Momentum Results Are for 1 Frequency Point

Model	Simulation Time
HP-MDS, *MSVIA*	0.37 sec
HP-Momentum	12.48 min
NET1 (EM-ANN Model)	0.54 sec

This section reviews the EM-ANN models developed recently [6, 7] for CPW transmission lines (frequency dependent Z_o and ϵ_{re}), 90° bends, short circuit stubs, open circuit stubs, step-in-width discontinuities, and symmetric T-junctions. Air-bridges are included where needed to suppress the unwanted slot-line mode [63]. All models have been developed using HP-Momentum [2] for EM simulation and its adaptive frequency sampling (AFS) feature. Common parameters for all models are the substrate parameters ($\epsilon_r = 12.9$, $H_{sub} = 625$ μm, $\tan\delta = 0.0005$) and air-bridge parameters ($H_a = 3$ μm and $W_a = 40$ μm). Once developed, these EM-ANN models are linked to a commercial microwave circuit simulator, where they provide accuracy approaching that of the EM simulation tool used for characterization of the CPW components over the entire range of the model input variables. In addition, the developed models allow for very fast, accurate electromagnetic/circuit optimization within the framework of the circuit simulator environment.

5.3.1 EM-ANN Modeling of CPW Transmission Lines

An EM-ANN model has been developed for frequency-dependent characteristics of CPW transmission lines. Variable input parameters and corresponding ranges are given in Table 5.12. Model outputs are frequency-dependent Z_o and ϵ_{re}. The EM-ANN model has been trained using PCAAMT [62], a program that provides full-wave solutions for printed transmission lines and general multilayer geometries. The training/test dataset consisted of 265 examples, while the verification dataset contained 51 examples. Twenty neurons were used in the hidden layer. Error results between the developed EM-ANN model and the full-wave solution for the training/test and verification datasets are shown in Table 5.13.

5.3.2 Modeling of CPW Bends

EM-ANN Modeling of Chamfered CPW 90° Bends

In this section, EM-ANN models are presented for two different chamfered CPW bend structures, as shown in Figure 5.8. Air-bridges are placed near

Table 5.12
Variable Parameter Ranges for CPW Components

	Min.	Max.
Frequency	1 GHz	50 GHz
W	20 μm	120 μm
G	20 μm	60 μm

Table 5.13
Error Results (Average and Standard Deviation) Between the EM-ANN Model and EM Simulation for the CPW Transmission Line

	Z_o (Ω)	Z_o (%)	ϵ_{re}
Train/test dataset			
Average error	0.105	0.234	0.00283
Standard dev.	0.098	0.224	0.00248
Verification dataset			
Average error	0.111	0.243	0.00341
Standard dev.	0.112	0.238	0.00240

CPW bends in order to reduce the unwanted slot-line mode, which tends to radiate [58]. The slot-line mode is generated in CPW bends due to the path length difference for the two slots. The inclusion of air-bridges, however, adds unwanted capacitance, which can degrade the performance of the bend. In addition to compensating for the reactances associated with the bend, chamfering provides a simple way to partially compensate for the effects of the air-bridges. This section investigates the effects of chamfering on the S-parameters for the CPW 90° bends shown in Figure 5.8

CPW bends of the type shown in Figure 5.8a have been studied in [30], for two values of the chamfer b = 0 and b = W + G. ANN modeling has been used for extending the work reported in [30] by determining the optimal chamfer for the bend structure. ANN has also been used to develop the compensated CPW bend shown in Figure 5.8b, which has been shown to improve upon the return loss and the insertion loss of the conventional chamfered bend of Figure 5.8a. In this study, all air-bridge parameters were held constant in order to concentrate on the effects caused by the chamfering of the bends.

Optimally Chamfered Conventional CPW Bend

An EM-ANN model has been developed for the CPW bend structure shown in Figure 5.8a. Variable inputs for the EM-ANN model are W, G, b/b_{max}, and frequency. Model outputs are S-parameters. Substrate material used is GaAs (ϵ_r = 12.9), and the thickness is 625 μm for this modeling effort. Air-bridges are 40 μm wide (W_a), 3 μm (H_a) above the GaAs surface, and are positioned at the bend discontinuity as shown.

EM simulations were performed on 17 bend structures, included within the range of parameters given in Table 5.12. Characteristic impedances used for port terminations have been determined from *linecalc* [27]. Five different

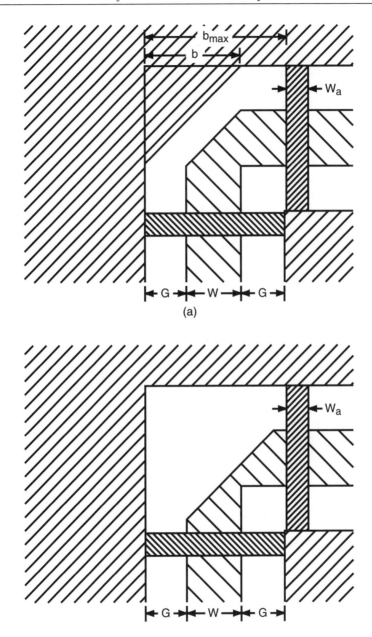

Figure 5.8 CPW 90° bend structures with $W_a = 40$ μm, $H_a = 3$ μm, $H_{sub} = 625$ μm, $\epsilon_r = 12.9$, and $\tan\delta = 0.0005$. (a) Conventional chamfered bend and (b) compensated bend (H_a is height of air-bridge above the substrate, W_a is the width of the air-bridge, and H_{sub} is the substrate thickness). (From [7], © 1997 IEEE, reprinted with permission.)

values of chamfer were simulated for each bend, with b/b_{max} ranging from 0 to 1. The strip corner was chamfered by a proportional amount given by $bW/(W + G)$. EM simulation data was separated into training/test (165 examples) and verification (300 examples) datasets for model development. Ten neurons were used in the hidden layer. Error results (EM-ANN model compared to EM simulation) are shown in Table 5.14.

It was determined that there indeed exists an optimal chamfer for each structure, especially when analyzing the return loss, $20\log_{10}|S_{11}|$. The developed EM-ANN model can reproduce the trends in S-parameters determined by the changes in the physical structure of the CPW bend. Using this model, the optimal chamfer for each bend was determined. Figure 5.9 shows the optimal chamfer, b/b_{max}, versus W/G for minimum return loss, as determined by the EM-ANN model. The optimal chamfer as a function of W/G is given by

Table 5.14
Error Results (Average and Standard Deviation) Between the EM-ANN Model and Full-Wave Simulation for the Optimally Chamfered CPW Bend
(From [7], © 1997 IEEE, Reprinted with Permission)

	$|S_{11}|$	$\angle S_{11}$ (°)	$|S_{21}|$	$\angle S_{21}$ (°)
Train/test dataset				
Average error	0.000701	0.102	0.000293	0.034
Standard dev.	0.000632	0.104	0.000270	0.029
Verification dataset				
Average error	0.000966	0.124	0.000354	0.048
Standard dev.	0.001058	0.141	0.000339	0.039

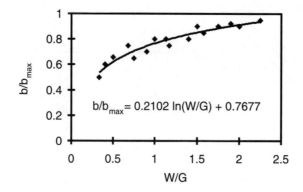

Figure 5.9 Optimal chamfer for return loss versus W/G for the conventional CPW 90° bend (Figure 5.8a) (from [7], © 1997 IEEE, reprinted with permission).

$$b/b_{max} = 0.2102 \ln(W/G) + 0.7677 \quad \text{for } 0 \leq W/G \leq 2 \qquad (5.1)$$
$$= 1 \quad \text{for } W/G > 2.5$$

Compensated CPW Bend

An EM-ANN model has also been developed for the compensated CPW bend structure shown in Figure 5.8b. This compensated bend is capable of improving upon the performance of the already discussed optimally chamfered CPW bend. Variable input parameters for the model are W, G, and frequency. Model outputs are S-parameters. For this bend geometry, the optimum chamfer for the strip is found to be the maximum allowable by air-bridge placement. EM simulations were performed on 17 bend structures—included within the range of parameters given in Table 5.12—to provide training/test (45 examples) and verification (35 examples) datasets for model development. Ten neurons were used in the hidden layer. Error results for the developed EM-ANN model are shown in Table 5.15.

CPW Bend Comparisons

The compensated CPW bend structure is found to improve the return loss over the optimally chamfered conventional bend, as shown in Table 5.16. This is believed to be due to a decrease in capacitance as the slot width at the corner is increased, thereby compensating for the increase in capacitance due to the air-bridges. Improvements are also seen in the insertion loss. Note that all air-bridge parameters remain the same as in the case for the optimally chamfered CPW bend.

Comparisons between unchamfered CPW corner, optimally chamfered conventional bend, and the compensated bends are shown in Figure 5.10 for

Table 5.15
Error Results (Average and Standard Deviation) Between the EM-ANN Model and Full-Wave Simulation for the Compensated Bend
(From [7], © 1997 IEEE, Reprinted with Permission)

	$|S_{11}|$	$\angle S_{11}$ (°)	$|S_{21}|$	$\angle S_{21}$ (°)
Train/test dataset				
Average error	0.000784	0.680	0.000584	0.363
Standard dev.	0.000518	0.513	0.000390	0.261
Verification dataset				
Average error	0.001390	1.022	0.000705	0.447
Standard dev.	0.000861	0.759	0.000365	0.380

Table 5.16
Comparison of Return Loss for the Conventional Optimally Chamfered Bend and the Novel Compensated Bend for Several Structures (Frequency = 50 GHz)
(From [7], © 1997 IEEE, Reprinted with Permission)

W (μm)	G (μm)	Optimally Chamfered Bend Return Loss (dB)	Novel Bend Return Loss (dB)	Improvement (dB)
70	60	−13.89	−20.92	7.03
120	40	−12.96	−16.36	3.40
70	20	−21.51	−26.94	5.43
35	55	−17.27	−20.18	2.91
105	55	−12.50	−17.02	4.52
105	25	−15.76	−19.33	3.57
70	40	−16.36	−21.41	5.05

CPW bend structures with W = 70 μm and G = 20 μm, corresponding to Z_o = 35 Ω. Note that all air-bridges and reference planes are in the same positions. Based on the results shown in Figure 5.10 and in Table 5.16, the compensated bend provides significant improvements in return and insertion loss over the other CPW bend structures. Improvements are also seen when comparing the optimally chamfered bend results to the corner (unchamfered) bend.

5.3.3 EM-ANN Models for CPW Opens and Shorts

Open and short circuit stubs are important components for many circuit designs, such as filters and impedance matching networks. The geometries of the open and short components considered are shown in Figure 5.11 and in Figure 5.12, respectively. Variable input parameters and corresponding ranges for model development were given earlier in Table 5.12. Model outputs are S-parameters. Typical output responses of open and short circuits with reference planes at the discontinuities are shown in Figure 5.13. For each model, EM simulations have been performed on 17 structures over the 1 GHz to 50 GHz frequency range. For the short circuit model development, 71 examples were used in the training/test dataset and 46 examples for the verification dataset, requiring 5 neurons in the hidden layer. Open circuit model development was accomplished using 95 examples in the training/test dataset, 53 examples for the verification dataset, and 6 neurons in the hidden layer. Error results for the training and verification datasets are given in Table 5.17 for the short circuit and in Table 5.18 for the open circuit.

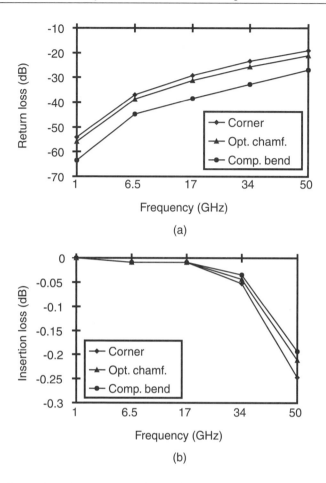

Figure 5.10 Comparison of unchamfered (Corner), conventional (Opt. Chamf.), and novel (Comp. Bend) CPW bends. W = 70 μm, G = 20 μm, ϵ_r = 12.9, and H_{sub} = 625 μm (from [7], © 1997 IEEE, reprinted with permission).

5.3.4 EM-ANN Modeling of CPW Step-in-Width

CPW step-in-width discontinuities are used extensively in circuit design for introducing impedance changes. The geometry of the step-in-width, for which an EM-ANN model has been developed, is shown in Figure 5.14. Variable input parameters for the model are frequency, W_1, W_2, and G. Also, only structures with $W_1 < W_2$ were used for model development. Parameter ranges are given in Table 5.12. Model outputs are S-parameters. A typical output response is shown in Figure 5.15. EM simulations have been performed on 30 structures over the 1 GHz to 50 GHz frequency range, providing 95 training/test examples and 55 verification examples. Eight neurons were used

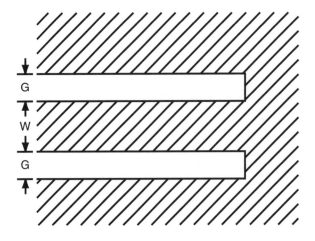

Figure 5.11 CPW short circuit geometry.

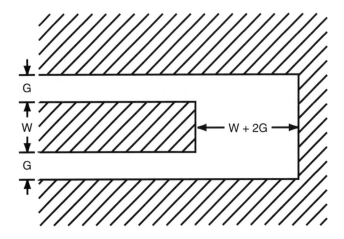

Figure 5.12 CPW open circuit geometry.

in the hidden layer. Residual errors are shown in Table 5.19 and are negligible for design applications.

5.3.5 EM-ANN Modeling of CPW Symmetric T-Junctions

The symmetric T-junction under consideration is shown in Figure 5.16. Variable model input parameters are frequency and the physical dimensions W_{in}, G_{in}, W_{out}, and G_{out}. Parameter ranges are given in Table 5.12. Model outputs are S-parameters. A typical output response is shown in Figure 5.17. EM simulations were performed on 25 structures over the 1 GHz to 50 GHz

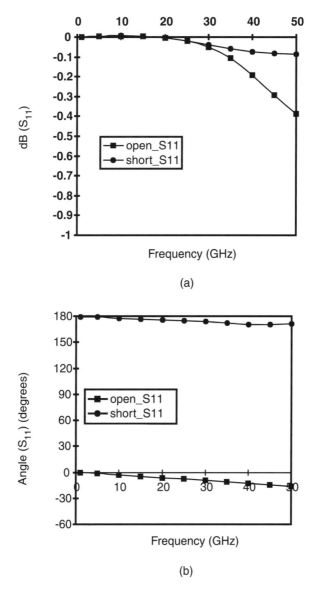

Figure 5.13 S-parameter response for CPW open and short circuits. W = 70 μm, G = 60 μm, and reference planes at the discontinuities (from [7], © 1997 IEEE, reprinted with permission).

Table 5.17
Error Results (Average and Standard Deviation) Between the EM-ANN Model and
Full-Wave Simulation for the CPW Short Circuit
(From [7], © 1997 IEEE, Reprinted with Permission)

	$|S_{11}|$	$\angle S_{11}$ (°)
Train/test dataset		
Average error	0.000248	0.396
Standard dev.	0.000350	0.271
Verification dataset		
Average error	0.000381	0.964
Standard dev.	0.000412	0.867

Table 5.18
Error Results (Average and Standard Deviation) Between the EM-ANN Model and
Full-Wave Simulation for the CPW Open Circuit
(From [7], © 1997 IEEE, Reprinted with Permission)

	$|S_{11}|$	$\angle S_{11}$ (°)
Train/test dataset		
Average error	0.000481	0.332
Standard dev.	0.000633	0.373
Verification dataset		
Average error	0.000520	0.634
Standard dev.	0.000892	0.676

frequency range, providing 155 training/test examples and 131 verification examples. Fifteen neurons were used in the hidden layer. Model error results are shown in Table 5.20 for both the training/test and verification datasets.

5.4 Other Passive Components' Models

In addition to the modeling of interconnect vias and CPW components discussed in Sections 5.2 and 5.3 above, the neural network modeling approach has been used for a variety of other RF and microwave passive components. These include spiral inductors [65, 66], multiconductor transmission lines [67–69], microstrip patch antennas [70–72], and coupled cavity waveguide filter components [73, 74]. These models are described briefly in this section.

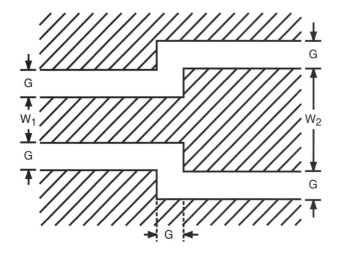

Figure 5.14 CPW step-in-width geometry.

5.4.1 Spiral Inductors

This modeling approach is similar to those discussed in the previous sections of this chapter. Inputs to the neural network model are the physical dimensions of the inductor and the desired frequency. The outputs are the S-parameter of that inductor at the respective frequency points. The neural net was trained with S-parameters of 70 distinct square spiral inductors. This resulted in 350 training sets for the C-band and X-band models and 630 for a combined C-X-band model. EM simulation data used for training was obtained with *em* software from Sonnet Software [75]. The layout of a typical inductor considered is shown in Figure 5.18. Test cases for EM simulation were selected by the fractional-factorial experimental design procedure—a technique similar to traditional DOE [66]. The correlations between neural model computed and EM-simulated results were greater than 0.98 for each modeled parameter.

5.4.2 Multiconductor Transmission Lines

Multiconductor striplines and microstrip lines are important components in multilayer RF and microwave circuits as well as for interconnects in VLSI digital circuits. Use of neural networks for modeling and optimization of high speed IC interconnects is discussed in Chapter 6. In this subsection, we briefly review modeling for multiconductor transmission lines as needed in multilayer RF and microwave circuits.

Multilayer transmission lines are used in the design of several multilayered RF and microwave circuits, such as directional couplers, baluns, filters, and so

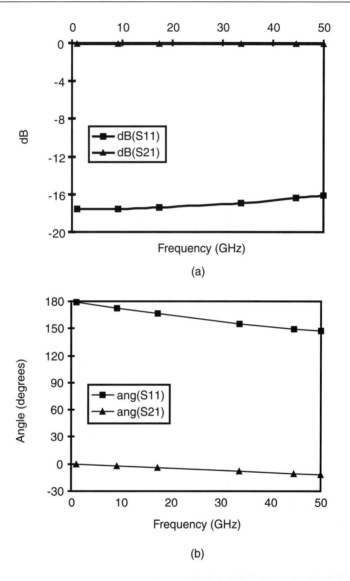

Figure 5.15 S-parameter response for a 71 Ω to 50 Ω CPW step-in-width transition. W_1 = 20 μm, W_2 = 70 μm, G = 60 μm, and reference planes at the discontinuity (from [7], © 1997 IEEE, reprinted with permission).

forth. A methodology for the design of multilayer asymmetric coupled line circuits was proposed recently [76]. In the proposed approach, network modeling and synthesis procedures are used to derive normal mode parameters (NMPs) for various multilayer coupled-line sections, parts of the circuit to be designed. The evaluation of physical geometry to realize the NMPs is not

Table 5.19
Error Results (Average and Standard Deviation) Between the EM-ANN Model and
Full-Wave Simulation for the Step-in-Width
(From [7], © 1997 IEEE, Reprinted with Permission)

	$|S_{11}|$	$\angle S_{11}$ (°)	$|S_{21}|$	$\angle S_{21}$ (°)
Train/test dataset				
Average error	0.000864	0.435	0.000792	0.321
Standard dev.	0.000721	0.454	0.000840	0.286
Verification dataset				
Average error	0.001214	0.553	0.000986	0.457
Standard dev.	0.001023	0.579	0.001032	0.390

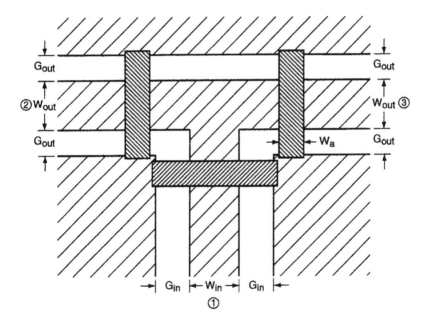

Figure 5.16 CPW symmetric T-junction geometry.

straightforward. The ANN modeling approach has been recommended for this purpose. ANN models have been developed for both synthesis (NMPs to the physical geometry) and analysis (physical geometry to NMPs). An example of a multilayer coupled-line configuration is shown in Figure 5.19. It consists of an asymmetrical coupled-line section in an inhomogeneous dielectric medium. Both layer-to-layer coupling and same layer coupling geometries are shown.

The development of the analysis model is straightforward. Models are developed for each type of coupled line section using the methodology discussed

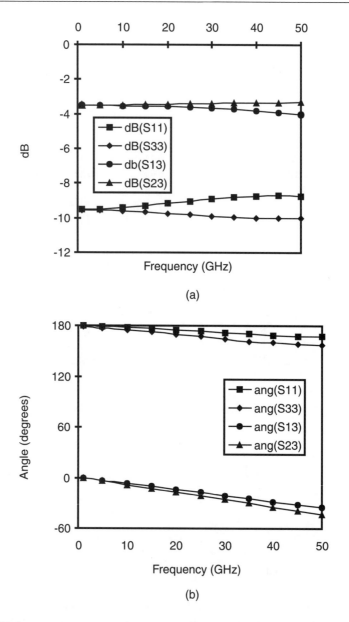

Figure 5.17 S-parameter response for a typical CPW symmetric T-junction. $W_{in} = W_{out} = 70\ \mu m$, $G_{in} = G_{out} = 60\ \mu m$, and reference planes at the air-bridge locations (from [7], © 1997 IEEE, reprinted with permission).

Table 5.20
Error Results Between the EM-ANN Model and EM Simulation for the CPW Symmetric T-Junction (Average and Standard Deviation). Input Branchline Port is Port 1, and the Output Ports on the Main Line are Ports 2 and 3 (From [7], © 1997 IEEE, Reprinted with Permission)

	$|S_{11}|$	$\angle S_{11}$ (°)	$|S_{13}|$	$\angle S_{13}$ (°)	$|S_{23}|$	$\angle S_{23}$ (°)	$|S_{33}|$	$\angle S_{33}$ (°)
Train Avg.	0.00150	0.754	0.00071	0.176	0.00084	0.246	0.00106	0.633
St.dev.	0.00128	0.696	0.00058	0.172	0.00097	0.237	0.00109	0.546
Verify Avg.	0.00345	0.782	0.00088	0.141	0.00126	0.177	0.00083	0.838
St.dev.	0.00337	0.674	0.00085	0.125	0.00105	0.129	0.00068	0.717

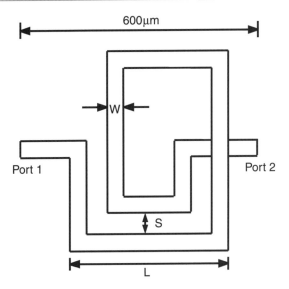

Figure 5.18 Example layout of square spiral inductor with 1.5 turns. The other factors used in the modeling are depicted as L, S, and W; which represent length, spacing, and width, respectively.

earlier. Physical parameters—such as the width of each line and the spacing between their edges—are used as ANN model inputs, as shown in Figure 5.20. The outputs are the elements of the L and C matrices corresponding to the coupled-line configuration. Once the L and C matrices have been obtained, they can be used to analytically determine the NMPs and consequently the S-parameters of a given coupled-line section. Analysis models may be linked to commercial microwave circuit simulators for circuit analysis and optimization.

The synthesis model shown in Figure 5.21 is developed by using normal mode parameters as inputs to the ANN model. The desired physical parameters

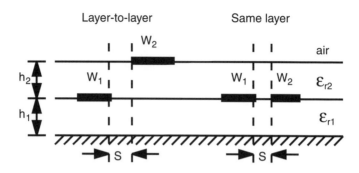

Figure 5.19 Example of multiconductor multilayer coupled-line geometries in an inhomogeneous medium. Both layer-to-layer and same-layer coupling sections are shown.

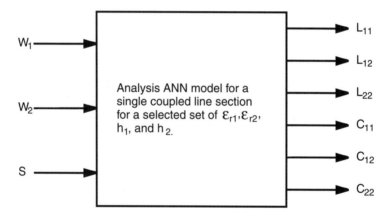

Figure 5.20 Analysis ANN model for asymmetric multilayer coupled line sections.

for the coupled line section are the outputs. This is known as an inverse modeling problem because the input and output variables are interchanged from those for characterization or the analysis model. For such problems there exists a well-defined forward (analysis) problem characterized by a single-valued mapping. The mapping for inverse problems, however, can often be multivalued. In such cases, if a least-squares error approach is used, the neural net tends to approximate the average of the desired target data [77]. This can lead to poor network performance since the average of several solutions is not necessarily itself a solution.

Another problem associated with inverse modeling is the coverage of the input variables' space by the selected training data. For the forward problem, it is possible to select points for ANN training which characterize the entire

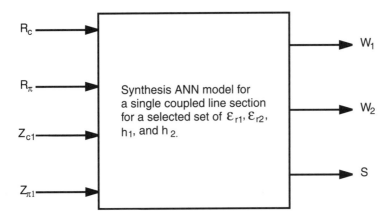

Figure 5.21 Synthesis ANN model for asymmetric multilayer coupled-line sections.

space spanned by the input variables. However, when the roles of the input and output variables are interchanged for inverse modeling, full characterization of the new input space is most likely not achieved. Therefore there can be valid input vectors (within the input space) for the ANN model, which produce incorrect results due to the absence of training data for particular regions of input space.

The problem for the asymmetrical coupled-line design case, as mentioned previously, is the fact that the mapping from physical parameters to NMPs is single-valued, whereas the inverse mapping from NMPs to physical parameters can be multivalued. Therefore a method is needed for determining the accuracy of the model outputs. The method for verifying the accuracy of the synthesis model outputs is shown in Figure 5.22. First, the desired NMPs are used as inputs for the synthesis ANN model, giving physical parameters as outputs. The outputs of the synthesis model are then used as inputs to the analysis model—which is a single-valued mapping—to determine the L and C matrices for the geometry obtained from the synthesis model. NMPs are calculated from the L and C matrices, and transformed into 4-port S-parameters [78]. These 4-port S-parameters are then compared to the 4-port S-parameters obtained from the NMPs, which have been used as inputs to the synthesis model. A determination can be made in this way as to the accuracy of the synthesis model for a given region of input space. If the model is not accurate for a given set of NMPs (input variables), the NMPs are altered by changing design parameters. The design parameter 'a' is related to the normal mode parameters of the asymmetric coupled line section [78, 81].

The comments made above for developing ANN models for synthesis are quite generic in nature. Any synthesis model needs to be used in conjunction

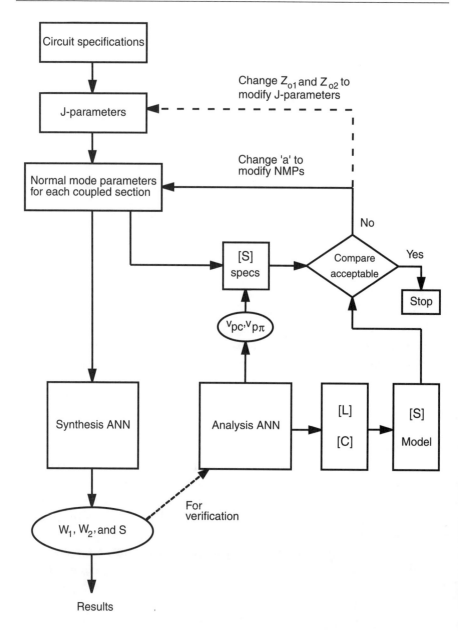

Figure 5.22 Synthesis procedure for asymmetric multilayer coupled line sections using ANN models (from [67], Watson, P. M., C. Cho, and K. C. Gupta, "EM-ANN Model for Synthesis of Physical Dimensions for Multilayer Asymmetric Coupled Transmission Line Structures," *Int. J. RF and Microwave CAE*, pp. 175–186, © 1999, John Wiley and Sons. Reprinted with permission from John Wiley and Sons, Inc.).

with the corresponding analysis model in order to assure the correctness of synthesis results.

5.4.3 Microstrip Patch Antennas

Microstrip patch antennas are the most common printed radiating elements at RF and microwave frequencies [79]. Geometry of a simple patch fed by a coaxial line is shown in Figure 5.23. There are several other feeding configurations, including a microstrip line feed network printed on the same substrate.

Although there are several simple models available for the analysis of microstrip patches, an accurate design requires the use of electromagnetic simulators. ANN modeling can provide an efficient design procedure for these antennas. So far, neural networks have been used for this application only to a limited extent. A neural network model has been developed for the calculation of resonance frequencies for an equilateral triangular microstrip antenna [70]. Training and test datasets used for the development of this model were obtained from previous works reported in the literature. Some of these results were measured and others were calculations based on several other models. A similar modeling has also been carried out for the calculation of resonance frequencies of a circular microstrip antenna [71]. In that case, results available in literature were used for training the neural network model as well. And a neural network model is also available for the calculation of bandwidth for electrically thin

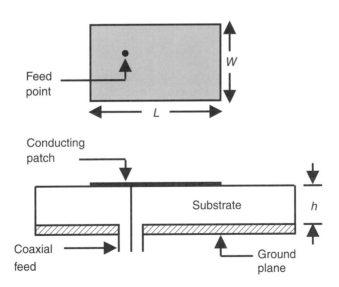

Figure 5.23 Schematic of a simple microstrip patch antenna.

and thick microstrip patch antennas [72]. This latter model was developed using experimentally measured data for neural network training.

The neural network modeling approach has also been used for design of other printed antennas like CPW patch antennas [80] discussed in Chapter 8.

5.4.4 Waveguide Filter Components

In addition to the design of planar RF and microwave components, ANN modeling has also been used for the design of waveguide components. Good examples of this application include E-plane metal insert waveguide filters [73], and direct-coupled cavity waveguide filters [74]. Generic coupled cavity waveguide filter geometry is shown in Figure 5.24. Finite element EM simulation results were used as ANN training data in this case. ANN model development procedure is similar to that discussed in other cases earlier and is summarized in Figure 5.25.

The geometry of E-plane metal insert filters for which ANN models have been developed is shown in Figure 5.26. The goal of this effort has been the development of a design procedure that takes manufacturing errors into account. Availability of a flexible and accurate model of the filter opens up several perspectives:

- A further improvement of neural model accuracy by means of a preliminary neural clustering of filter curves;

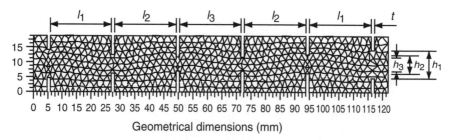

Figure 5.24 An example of a fifth-order filter with rectangular irises in a WR75 waveguide. The inner region is meshed to perform the FEM analysis. The geometrical parameters characterizing the filtering device—the neural network outputs—are the length of each cavity and the opening length of each iris. The thickness of the irises is set to t = 1 mm (from [73], Fedi, G., et al., "Direct-Coupled Cavity Filter Design Using a Hybrid Feedforward Neural Network-Finite Element Procedure," *Int. J. RF Microwave CAE*, pp. 287–296, © 1999, John Wiley and Sons. Reprinted with permission from John Wiley and Sons, Inc.).

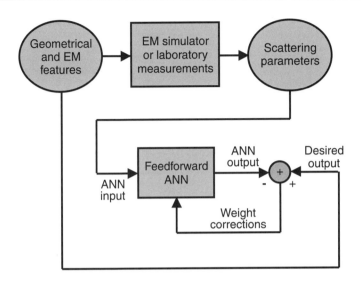

Figure 5.25 Procedure used for the development of ANN models for direct-coupled waveguide filters (from [73], Fedi, G., et al., "Direct-Coupled Cavity Filter Design Using a Hybrid Feedforward Neural Network-Finite Element Procedure," *Int. J. RF Microwave CAE*, pp. 287–296, © 1999, John Wiley and Sons. Reprinted with permission from John Wiley and Sons, Inc.).

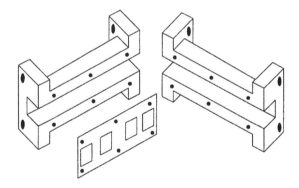

Figure 5.26 E-plane metal insert filter configuration for which ANN models have been developed (from [74], © 1998 IEEE, reprinted with permission).

- The automatic estimate of filter response sensitivity to inaccuracies of geometrical dimensions—useful in obtaining information about the manufacturing accuracy needed in order to maintain the overall filter response within a desired tolerance;
- Inverting the neural model in order to obtain a new set of geometrical dimensions which take manufacturing tolerances into account.

References

[1] Liao, S. Y., *Microwave Circuit Analysis and Amplifier Design*, NJ: Prentice-Hall, Inc., 1987.

[2] HP-Momentum, ver. A.02.51, Hewlett-Packard Co. (now Agilent Technologies, Inc.), Santa Rosa, CA, 1996 (and other versions).

[3] Montgomery, D. C., *Design and Analysis of Experiments*, NY: John Wiley and Sons, Inc., 1991.

[4] Watson, P. M., and K. C. Gupta, "EM-ANN Models for Via Interconnects in Microstrip Circuits," *1996 IEEE MTT-S Symposium Digest*, pp. 1819–1822.

[5] Watson, P. M., and K. C. Gupta, "EM-ANN Models for Microstrip Vias and Interconnects in Multilayer Circuits," *IEEE Trans. Microwave Theory and Techniques*, Vol. 44, No. 12, Dec. 1996, pp. 2495–2503.

[6] Watson, P. M., and K. C. Gupta, "EM-ANN Modeling and Optimum Chamfering of 90 Degree CPW Bends with Air Bridges," *IEEE Intn. Microwave Symp. Digest*, Denver, CO, June 1997, pp. 1603–1606.

[7] Watson, P. M., and K. C. Gupta, "Design and Optimization of CPW Circuits Using EM-ANN Models for CPW Components," *IEEE Trans. Microwave Theory and Techniques*, Vol. 45, Dec. 1997, pp. 2515–2523.

[8] HP-MDS, ver. 7.00.00, Hewlett-Packard Co., Santa Rosa, CA, 1996 (and other versions).

[9] Watson, P. M., "Artificial Neural Network Modeling for Computer-Aided Design of Microwave and Millimeter-Wave Circuits," Ph.D. thesis (Advisor: Prof. K. C. Gupta), ECE Dept., University of Colorado at Boulder, 1998.

[10] Cerri, G., M. Mongiardo, and T. Rozzi, "Radiation from Via-Hole Grounds in Microstrip Lines," *IEEE MTT-S Int. Microwave Symp. Dig.*, 1994, pp. 341–344.

[11] Wang, T., R. F. Harrington, and J. Mautz, "Quasi-Static Analysis of a Microstrip Via through a Hole in a Ground Plane," *IEEE Trans. Microwave Theory and Techniques*, Vol. 36, No. 6, June 1988, pp. 1008–1013.

[12] Goldfarb, M. E., and R. A Pucel, "Modeling Via Hole Grounds in Microstrip," *IEEE Microwave and Guided Wave Letters*, Vol. 1, No. 6, June 1991, pp. 135–137.

[13] Finch, K. L., and N. G. Alexopoulos, "Shunt Posts in Microstrip Transmission Lines," *IEEE Trans. Microwave Theory and Techniques*, Vol. MTT-38, No. 11, Nov. 1990, pp. 1585–1594.

[14] Jansen, R. H., "A Full-wave Electromagnetic Model of Cylindrical and Conical Via Hole Grounds for use in Interactive MIC/MMIC Design," *IEEE MTT-S Microwave Symp. Digest*, 1992, pp. 1233–1236.

[15] Maeda, S., T. Kashiwa, and I. Fukai, "Full-wave Analysis and Propagation Characteristics of a Through Hole Using the Finite Difference Time-Domain Method," *IEEE Trans. Microwave Theory and Techniques*, Vol. MTT-39, No. 12, Dec. 1991, pp. 2154–2159.

[16] Becker, W. D., P. Harms, and R. Mittra, "Time Domain Electromagnetic Analysis of a Via in a Multilayer Computer Chip Package," *IEEE MTT-S Microwave Symp. Digest*, 1992, pp. 1229–1232.

[17] Visan, S., O. Picon, V. F. Hanna, "3D Characterization of Air Bridges and Via Holes in Conductor Backed Coplanar Waveguides for MMIC Applications," *IEEE MTT-S Microwave Symp. Digest*, 1993, pp. 709–712.

[18] Eswarappa, C., and W. J. R. Hoefer, "Time-Domain Analysis of Shorting Pins in a Microstrip using 3-D SCN TLM," *IEEE MTT-S Digest*, 1993, pp. 917–920.

[19] Tsai, W. J., and J. T. Aberle, "Analysis of a Microstrip Line Terminated with a Shorting Pin," *IEEE Trans. Microwave Theory and Techniques*, Vol. MTT-40, No. 4, 1992, pp. 645–651.

[20] Cerri, G., M. Mongiardo, and T. Rozzi, "Full-wave Equivalent Circuit of Via Hole Grounds in Microstrip," *Proc. 23rd European Microwave Conference*, 1993, pp. 207–208.

[21] Daigle, R. C., G. W. Bull, and D. J. Doyle, "Multilayer Microwave Boards: Manufacturing and Design," *Microwave Journal*, Apr. 1993, pp. 87–97.

[22] Sorrentino, R., et al., "Full-Wave Modeling of Via-Hole Grounds in Microstrip by Three-dimensional Mode Matching technique," *IEEE Trans. Microwave Theory and Techniques*, Vol. MTT-40, No. 12, Dec. 1992.

[23] Becker, W., P. Harms, and R. Mittra, "Time-Domain Electromagnetic Analysis of Interconnects in a Computer Package," *IEEE Trans. Microwave Theory and Techniques*, Vol. MTT-40, No. 12, Dec. 1992, pp. 2155–2163.

[24] Harms, P., J. Lee, and R. Mittra, "Characterizing the Cylindrical via Discontinuity," *IEEE Trans. Microwave Theory and Techniques*, Vol. MTT-41, No. 1, Jan. 1993, pp. 153–156.

[25] Gu, O., E. Yang, and M. Tassoudji, "Modeling and Analysis of Vias in Multilayered Integrated Circuits," *IEEE Trans. Microwave Theory and Techniques*, Vol. MTT-41, No. 2, Feb. 1993, pp. 206–214.

[26] Swanson, D., " Grounding Microstrip Lines with Via Holes," *IEEE Trans. Microwave Theory and Techniques*, Vol. MTT-40, No. 8, Aug. 1992, pp. 1719–1721.

[27] linecalc, ver. 5.0, Hewlett-Packard Co. (now Agilent Technologies, Inc.), Santa Rosa, CA, 1994.

[28] Wu, M., et al., "Full-Wave Characterization of the Mode Conversion in a CPW Right-Angled Bend," *IEEE Trans. on Microwave Theory and Techniques*, Vol. 43, No. 11, Nov. 1995, pp. 2532–2538.

[29] Becks, T., and I. Wolff, "Full-Wave Analysis of Various CPW Bends and T-junctions with Respect to Different Types of Air-Bridges," *MTT-S Int. Microwave Symp. Dig.*, 1993, pp. 697–700.

[30] Omar, A. A., et al., "Effects of Air-Bridges and Mitering on CPW 90° Bends: Theory and Experiment," *MTT-S Int. Microwave Symp. Dig.*, 1993, pp. 823–826.

[31] Sinclair, C., et al., "Closed-Form Expressions for CPW Discontinuities," *Conference Proceedings: Military Microwaves 92*, pp. 227–229.

[32] Alexandrou, S., et al., "Time-Domain Characterization of Bent CPWs," *IEEE Journal of Quantum Electronics*, Vol. 28, No. 10, Oct. 1992, pp. 2325–32.

[33] Bromme, R., and R. H. Jansen, "Systematic Investigation of CPW MIC/MMIC Structures using a Unified Strip/Slot 3D EM Simulator," *MTT-S Int. Microwave Symp. Dig.*, 1991, pp. 1081–4.

[34] Sawasa, H., et al., "Transmission Characteristics of CPW Bends for Various Curvatures," *IEICE Transactions on Electronics*, Vol. E77-C, No. 6, June 1994, pp. 949–51.

[35] Sawasa, H., et al, "Radiation from Bend in CPW Angular and Circular Bends," *1994 Int. Symp. on EM Comp.*, p. 325.

[36] Mirshekar-Syahkal, D., "Computation of Equivalent Circuits of CPW Discontinuities Using Quasi-Static Spectral Domain Method," *IEEE Trans. on Microwave Theory and Techniques*, Vol. 44, No. 6, June 1996, pp. 979–984.

[37] Nadarassin, M., et al., "Analysis of Planar Structures by an Integral Approach Using Entire Domain Trial Functions," *IEEE Trans. on Microwave Theory and Techniques*, Vol. 43, No. 10, Oct. 1995, pp. 2492–2495.

[38] Mernyei, F., et al., "New Cross-T Junction for CPW Stub-Filters on MMIC's," *IEEE Microwave and Guided Wave Letters*, Vol. 5, No. 5, May 1995, pp. 139–141.

[39] Naghed, M., et al., "A New Method for the Calculation of the Equivalent Inductances of Coplanar Waveguide Discontinuities," *MTT-S Int. Microwave Symp. Dig.*, 1991, pp. 747–750.

[40] Omar, A., et al., "A Versatile Moment Method Solution of the Conventional and Modified Coplanar Waveguide T-Junctions," *IEEE Trans. on Microwave Theory and Techniques*, Vol. 41, No. 4, April 1993, pp. 687–692.

[41] Simons, R., et al., "Modeling of Some Coplanar Waveguide Discontinuities," *IEEE Trans. on Microwave Theory and Techniques*, Vol. 36, No. 12, Dec. 1988, pp. 1796–1803.

[42] Naghed, M., et al., "Equivalent Capacitances of Coplanar Waveguide Discontinuities and Interdigitated Capacitors Using a Three-Dimensional Finite Difference Method," *IEEE Trans. on Microwave Theory and Techniques*, Vol. 38, No. 12, Dec. 1990, pp. 1808–1815.

[43] Drissi, M., et al., "Analysis of Coplanar Waveguide Radiating End Effects Using the Integral Equation Technique," *IEEE Trans. on Microwave Theory and Techniques*, Vol. 39, No. 1, Jan. 1991, pp. 112–116.

[44] Isele, B., and P. Russer, "The Modeling of Coplanar Circuits in a Parallel Computing Environment," *MTT-S Int. Microwave Symp. Dig.*, 1996, pp. 1035–1038.

[45] Dib, N. I., et al., "A Comprehensive Theoretical and Experimental Study of Coplanar Waveguide Shunt Stubs," *MTT-S Int. Microwave Symp. Dig.*, 1992, pp. 947–950.

[46] Yu, M., et al., "Analysis of Planar Circuit Discontinuities Using the Quasi-Static Space-Spectral Domain Approach," *MTT-S Int. Microwave Symp. Dig.*, 1992, pp. 845–848.

[47] Doerner, R., et al., "Modeling of Passive Elements for Coplanar SiGe MMICs," *MTT-S Int. Microwave Symp. Dig.*, 1995, pp. 1187–1190.

[48] Jin, H., and R. Vahldieck, "Full-Wave Analysis of Coplanar Waveguide Discontinuities Using the Frequency Domain TLM Method," *IEEE Trans. on Microwave Theory and Techniques*, Vol. 41, No. 9, Sept. 1993, pp. 1538–1542.

[49] Yu, M., et al., "Theoretical and Experimental Characterization of Coplanar Waveguide Discontinuities," *IEEE Trans. on Microwave Theory and Techniques*, Vol. 41, No. 9, Sept. 1993, pp. 1638–1640.

[50] Alessandri, F., et al., "A 3-D Matching Technique for the Efficient Analysis of Coplanar MMIC Discontinuities with Finite Metallization Thickness," *IEEE Trans. on Microwave Theory and Techniques*, Vol. 41, No. 9, Sept. 1993, pp. 1625–1629.

[51] Tran, A., and T. Itoh, "Full-Wave Modeling of Coplanar Waveguide Discontinuities with Finite Conductor Thickness," *IEEE Trans. on Microwave Theory and Techniques*, Vol. 41, No. 9, pp. 1611–1615, Sep. 1993.

[52] Chiu, C., and R. Wu, "A Moment Method Analysis for Coplanar Waveguide Discontinuity Inductances," *IEEE Trans. on Microwave Theory and Techniques*, Vol. 41, No. 9, Sept. 1993, pp. 1511–1514.

[53] Dib, N. I., et al., "Characterization of Asymmetric Coplanar Waveguide Discontinuities," *IEEE Trans. on Microwave Theory and Techniques*, Vol. 41, No. 9, Sept. 1993, pp. 1549–1557.

[54] Dib, N. I., et al., "A Theoretical and Experimental Study of Coplanar Waveguide Shunt Stubs," *IEEE Trans. on Microwave Theory and Techniques*, Vol. 41, No. 1, Jan. 1993, pp. 38–44.

[55] Klingbeil, H., et al., "FDFD Full-Wave Analysis and Modeling of Dielectric and Metallic Losses of CPW Short Circuits," *IEEE Trans. on Microwave Theory and Techniques*, Vol. 44, No. 3, March 1996, pp. 485–487.

[56] Beilenhoff, K., et al., "Open and Short Circuits in Coplanar MMIC's," *IEEE Trans. on Microwave Theory and Techniques*, Vol. 41, No. 9, Sept. 1993, pp. 1534–1537.

[57] Mao, M., et al., "Characterization of Coplanar Waveguide Open End Capacitance—Theory and Experiment," *IEEE Trans. on Microwave Theory and Techniques*, Vol. 42, June 1994, No. 6, pp. 1016–1024.

[58] Harokopus, W., Jr., and P. B. Katehi, "Radiation Loss from Open Coplanar Waveguide Discontinuities," *MTT-S Int. Microwave Symp. Dig.*, 1991, pp. 743–746.

[59] Wguyen, C., et al., "Analysis of Coplanar Waveguide Multiple-Step Discontinuities," *IEEE AP-S Symposium*, 1993, pp. 185–188.

[60] Lin, Y., "Characterization of Coplanar Waveguide Open-End and Short-End Discontinuities by the Integral Equation Method," *Asia-Pacific Microwave Conf. Digest*, 1993, pp. 29–32.

[61] Pogatzki, P., et al., "A Comprehensive Evaluation of Quasi-Static 3D-FD Calculations for more than 14 CPW Structures—Lines, Discontinuities, and Lumped Elements," *MTT-S Int. Microwave Symp. Dig.*, 1994, pp. 1289–1292.

[62] Pogatzki, P., and O. Kramer, "A Coplanar Element Library for the Accurate CAD of (M)MICs," *Microwave Engineering Europe*, Dec./Jan. 1994.

[63] Riaziat, M., et al., "Coplanar Waveguides used in 2–18 GHz Distributed Amplifier," *MTT-S Int. Microwave Symp. Dig.*, 1986, pp. 337–338.

[64] PCAAMT, ver. 2.0, EM-CAD Laboratory, Brooklyn Polytechnic University, 1990.

[65] Creech, G. L., et al., "Artificial Neural Networks for Accurate Microwave CAD Applications," *IEEE MTT-S Int. Microwave Symp. Dig.*, 1996, pp. 733–736.

[66] Creech, G. L., et. al., "Artificial Neural Networks for Fast and Accurate EM-CAD of Microwave Circuits," *IEEE Trans. on Microwave Theory and Techniques*, Vol. 45, No. 5, pt. 2, May 1997, pp. 794–802.

[67] Watson, P. M., C. Cho, and K. C. Gupta, "EM-ANN Model for Synthesis of Physical Dimensions for Multilayer Asymmetric Coupled Transmission Line Structures," *Int. J. of RF and Microwave CAE*, Vol. 9, 1999, pp. 175–186.

[68] Zhang, Q. J., F. Wang, and M. S. Nakhla, "Optimization of High-Speed VLSI Interconnects: A Review," *Int. J. of Microwave and Millimeter-Wave CAE*, Vol. 7, 1997, pp. 83–107.

[69] Wang, F., et al., "A Hierarchical Neural Network Approach to the Development of Library of Neural Models for Microwave Design," *IEEE MTT-S Int. Microwave Symp.*, 1998, pp. 1967–1770.

[70] Sagiroglu, S., K. Guney, and M. Erler, "Calculation of Bandwidth for Electrically Thin and Thick Rectangular Microstrip Antennas with the Use of Multilayered Perceptrons," *Int. J. of RF and Microwave CAE*, Vol. 9, 1999, pp. 277–286.

[71] Sagiroglu, S., and K. Guney, "Calculation of Resonant Frequency for An Equilateral Triangular Microstrip Antenna Using Artificial Neural Networks," *Microwave Opt. Technology Lett.*, 14, 1997, pp. 89–93.

[72] Sagiroglu, S., K. Guney, and M. Erier, "Resonant Frequency Calculation for Circular Microstrip Antennas Using Artificial Neural Networks," *Int. J. RF Microwave Computer-Aided Eng.*, 8, 1998, pp. 270–277.

[73] Fedi, G., et al., "Direct-Coupled Cavity Filter Design Using a Hybrid Feedforward Neural Network-Finite Element Procedure," *Int. J. RF Microwave Computer-Aided Eng.*, 9, 1999, pp. 287–296.

[74] Burrascano, P., M. Dionigi, C. Fancelli, and M. Mongiardo, "A Neural Network Model for CAD and Optoimization of Microwave Filters," *IEEE MTT-S Int. Microwave Symp.*, 1998, pp. 13–16.

[75] em Simulator, Version 2.4, Sonnet Software, Liverpool, NY, 1993.

[76] Cho, C., and K. C. Gupta, "Design Methodology for Multilayer Coupled Line Filters," *IEEE Int. Microwave Symp. Digest*, June 1997, pp. 4603–4606.

[77] Bishop, C. M., *Neural Networks for Pattern Recognition*, New York: Oxford University Press, 1996.

[78] Tsai, C. M., "Field Analysis, Network Modeling, and Circuit Applications of Inhomogeneous Multi-Conductor Transmission Lines," Ph.D. Dissertation (Advisor: Prof. K. C. Gupta), University of Colorado at Boulder, Dept. of Electrical and Computer Eng., 1993.

[79] James, J. R., and P. S. Hall, eds., "Handbook of Microstrip Antennas," London, UK: Peter Peregrinus, 1989.

[80] Watson, P. M., G. L. Creech, and K. C. Gupta, "Knowledge Based EM-ANN Models for the Design of Wide Bandwidth CPW Patch/Slot Antennas," *1999 IEEE/AP-S Int. Symposium on Antennas and Propagation*, Orlando, pp. 2588–2591.

[81] Tsai, C.-M., and K. C. Gupta, "A Generalized Model for Coupled Lines and its Applications to Two-Layer Planar Circuits," *IEEE Trans. on Microwave Theory and Techniques*, Vol. 40, No. 12, Dec. 1992, pp. 2190–2199.

6

Modeling of High-Speed IC Interconnects

This chapter[1] presents applications of neural networks for modeling and optimization of high-speed VLSI interconnects in digital systems. The neural network approach is compared with traditional techniques for high-speed interconnect modeling and signal integrity analysis. Applications and numerical examples presented include: three parallel coupled interconnects, asymmetric two-line interconnects, an eight-bit digital bus configuration, a transmission line circuit with nonlinear termination, and delay and crosstalk optimization.

6.1 Introduction

VLSI interconnect modeling and optimization is an important step toward high-speed electronic system design. With rapid developments in VLSI technology, design, and CAD techniques—at both chip and package levels—the central processor cycle times are reaching the vicinity of 1 ns. The communication switches are now being designed to transmit data at bit rates faster than 1 Gbps. At such high signal speeds, previously negligible interconnect effects—such as signal delay, crosstalk, ground noise, ringing, and distortion—become significant [1–5]. These interconnect effects—also known as *signal integrity* effects (described by analog analysis techniques)—must be taken into account during digital circuit design. In addition, the drive within the VLSI industry

---

1. Sections 3 and 4 of this chapter are based upon A. Veluswami's thesis [42] at Carleton University.

toward miniature designs, low power consumption, and increased integration of analog circuits with digital blocks, has further complicated the issue of signal integrity analysis. The scaling of chip dimensions to achieve miniaturization and high signal speed affects the global interconnect delay. As shown in Figure 6.1, the global interconnect delay grows as a cubic power of the scaling factor ($\gg$ 1) [1, 6]. It is predicted that interconnects will be responsible for nearly 70–80% of the signal delay in high-speed systems. As such, there is a tremendous effort in the VLSI CAD research area to develop accurate methods for signal integrity analysis and design of VLSI interconnects.

High-speed interconnect analysis is an expensive computational task, characterized by long run-times and huge memory requirements, and this is due to two main reasons. First of all, the use of an interconnect model in a signal integrity design could be massive, since the number of interconnects in any VLSI system is extremely large. Second, most CAD/optimization methods currently use iterative techniques, in which a given interconnect network is evaluated online repeatedly until an optimal solution is obtained. Examples of such techniques [7–13] include simulated annealing, l_p-based optimization, gradient-based methods, Monte Carlo techniques, statistical/yield analysis, and so forth. It is apparent that the online time required for interconnect analysis

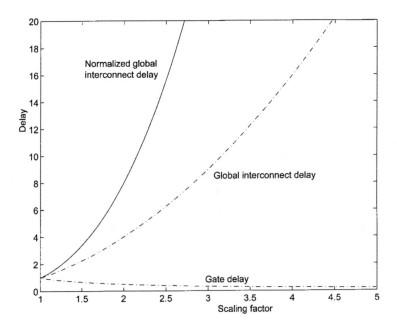

Figure 6.1 Impact of scaling on signal delay in high-speed systems (from [1], © 2000 CRC Press, reprinted with permission).

using these techniques depends upon the efficiency of the interconnect models employed.

In this chapter, the neural network approach described earlier is applied to the simulation and optimization of high-speed interconnects and interconnect circuitry. We present a description of an interconnect analysis problem, followed by neural network techniques and their application to the modeling of interconnects at the component level, as well as the modeling of signal integrity characteristics of interconnect networks at the circuit level. Signal integrity optimization examples are also presented.

6.2 High-Speed Interconnect Modeling and Signal Integrity Analysis

High-speed interconnects can exist at various levels of a system hierarchy, for example on chips, package structures, multichip modules, printed circuit boards, and backplanes [1, 3], as shown in Figure 6.2. As illustrated in Figure 6.3, a VLSI interconnect delivers a signal from a driver to a receiver. The equivalent circuit that simulates the interconnect effects includes interconnect models, driver/receiver models, and possibly additional models of the terminating components. The drivers/receivers can be modeled by lumped linear or nonlinear components, representing the transistors in driver/receiver circuits. The interconnects are, in general, coupled multiconductor transmission lines. The overall objective is to analyze signal integrity for signal delay, crosstalk, etc. from interconnect and termination parameters. There are two tasks—one at the individual interconnect component level, and the other at the circuit simulation level.

6.2.1 Traditional Techniques

In this section, we describe the existing interconnect modeling approaches, and the circuit-level interconnect simulation and optimization techniques. The situations under which the existing techniques become cumbersome are discussed.

There are several types of traditional interconnect models—namely, the lumped equivalent model, the distributed model, and the full-wave 3-D EM model. The lumped models (e.g., RLC networks) are often used to model chip level interconnects. The distributed model (e.g., lossy coupled multiconductor transmission line described by Telegrapher's equations [14] with cross-sectional per unit length RLCG parameters) are suitable for board or multichip module level interconnects. The full-wave 3-D EM models are free from assumptions

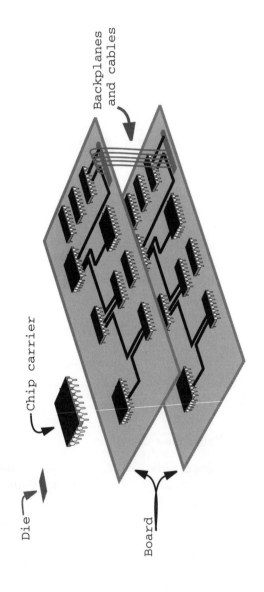

Figure 6.2 VLSI interconnect hierarchy (from [1], © 2000 CRC Press, reprinted with permission).

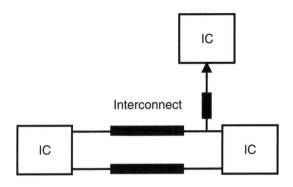

Figure 6.3 An example of interconnects delivering signals from drivers to receivers.

and approximations, and give very accurate results by accommodating EM effects at very high frequencies [15, 16]. However, the 3-D EM models are highly CPU-intensive and thus not suitable for online use in large-scale CAD and optimization. Extensively-used models are based on RLGC, the lumped or distributed interconnect models, as described in [17–21].

At the circuit level, standard approaches for analyzing signal integrity in a network of interconnects include circuit simulation techniques—such as SPICE integration technique, asymptotic waveform evaluation (AWE) [22], complex frequency hopping (CFH) [23], Padé via Lanczos (PVL) [24], and numerical inversion of Laplace transforms (NILT) [25]. In order to perform signal integrity optimization, repeated simulation of interconnect networks becomes necessary. Yield optimization requiring statistical variations in interconnects can be even more computationally expensive. Several modeling approaches for speeding up circuit optimization have been developed. The purpose of these approaches is to build a macromodel to replace the original circuit partially or completely during optimization. Examples of such models include the conventional polynomial [26], spline [27], and table lookup models [28], as well as macromodeling based on Padé approximation [29].

The polynomial models (e.g., quadratic models) are popular models for optimization but usually have limited nonlinearity and require rebuilding and updating during optimization. The table lookup models can represent more nonlinearity and can be used without being updated during optimization. They are thus more suitable for interactive circuit design and optimization, since most of the online computation is shifted to off-line model building. However, for high-dimensional problems with many input variables to the model (i.e., n is very large), the table lookup models become prohibitively large.

Recently, macromodeling techniques based on model-reduction—using moment-matching techniques (MMT), Padé approximation [22, 23, 30–34], and Krylov-space methods [35–40]—have been suggested. These macromodeling techniques can be used to model the time or frequency responses of a large linear subnetwork. Such models are derived through an analytical process and do not rely heavily on data samples from original simulations. Model building can therefore be much faster than table lookup models, and the structure of the model may have a meaningful relationship with the original problem. Such macromodels do not usually include optimization variables, however, and therefore may need repetitive model rebuilding during optimization. These models are most powerful in iterative solutions of nonlinear networks where the linear part is large and represented by a macromodel, and in optimization where optimization variables exist outside the macromodel.

6.2.2 Neural Network Approach

The objective is to develop a neural network interconnect model that can be used online in iterative CAD/optimization routines. The model is expected to capture the relationship between the set of parameters defining the physical configuration of an interconnect structure or group of interconnects in a VLSI system, and its operational characteristics including the signal integrity characteristics of the system. The model can be mathematically represented as

$$y = f(x) \tag{6.1}$$

where y is an m-dimensional output vector representing the parameters to be modeled, such as RLCG matrices of interconnects, signal propagation delays at the terminations, crosstalk, and level of ground-bounce noise of interconnect networks. x is an n-dimensional input vector containing the interconnect parameters necessary to compute y. The input parameters can include physical dimensions and termination impedance of interconnects, dielectric substrate characteristics, topology of the interconnect network, input voltage signal, frequency of operation (or rise time in the case of a digital signal), termination impedance of the interconnects, and so forth, as shown in Figure 6.4.

In the neural network approach, the function f is modeled using a neural network. Feedforward neural network structures described in Chapter 3 can be considered as mathematical tools that are capable of nonlinear mapping in high dimensions [41]. We consider the use of neural networks at two levels of interconnect design, namely the physical/EM level and the circuit level.

Physical/EM Level

At this level, neural network models are trained to capture the relationship between the physical and geometrical parameters of the interconnects and the

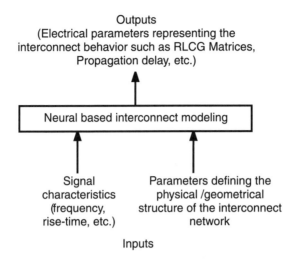

Figure 6.4 Typical input-output parameters of neural models for interconnect analysis (from [8], © 1997 IEEE, reprinted with permission).

equivalent electrical parameters [42]. As an example, in the interconnect neural model for the 4-conductor interconnect structure in Figure 6.5, the inputs x consist of length and width of conductors, height of substrate, and separation between conductors; and the outputs y could be the RLCG parameters. Such a relationship would initially require quasi-TEM solutions of EM problems [19] that are slow. Since the models would be used massively and repetitively in the overall system design, fast RLGC evaluation using neural models is desirable.

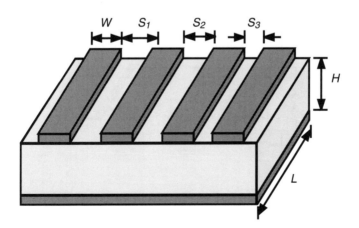

Figure 6.5 A coupled 4-conductor transmission line for VLSI interconnect modeling.

Circuit Level

At this level, the circuit may be partly or completely replaced by a neural network model [43, 44]. The inputs x consist of different circuit parameters, and the predictions y from the neural model can be signal delay, crosstalk, and so forth. Figure 6.6 shows an example circuit, which is a network of interconnects. Signal delays at the terminations are functions of individual interconnect parameters and terminations. Instead of solving the time-consuming original circuit equations, we use neural network models to predict signal delay. On a typical printed circuit board (PCB), there are a large number of such interconnect tree structures. As such, once a neural network is trained to represent this network, it can be used massively and repetitively in an overall chip or PCB design.

The neural network learns the interconnect relationships from measured/ simulated input-output sample pairs through a training process described in Chapter 4. After training, the neural model is ready to be integrated into circuit simulators for subsequent use in simulation and optimization. The model can predict the outputs y for given inputs x that may not have been seen during training. The neural model requires less computation than the original physics/ EM/circuit equations.

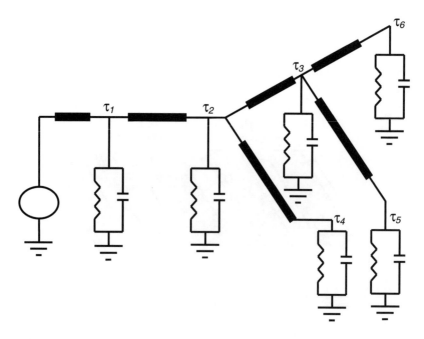

Figure 6.6 A VLSI interconnect tree representing signal path from driver to receivers.

When outputs y represent interconnect parameters at the component level, the neural model needs to be formulated into a modified nodal admittance (MNA) stamp and then presented to the circuit simulator [43]. A circuit level neural model can directly represent circuit responses such as signal delay and crosstalk. In this case, the error functions for signal integrity optimization are directly formulated from neural network outputs. A powerful feature of the neural network approach is the fact that the input parameters x of the neural model can be the optimization variables, thereby replacing expensive repetitive simulation of the original problem with repetitive neural network feedforward.

6.3 Application Examples

6.3.1 Example A: Three Parallel Coupled Interconnects

This example is from [8], where a configuration of the three parallel interconnects shown in Figure 6.7 is modeled. During design optimization, each individual configuration would vary in terms of width W and thickness t of the interconnects, separation between the conductors s, height of the dielectric h, and the frequency of operation f. Hence,

$$x = [W \quad t \quad s \quad h \quad f]^T$$

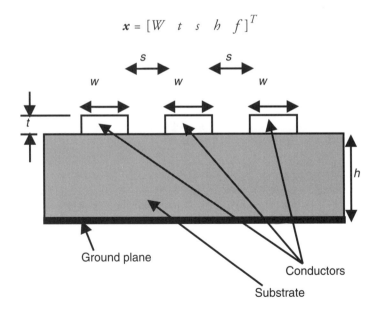

Figure 6.7 Cross-section of three parallel interconnects showing parameters for neural modeling (from [8], © 1997 IEEE, reprinted with permission).

The ranges of the input variables are given in Table 6.1. The outputs of the model are the elements of the L and C matrices

$$L = \begin{bmatrix} L_{11} & L_{12} & L_{13} \\ L_{21} & L_{22} & L_{23} \\ L_{31} & L_{32} & L_{33} \end{bmatrix}$$

and

$$C = \begin{bmatrix} C_{11} & C_{12} & C_{13} \\ C_{21} & C_{22} & C_{23} \\ C_{31} & C_{32} & C_{33} \end{bmatrix}.$$

Since the conductors have identical dimensions and are symmetrically arranged, $L_{ij} = L_{4-j,4-i}$ and $L_{ij} = L_{ji}$. Elements in the capacitor matrix will have similar relations. Thus elements L_{11}, L_{12}, L_{13}, C_{11}, C_{12} and C_{13} are selected as independent parameters to construct the L and C matrices, that is,

$$y = [L_{11}, L_{12}, L_{13}, C_{11}, C_{12}, C_{13}]^T$$

The implemented neural network model has the features shown in Table 6.2. The other per-unit length parameters of the structure, R and G, would be modeled in a manner identical to that of L and C.

Data was generated off-line using Nortel Network's SALI EM field solver [45]. Training was carried out on a Sun SPARCstation 10. The average test error was 0.0174 at the end of training. The neural model was found to estimate the L and C parameters with accuracy comparable to the field solver

Table 6.1
Neural Network Input Parameters and Their Range for Example A
(From [8], © 1997 IEEE, Reprinted with Permission)

Parameter	Symbol	Minimum	Maximum
Conductor width	w	5 mil	11 mil
Conductor thickness	t	0.7 mil	2.8 mil
Separation	s	1 mil	16 mil
Substrate height	h	5 mil	10 mil
Frequency	f	1 MHz	8 GHz

Table 6.2
Features of the Neural Network Model for Example A
(From [8], © 1997 IEEE, Reprinted with Permission)

Feature	Value
Number of inputs	5
Number of outputs in the overall model	18
Number of output neurons in neural network	6
Number of neurons in the hidden layer	10
Size of model (number of weights in w)	126
Size of training set	500
Size of test set	500
Data generation technique	SALI simulator
Average training error	0.0162
Average test error	0.0174

used. The average test errors of the neural model for 100 random circuit configurations from the test set are shown in Figure 6.8. Figure 6.9 compares the variation of one of the outputs, C_{13}, against the width of separation, as estimated by a neural model and computed from an EM-field solver.

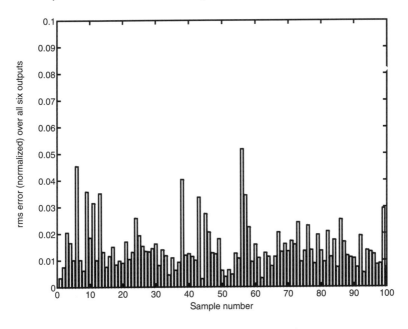

Figure 6.8 Test error for 100 random inputs from the test set (Example A) (from [8], © 1997 IEEE, reprinted with permission).

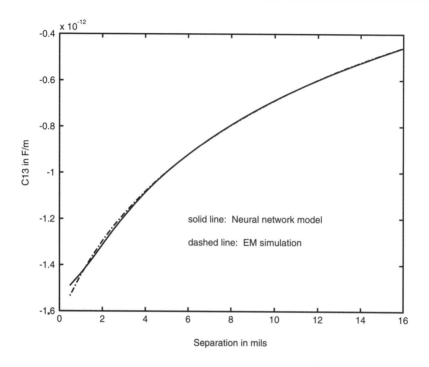

Figure 6.9 Matrix element C_{13} as a function of separation, keeping other parameters constant (Example A) (from [8], © 1997 IEEE, reprinted with permission).

6.3.2 Example B: Two Asymmetric Interconnects

This example is from [8], where the L and C parameters of an asymmetric configuration of two parallel interconnects shown in Figure 6.10 are modeled. The model inputs are identical to those in Example A, except that the width

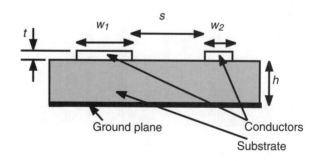

Figure 6.10 Two asymmetric microstrip lines for VLSI interconnects (Example B) (from [8], © 1997 IEEE, reprinted with permission).

of the two conductors must be treated separately. The inputs to the neural network are

$$x = [W_1 \quad W_2 \quad t \quad s \quad h \quad f]^T$$

The ranges of inputs are the same as those in Table 6.1. Data is generated off-line using EM simulation, in order to obtain the L and C parameters for different values of input parameters. There are eight outputs,

$$y = [L_{11} \; L_{12} \; L_{21} \; L_{22} \; C_{11} \; C_{12} \; C_{21} \; C_{22}]^T$$

Table 6.3 shows the features of the neural network model developed. The accuracy of the neural network model is close to that of the EM simulator.

6.3.3 Example C: An Eight-Bit Digital Bus Configuration

This example demonstrates the neural modeling of an 8-bit bus configuration [8]. Figure 6.11 illustrates the bus structure—whose cross-sectional view is shown in Figure 6.12. In the region between the I/O pads and IC pins to which the inputs are connected, all eight metallic interconnects run parallel to each other. As such, this is a region of high crosstalk and noise, and it is critical in layout design.

Table 6.3
Features of Neural Network Model for Example B
(From [8], © 1997 IEEE, Reprinted with Permission)

Feature	Value
Number of inputs	6
Number of outputs in the overall model	8
Number of output neurons in neural network	8
Number of neurons in the hidden layer	10
Size of model (number of weights in w)	158
Size of training set	500
Size of test set	500
Data generation technique	SALI simulator
Average training error	0.0233
Average test error	0.0253

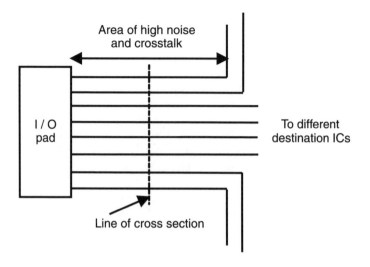

Figure 6.11 Eight-bit bus on a printed circuit board (PCB) showing region of high crosstalk and signal noise (from [8], © 1997 IEEE, reprinted with permission).

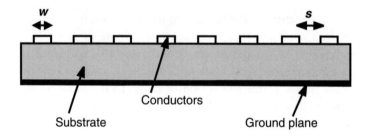

Figure 6.12 Cross sectional view of a 8-bit digital bus (from [8], © 1997 IEEE, reprinted with permission).

The inputs to the model and their ranges are the same as those of Example A. The outputs modeled are the elements of the L and C matrices, each of which contains 64 elements. However, because the eight interconnects are equally spaced, identical in physical dimensions, and have symmetrical cross sections, the 8×8 matrices are symmetric about both diagonals. Thus the entire L matrix can be constructed from L_{11} and 16 other values, and the same goes for the C matrix. Furthermore, in the case of the inductance matrix, the values of the mutual inductances between the conductors that are apart (i.e., L_{18}, L_{17}, L_{28}, and L_{27}) are extremely small, and can therefore be set to a nominally low average value. The L matrix is represented as

$$L = \begin{bmatrix} L_{11} & L_{12} & L_{13} & L_{14} & L_{15} & L_{16} & \overline{L_{17}} & \overline{L_{18}} \\ L_{12} & L_{11} & L_{23} & L_{24} & L_{25} & \overline{L_{26}} & \overline{L_{27}} & \overline{L_{28}} \\ L_{13} & L_{23} & L_{11} & L_{34} & L_{35} & L_{36} & \overline{L_{37}} & L_{16} \\ L_{14} & L_{24} & L_{34} & L_{11} & L_{45} & L_{35} & L_{25} & L_{15} \\ L_{15} & L_{25} & L_{35} & L_{45} & L_{11} & L_{34} & L_{24} & L_{14} \\ L_{16} & \overline{L_{62}} & L_{36} & L_{35} & L_{34} & L_{11} & L_{23} & L_{13} \\ \overline{L_{71}} & \overline{L_{72}} & \overline{L_{73}} & L_{25} & L_{24} & L_{23} & L_{11} & L_{12} \\ \overline{L_{81}} & \overline{L_{82}} & L_{16} & L_{15} & L_{14} & L_{13} & L_{12} & L_{11} \end{bmatrix}$$

where the elements with bars are fixed at nominal average values and are not modeled. Hence, the number of outputs for the neural network is 30, resulting from 13 L elements and 17 C elements.

The main features of the developed neural model are tabulated in Table 6.4. Figure 6.13 shows the values of the mutual capacitances in a given configuration as one moves away from the first conductor—that is, the values on the first row of the C matrix. The test errors on 100 random input points are shown in Figure 6.14.

6.3.4 Example D: Interconnect Circuit with Nonlinear Terminations

In this example, a different type of neural network model is demonstrated. Instead of modeling a device or a circuit element, we model the responses of the entire circuit [43]. Figure 6.15 represents a high-speed VLSI interconnect network modeled by seven transmission lines and five nonlinear driver/receivers.

Table 6.4
Features of Neural Network Model for Example C
(From [8], © 1997 IEEE, Reprinted with Permission)

Feature	Value
Number of inputs	5
Number of outputs in the overall model	128
Number of output neurons in neural network	30
Number of neurons in the hidden layer	10
Size of model (number of weights in w)	390
Size of training set	500
Size of test set	500
Data generation technique	SALI simulator
Average training error	0.0348
Average test error	0.0351

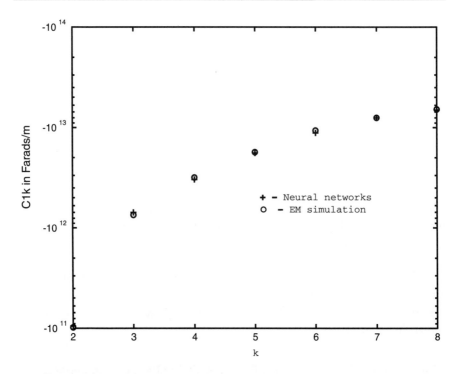

Figure 6.13 The values of capacitances between conductors 1 and k (Example C) (from [8], © 1997 IEEE, reprinted with permission).

Signal delay through such interconnect networks is an important criterion in high-speed VLSI system design. However, repeated signal delay analysis of this circuit using conventional circuit simulators like HSPICE is CPU intensive. The inputs to the neural model x contain six termination variables including capacitors, inductors, and resistors at the four terminations. The signal integrity characteristics y include the signal propagation delays of V_{out1} through V_{out4}. The number of hidden neurons in the neural network is 30. A total of 350 training samples are generated using HSPICE simulation. After training, the error between neural model and training data is mostly below 0.4%, as shown in Figure 6.16. The four signal integrity responses predicted by the trained neural network are compared with those from the HSPICE, for 100 randomly generated test samples not used for training. The result of such a comparison can be seen in Figure 6.17, and the errors are within ±0.4%.

6.3.5 Example E: Signal Integrity Optimization

This example illustrates the use of neural networks in signal integrity optimization [46]. Signal crosstalk is observed between adjacent interconnects at high

Modeling of High-Speed IC Interconnects

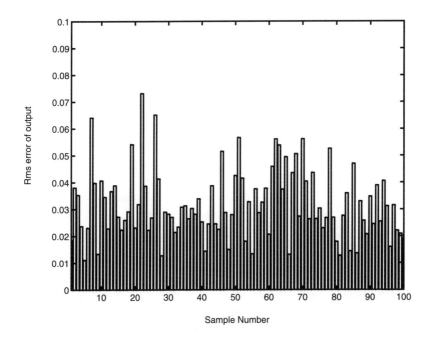

Figure 6.14 Test error for 100 random inputs from the test set for Example C (from [8], © 1997 IEEE, reprinted with permission).

signal speeds. Such a phenonmenon in a coupled interconnect network containing three 2-conductor transmission lines is shown in Figure 6.18. The signal integrity analysis of this circuit involves eigen-value solutions of the transmission line matrices and mixed time- and frequency-domain analysis, such as FFT or numerical inversion of Laplace transform (NILT) [25]. The inputs to the neural model x contain six variables—namely, the length (l_i) and separation (s_i) between coupled conductors of each transmission line. The signal integrity responses y include the signal propagation delays of V_{out1} and V_{out2}, and the magnitude of crosstalks V_{cross1} and V_{cross2}. The neural network structure used in this example is shown in Figure 6.19. A training dataset of 500 samples was generated using a NILT simulator. After training, the neural network can represent the behavior of the entire circuit—that is, for given length and spacing values, the neural model can predict delay and crosstalk. This model is very useful for signal-integrity-driven layout, where the interconnect length and spacing are varied during optimization.

To take manufacturing tolerances into account, all six input parameters are allowed to be statistical variables following Gaussian distribution with 5% standard deviation. We then perform a Monte Carlo analysis with 500 interconnect circuits, each with randomly generated values of interconnect

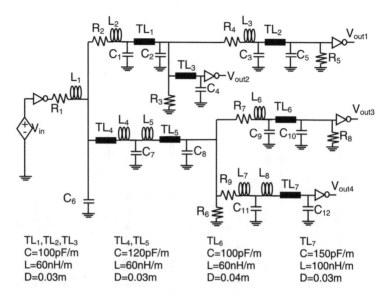

Figure 6.15 A high-speed VLSI interconnect network represented by a seven transmission line circuit with nonlinear terminations (from [43], © 1995 IEEE, reprinted with permission).

lengths and spacings. The design specifications for the four responses are 1.8 ns, 1.72 ns, 0.02V, and 0.02V respectively. The specifications were severely violated before optimization and the yield was 0%.

In order to improve the yield, a yield optimization was performed using the neural network model. During optimization, the neural model is used to estimate the yield for a given circuit configuration, using 100 random circuit configurations distributed around the given configuration. The nominal values of the interconnect parameters (lengths and spacings) are then changed— effecting a change in all 100 circuit configurations—and the yield is reestimated. During this process, the weights of the neural network model were fixed, and the outputs forced to match or exceed design specifications by adjusting the neural model inputs.

The yield optimization solution consists of the final nominal values of the neural model inputs (i.e., lengths and spacings of the transmission lines). After yield optimization, Monte Carlo analysis with 500 randomly-generated circuit configurations was performed, and the final yield of 98% achieved. The histograms for delay of V_{out2} before and after optimization are shown in Figure 6.20(a) and Figure 6.20(b), respectively. The highly repetitive use of the neural network model instead of the original NILT simulation allowed for fast yield optimization.

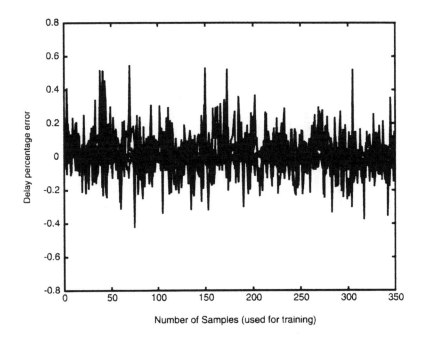

Figure 6.16 The seven transmission line with nonlinear termination example. Percentage errors between signal delays predicted from the neural network model and that from training data for 350 randomly generated sets of samples (from [43], © 1995 IEEE, reprinted with permission).

6.3.6 Example F: Neural Networks for Interconnects on a Printed Circuit Board

In this example, a neural model is developed to represent the delays of signal propagation in a network of interconnects on a printed circuit board (PCB) [8, 42]. A typical section of such a network is shown in Figure 6.21. One IC pin acts as a source (driver) from which a digital signal is transmitted to several other receiver IC pins. The number of pins connected to a given source is related to the term *fan-out* in a digital IC. The neural model developed here treats drivers connected to four pins. The equivalent circuit of the interconnects of Figure 6.21 is shown in Figure 6.22.

A PCB could consist of several hundred of such configurations. During layout optimization of such a PCB, each individual interconnect network would vary in terms of its driver (source) characteristics, receiver pin load characteristics, lengths of interconnects, and network topology. The input variables are $\boldsymbol{x} = [l_i, R_i, C_i, R_s, t_R, V_p, \epsilon_j]^T$, $i = 1, 2, 3, 4$, and $j = 1, 2, 3$. Here, l_i is length of the ith interconnect, R_i and C_i are ith

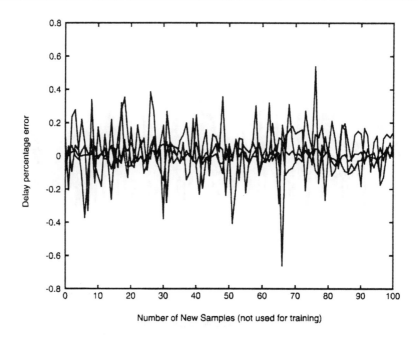

Figure 6.17 The seven transmission line with nonlinear termination example. Percentage errors between signal delays predicted from the neural network model and that from exact simulation for 100 randomly generated sets of samples not used for training.

interconnect terminations, R_s is the source impedance, t_R and V_p are rise time and peak value of the source signal.

The input parameter ϵ_j identifies the interconnect network topology, which is an important design variable available to a circuit layout designer. In order to use a neural network approach for design, an explicit parameterization of the topology is required. Given a certain number of interconnects in a network, there are a number of configurations into which they can be driven by the source node. The problem here is to identify a unique numbering scheme to differentiate possible topologies, and then to convert those numbers into a new neural network input. To achieve this, the graph theory was used to number the source node as the root of a set of nodes to be connected as a directed graph. Each possible circuit configuration thus forms a rooted tree, and the network topology becomes one of the possible rooted trees for a given number of nodes. The number of nodes in the tree represents the number of interconnects in the circuit. The following steps are used to number the nodes, as illustrated in Figure 6.23 [42].

Step 1: The source is numbered 1;

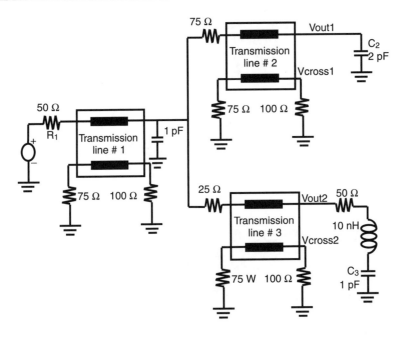

Figure 6.18 A coupled interconnect network with three coupled transmission lines.

Step 2: The other nodes in the tree are numbered in a breadth-first manner, that is, nodes at the same layer (having the same distance to their common roots) are numbered consecutively;

Step 3: On a layer with more than one node, the node spawning the deepest descendant is numbered first;

Step 4: If there is a tie, the node with the larger number of children is numbered first.

The outputs of the neural model are the propagation delays at the four terminations, $y = [\tau_1\ \tau_2\ \tau_3\ \tau_4]^T$, where the propagation delays are defined as the time taken for the signal to reach 80% of its steady state value. The neural network structure is shown in Figure 6.24. The ranges of the input parameters are shown in Table 6.5.

The number of inputs, outputs, and hidden neurons for the neural network model are shown in Table 6.6. Due to the large dimensionality of the problem and the special nature of the input variables (particularly the topology variable), a relatively huge amount of training and test data samples are used, and the data is generated off-line using NILT [25]. Training is carried out on a Sun SPARCstation 10, and the test error decreased continuously until it reached a value of 0.04. The error distribution of the neural model

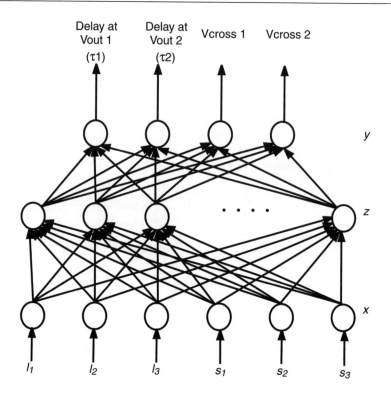

Figure 6.19 An example of a neural network model at the circuit level for the circuit of Figure 6.18. l_i and s_i are the length and separation between coupled conductors of the ith transmission line component.

prediction for 100 random interconnect network configurations is shown in Figure 6.25.

This example forms a basis for a neural network interconnect simulator that can be used in propagation delay analysis. Based on the methodology used in the example, neural network models for different-sized networks (with two to six branches) were built in [42], and subsequently used to construct an interconnect network simulator. Substantial speed-ups were obtained over the existing techniques, especially in an optimization environment where a large number of interconnect trees with different interconnect lengths, terminations, and tree topologies were needed to be simulated repetitively.

6.4 Discussion

6.4.1 Run-Time Comparison

In this section, online run-time requirements of the neural network models are compared with existing simulation techniques [8, 42]. During each iteration

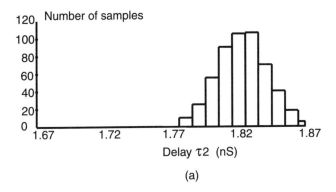

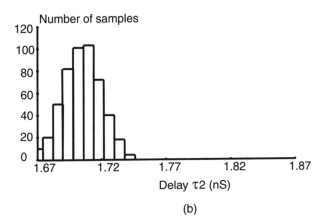

Figure 6.20 Histogram of the signal delay for V_{out2} of the coupled interconnect network in Figure 6.18 obtained from an exact circuit simulation of 500 random samples. Samples with a delay value of more than 1.72 ns violate specifications (a), before optimization, and (b), after optimization using the neural network approach (from [46], © 1994 IEEE, reprinted with permission).

of the optimization routine, the interconnect model is called repetitively for different configurations of the circuit (circuits with different interconnect lengths, terminations, and tree topologies). In practical IC design, the number of times an interconnect model is called during optimization in order to obtain a reasonably (if not globally) optimal solution could easily reach or exceed tens of thousands.

In Table 6.7, the run-times required for estimating the L and C matrices of 20,000 different interconnect structures using EM simulation and neural network techniques are compared. A substantial speed-up is observed in the

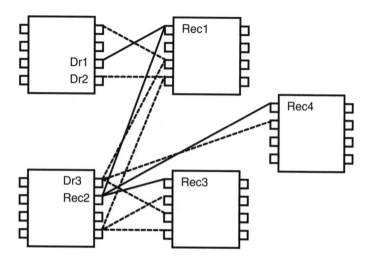

Figure 6.21 A typical interconnect network with IC pins connected by several interconnects.

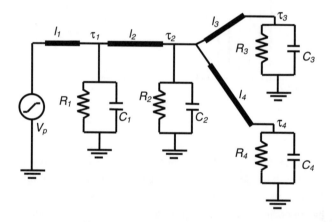

Figure 6.22 Electrical equivalent of an interconnect configuration showing the outputs of the model (Example F).

case of neural models. Table 6.8 shows a comparison of the run-time for the simulation of 20,000 PCB structures (Example F). The neural network technique is significantly faster compared to simulation techniques like AWE [29] and NILT. Comparisons were made on a Sun SPARCstation 10.

6.4.2 Performance Evaluation

It can be seen from the above examples that neural network techniques offer substantial computational speed-up over standard simulation techniques. This

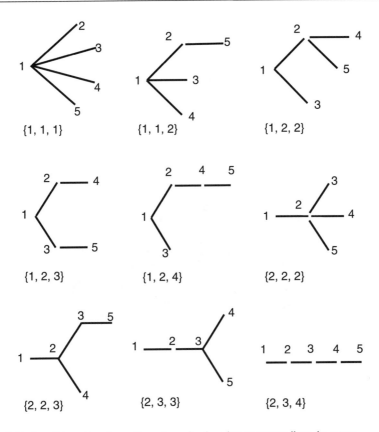

Figure 6.23 Possible network configurations for four interconnect lines in a tree interconnect network. The values of the neural network input variables $\{\epsilon_1, \epsilon_2, \epsilon_3\}$ are shown in curly brackets. Each combination of these three input variables define an interconnect topology.

allows designers to be much more liberal in defining the critical paths to be tested and optimized, and permits a large number of iterations to be performed in the optimization routine. Using neural networks as a fast online simulation tool, one can perform exhaustive optimization routines without too much concern for the convergence time, which is usually a major concern for existing simulation tools.

A neural network model can be developed for many interconnect patterns or networks that might be encountered during layout optimization, irrespective of size or number of conductors. Memory requirements do not grow exponentially with an increase in the number of model inputs/outputs. In the examples presented in the chapter, the number of training and test data samples required is relatively lower than the number of data samples required to construct table

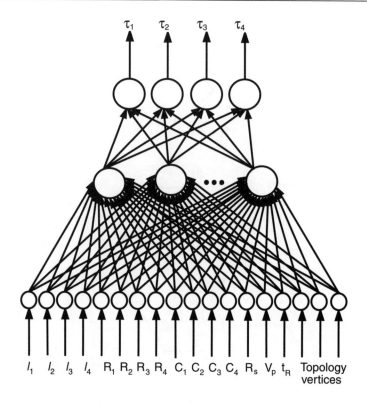

Figure 6.24 Neural network for a tree of four interconnects on a PCB.

Table 6.5
Input Variables and Their Ranges for Example F
(From [8], © 1997 IEEE, Reprinted with Permission)

Variable Name	Symbol	Range of Values
Interconnect length	l_i	1–15 cm
Termination resistance	R_i	100–100,000 Ohms
Termination capacitance	C_i	3.3–5 nF
Source resistance	R_s	13.3–45 Ohms
Input risetime	t_R	1.6–10 ns
Peak voltage	V_p	0.8–5 V
Topology vertices	ϵ_j	1–4

Table 6.6
Features of Neural Network Model for Example F
(From [8], © 1997 IEEE, Reprinted with Permission)

Feature	Value
Number of inputs	18
Number of outputs in the overall model	4
Number of output neurons in neural network	4
Number of neurons in the hidden layer	40
Size of model (in floating point numbers)	924
Size of training set	4500
Size of test set	4500
Data Generation Technique	NILT simulator
Average training error	0.0356
Average test error	0.04

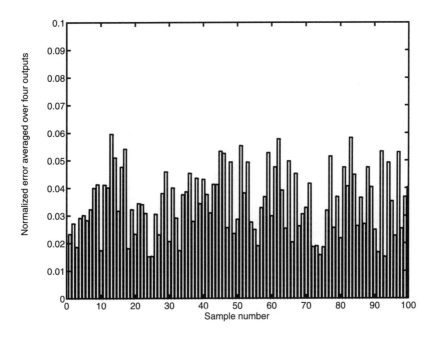

Figure 6.25 Error for 100 random circuit configurations for the four interconnect network.

look-up models. In the case of Example F, for instance, where there are 18 inputs, a large amount of data would be needed to build a table look-up model.

The large number of variables neural network models can handle makes them a viable alternative to polynomial curve-fitting techniques. In Example

Table 6.7
Comparison of Run-Time Requirements for the Interconnect Neural Network Model and EM Simulation (From [8], © 1997 IEEE, Reprinted with Permission)

Method	Run-Time for 20,000 Configuration
EM simulation	20–80 hours
Neural Network Model	40–130 seconds

Table 6.8
CPU Time Needed for Simulation of 20,000 Circuit Configurations for the Interconnect Tree Structure (From [8], © 1997 IEEE, Reprinted with Permission)

Method	CPU Time (Speed-Up)
NILT	34.43 hours (310)
AWE	9.56 hours (86)
Neural network	6.67 minutes (1)

C, 30 output variables were modeled simultaneously. In Example F, the number of inputs considered was 18. High-order polynomial curve-fitting with 18 independent input variables would prove to be significantly difficult to achieve. But on the other hand, low-order polynomial models (such as quadratic models with 18 variables) would have a much lower range of validity than neural models.

6.5 Conclusions

The development of fast, accurate, and reliable interconnect models continues to be a vital activity for high-speed circuit design and optimization. The objective is to provide faster evaluation of signal integrity characteristics so that the massive and highly-repetitive process of interconnect network analysis and design can be sped up. This chapter presented preliminary efforts in the area of interconnect modeling and optimization with neural network techniques.

In VLSI systems, the end-to-end signal travels through a hierarchy of interconnect paths—first from chip to package, then to board, then backplane, and then finally back to board and chip. Corresponding hierarchical signal integrity design involving interconnects at all the levels is a complicated task. In addition, today's VLSI design increasingly requires multidisciplinary considerations, including electrical, physical, thermal, EMI/EMC, reliability aspects, and many others. A conventional single-discipline optimization may lead to poor system performance in other disciplines prolonging design iterations between various disciplines. A concurrent multidisciplinary optimization approach can be very useful. Various techniques, including the neural network techniques, are continuously being developed to address these needs.

References

[1] Nakhla, M. S., and R. Achar, "Interconnect Modeling and Simulation," in *The VLSI Handbook*, Boca Raton, FL: CRC Press, 2000, pp. 17.1–17.29.

[2] Deutsch, A., "Electrical Characteristics of Interconnections for High-Performance Systems," *Proceedings of the IEEE*, Vol. 86, Feb. 1998, pp. 315–355.

[3] Nakhla, M. S., and Q. J. Zhang, eds., *Modeling and Simulation of High Speed VLSI Interconnects*, Norwell, MA: Kluwer, 1994.

[4] Dai, W. W. M., ed., "Special Issue on Simulation, Modeling, and Electrical Design of High Speed and High Density Interconnects," *IEEE Trans. on Circuit Systems I*, Vol. 39, Nov. 1992, pp. 857–982.

[5] Gao, D. S., A. T. Yang, and S. M. Kang, "Modeling and Simulation of Interconnection Delays and Crosstalk in High Speed Integrated Circuits," *IEEE Trans. on Circuit Systems I*, Vol. 37, Jan. 1990, pp. 1–9.

[6] Bakoglu, H. B., *Circuits, Interconnections and Packaging for VLSI*, Reading, MA: Addison-Wesley, 1990.

[7] Zhang, Q. J., F. Wang, and M. S. Nakhla, "Optimization of High Speed VLSI Interconnects: a Review," *Int. Journal on Microwave and Millimeter-wave Computer-Aided Engineering*, Special Issue on Optimization Oriented Microwave CAD, (invited), Vol. 7, 1997, pp. 83–107.

[8] Veluswami, A., M. S. Nakhla, and Q. J. Zhang, "The Application of Neural Networks to EM-Based Simulation and Optimization of Interconnects in High Speed VLSI Circuits," *IEEE Trans. Microwave Theory and Techniques*, Vol. 45, 1997, pp. 712–723.

[9] Zhang, Q. J., S. Lum, and M. S. Nakhla, "Minimization of Delay and Crosstalk in High-Speed VLSI Interconnects," *IEEE Trans. on Microwave Theory and Techniques*, Vol. 40, 1992, pp. 1555–1563.

[10] Mihan, K. K., et al., "Concurrent Thermal and Electrical Optimization of High Speed Packages and Systems," In *Int. Intersociety Elec. Packaging Conf.*, Kaanapali Beach, HI: Mar. 1995, pp. 221–227.

[11] Williamson, J., et al., "Ground Noise Minimization in Integrated Circuit Packages through Pin Assignment Optimization," *IEEE Trans. Components, Packaging and Manuf. Technology, Part B*, Vol. 19, 1996, pp. 361–371.

[12] Wei, Y. J., Q. J. Zhang, and M. S. Nakhla, "Multilevel Optimization of High-Speed VLSI Interconnect Networks by Decomposition," *IEEE Trans. on Microwave Theory and Techniques*, Vol. 42, 1994, pp. 1638–1650.

[13] Zhang, Q. J., and M. S. Nakhla, "Statistical Simulation and Optimization of High-Speed VLSI Interconnects," *International Journal on Analog Integrated Circuits and Signal Processing*, Special Issue on High-Speed Interconnects, Vol. 5, 1994, pp. 95–106.

[14] Djordjevic, A. R., T. K. Sarkar, and R. F. Harrington, "Time-Domain Response of Multiconductor Transmission Lines," *Proceedings of the IEEE*, Vol. 75, June 1987, pp. 743–764.

[15] Achar, R., M. S. Nakhla, and Q. J. Zhang, "Full-Wave Analysis of High-Speed Interconnects using Complex Frequency Hopping," *IEEE Trans. On Computer-Aided Design*, Vol. 17, 1998, pp. 997–1016.

[16] Wang, R., and O. Wing, "A Circuit Model of a System of VLSI Interconnects for Time Response Computation," *IEEE Trans. on Microwave Theory and Techniques*, Vol. 39, 1991, pp. 688–693.

[17] Paul, C. R., *Analysis of Multiconductor Transmission Lines*, New York: Wiley, 1994.

[18] Harrington, R. F., *Field Computation by Moment Methods*, New York: Macmillan, 1968.

[19] Djordjevic, A. R., R. F. Harrington, T. K. Sarkar, and M. Bazdar, *Matrix Parameters of Multiconductor Transmission Lines*. Norwood, MA: Artech House, 1989.

[20] Poltz, J., "Optimizing VLSI Interconnect Model for SPICE Simulation," *Int. J. Analog Integrated Circuits and Signal Processing*, Vol. 5, 1994, pp. 87–94.

[21] Costache, G. I., "Finite Element Method Applied to Skin-Effect Problems in Strip Transmission Lines," *IEEE Trans. on Microwave Theory and Techniques*, Vol. MTT-35, Nov. 1987, pp. 1009–1013.

[22] Pillage, L. T., and R. A. Rohrer, "Asymptotic Waveform Evaluation for Timing Analysis," *IEEE Trans. Computer-Aided Design*, Vol. 9, Apr. 1990, pp. 352–366.

[23] Chiprout, E., and M. S. Nakhla, "Analysis of Interconnect Networks using Complex Frequency Hopping (CFH)," *IEEE Trans. on Computer-Aided Design*, Vol. 14, Feb. 1995, pp. 186–199.

[24] Feldmann, P., and R. W. Freund, "Efficient Linear Circuit Analysis by Padé Via Lanczos Process," *IEEE Trans. Computer-Aided Design*, Vol. 14, May 1995, pp. 639–649.

[25] Griffith, R., and M. S. Nakhla, "Time-Domain Analysis of Lossy Coupled Transmission Lines," *IEEE Trans. Microwave Theory and Techniques*, Vol. 38, 1990, pp. 1480–1487.

[26] Biernacki, R., et al., "Efficient Quadratic Approximation for Statistical Design," *IEEE Trans. Circuits Syst.*, Vol. CAS-36, 1989, pp. 1449–1454.

[27] Barby, J. A., J. Vlach, and K. Singhal, "Polynomial Splines for MOSFET Model Approximation," *IEEE Trans. Computer-Aided Design*, Vol. 7, 1988, pp. 557–567.

[28] Meijer, P., "Fast and Smooth Highly Nonlinear Multidimensional Table Models for Device Modelling," *IEEE Trans. Circuits Syst.*, Vol. 37, 1990, pp. 335–346.

[29] Chiprout, E., and M. S. Nakhla, *Asymptotic Waveform Evaluation and Moment Matching for Interconnect Analysis*, Boston, MA: Kluwer, 1994.

[30] Bracken, J. E., V. Raghavan, and R. A. Rohrer, "Interconnect Simulation with Asymptotic Waveform Evaluation (AWE)," *IEEE Trans. Circuit Systems I*, Vol. 39. Nov. 1992, pp. 869–878.

[31] Sanaie, R., et al., "A Fast Method for Frequency and Time-Domain Simulation of High-Speed VLSI Interconnects," *IEEE Trans. Microwave Theory and Techniques*, Vol. 42, Dec. 1994, pp. 2562–2571.

[32] Achar, R., M. S. Nakhla, and E. Chiprout, "Block CFH: A Model-Reduction Technique for Distributed Interconnect Networks," *Proc. IEEE European Conf. On Circuits Theory and Design (ECCTD)*, Budapest, Hungary, Sept. 1997, pp. 396–401.

[33] Dounavis, A., et al., "Passive Closed-Form Transmission-Line Model for General-Purpose Circuit Simulators," *IEEE Trans. Microwave Theory and Techniques*, Vol. 47, Dec. 1999, pp. 2450–2459.

[34] Lin, S., and E. S. Kuh, "Transient Simulation of Lossy Interconnects Based on the Recursive Convolution Formulation," *IEEE Trans. on Circuit Systems*, Vol. 39, Nov. 1992, pp. 879–892.

[35] Silveria, M., M. Kamon, I. Elfadel, and J. White, "A Coordinate-Transformed Arnoldi Algorithm for Generating Guaranteed Stable Reduced-Order Models of RLC Circuits," *Proc. IEEE IC-CAD*, Nov. 1996, pp. 288–294.

[36] Kerns, K. J., and A. T. Yang, "Preservation of Passivity during RLC Network Reduction via Split Congruence Transformation," *IEEE Trans. on CAD*, Vol. 17, July 1998, pp. 582–591.

[37] Odabasioglu, A., M. Celik, and L. T. Pileggi, "PRIMA: Passive Reduced-Order Interconnect Macromodeling Algorithm," *Proc. of IC-CAD*, Nov. 1997, pp. 58–65.

[38] Achar, R., et al., "PIRAMID: Passive Multilevel Multipoint Interconnect Reduction Algorithm for Macromodleing of Multiport Networks Including Distributed Elements," *IEEE Trans. Circuit Systems*, (To appear).

[39] Elfadel, I. M., and D. D. Ling, "A Block Rational Arnoldi Algorithm for Multiport Passive Model-Order Reduction of Multiport RLC Networks," *Proc. of IC-CAD*, Nov. 1997, pp. 66–71.

[40] Yu, Q., J. M. Wang, and E. S. Kuh, "Multipoint Moment-Matching Model for Multiport Distributed Interconnect Networks," *Proc. of IC-CAD*, Nov. 1998, pp. 85–90.

[41] Haykin, S., *Neural Networks: A Comprehensive Foundation*, New York: IEEE Press, 1994.

[42] Veluswami, A., "*Neural Network-Based CAD for High-Speed Interconnects and Monolithic Inductors,*" Master's thesis, (Supervisors: M. S. Nakhla and Q. J. Zhang) Carleton University, Ottawa, Canada, November 1995.

[43] Zaabab, A. H., Q. J. Zhang, and M. S. Nakhla, "Neural Network Modeling Approach to Circuit Optimization and Statistical Design," *IEEE Trans. Microwave Theory and Techniques*, Vol. 43, 1995, pp. 1349–1358.

[44] Zaabab, A. H., Q. J. Zhang, and M. S. Nakhla, "Application of Neural Networks in Circuit Analysis," In *Proc. IEEE Int. Conf. Neural Networks*, Perth, Australia, Nov. 1995, pp. 423–426.

[45] SALI: *Structure Analysis for Lossy Interconnect*, Northern Telecom Ltd., Ottawa, Ont., Canada, 1995.

[46] Zhang, Q. J., and M. S. Nakhla, "Signal Integrity Analysis and Optimization of VLSI Interconnects using Neural Network Models," *Proc. IEEE Int. Symp. Circuits and Systems*, London, England, May 1994, pp. 459–462.

7

Active Component Modeling Using Neural Networks

This chapter introduces applications of neural networks to active component modeling. The description includes direct modeling of the device external behavior using neural networks for DC, small-signal, and large-signal device modeling. It also describes indirect neural modeling through a known equivalent circuit of the device, and its application in global modeling of microwave circuits. Discussions on the incorporation of the large-signal neural model into circuit simulations and time-varying Volterra kernel-based neural models for transistors are provided.

7.1 Introduction

Active device modeling is one of the most important areas of microwave CAD. Most frequently-used approaches in today's circuit design are based on lumped equivalent circuits, an example of which is shown in Figure 7.1. A large variety of such equivalent circuit models have been developed in the past, because no single equivalent model can represent all kinds of transistor behaviors. The specific equivalent circuit structure in a model optimized for one type of device becomes a limitation of the model for other devices. With the rapid technology change, new types of semi-conductor devices are constantly evolving, and development of models to represent the new transistor behaviors is a continuous activity. Developing new equivalent circuit models requires human experience and judgment. Often a time-consuming trial-and-error process is used for formulating new equivalent circuit topology and for creating formulas for

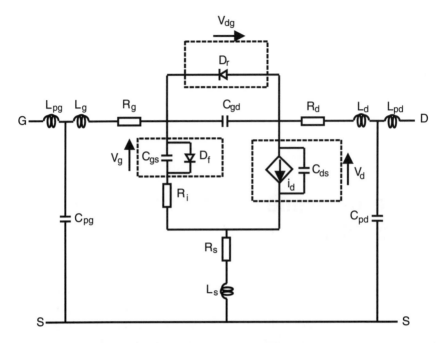

D_f, D_r, C_{gs} and i_d are non-linear elements. For example,

$$i_d = I_{dss}\left(1 - \frac{V_g}{V_p}\right)^2 \tanh\left(\frac{áV_d}{V_g - V_p}\right)$$

$$V_p = V_{po} + \gamma V_d$$

Figure 7.1 An example of a large signal equivalent circuit model of a field effect transistor (from [1], © 1985 IEEE, reprinted with permission).

nonlinear elements. Recently, researchers have started investigating neural network approaches for representing transistor DC, small-signal, and large-signal behaviors. Neural networks have the potential of modeling new transistor behaviors through computerized training, thus possibly making the process of model development more automated. Through learning from measurement data, neural models can also be developed for a new semiconductor device, even if the device theory/equations are still unavailable. In this chapter, we describe the recent activities in the area of application of neural networks to transistor modeling.

7.2 Direct Modeling Approach

In this approach, the component external behaviors are directly modeled by neural networks. There is no need to investigate the internal structure of transistors. The transistors are characterized by external behaviors, such as the terminal DC currents or bias-dependent S-parameters. The following subsections will describe various models for handling different cases of transistor modeling requirements.

7.2.1 Transistor DC Model

For transistors, formulation of a DC model is straightforward. The inputs to the neural models are device physical/process and bias parameters. The current flowing through the terminals of the transistor is the neural model output.

Example of a Physics-Based MESFET DC Neural Model.

In this example, a neural model to represent the DC characteristics of a physics-based MESFET is developed [2]. The inputs to the neural model include the transistor physical/process parameters (channel length L, channel width W, doping density N_d, channel thickness a), and terminal voltages (gate-source voltage V_{GS}, drain-source voltage V_{DS}). The drain current (I_D) is the neural network output, as shown in Figure 7.2. The original problem is physics based [3] and requires a slow numerical simulation. Training samples were obtained by simulating the original Khatibzadeh and Trew model [4] using the OSA90 simulator [5] at randomly chosen inputs. The neural network was trained on SPARC station 5 using a gradient-based l_2 optimization approach, and the CPU time was about five minutes.

Figure 7.3 shows the I-V curves predicted by the neural model, which was trained by 300 samples. The neural models are much faster than the

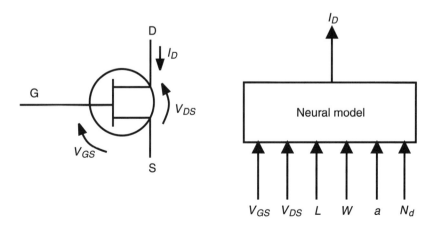

Figure 7.2 Neural network DC model of a MESFET.

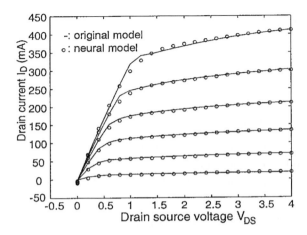

Figure 7.3 I-V curves from the neural network model of a MESFET.

original physics-based MESFET model. As a comparison, the time taken for 1,000 repetitive simulations in a Monte Carlo analysis with random values of device physical/process parameters was about four seconds for the neural model and 27 minutes for the original theory-based MESFET model [4, 5].

7.2.2 Small-Signal Models

Small-signal models of active components are very important for active circuit design. Neural networks can be trained to learn the nonlinear relationship between the small-signal transistor behavior and the physical/process/bias parameters. The inputs to the small-signal neural models can be device physical/process parameters and bias conditions. The outputs of the neural model are S-parameters.

Example of a Small-Signal HBT Neural Model

In this example, small-signal neural models of an HBT are developed [6]. S-parameters of an HBT are measured for various input combinations of frequency, bias current, and bias voltage. A separate neural model was developed for different transistor sizes. The neural network structure is shown in Figure 7.4. The neural model has three inputs, namely, the frequency f, the collector bias current I_C, and the base voltage V_{BE}. There are eight outputs, namely, MS_{ij} and PS_{ij} $i, j = 1, 2$, which are the magnitudes and the phases of S_{11}, S_{12}, S_{21}, and S_{22}, respectively. Three-layer perceptrons (MLP3) with different numbers of hidden neurons were trained for each transistor size. The trained neural models were tested by an independent set of measured data called the

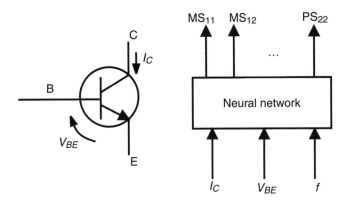

Figure 7.4 HBT and the corresponding neural model.

test data. Average test errors of the neural models for S-parameters are observed to be well below 2%. In Table 7.1, MLP 3-12-8 represents a three-layer perceptron with three inputs, 12 hidden neurons, and eight outputs. The neural model prediction of the S-parameter is very accurate, considering that the model is developed with measurement errors in data, as can be seen from Figure 7.5 and Figure 7.6.

S-parameters were then converted to equivalent Y-parameters. Three-layer perceptrons with different numbers of hidden neurons were trained for each transistor size. The outputs of the neural models are the real and imaginary parts of Y_{11}, Y_{12}, Y_{21}, and Y_{22}. It is observed that the neural models can estimate the Y-parameters with average test errors well below 3% as shown in Table 7.2.

This example illustrates the development of a transistor model simply by neural network training with measurement data, avoiding the trial-and-error process in developing new equivalent circuit topologies. The finished neural

Table 7.1
Average Test Errors of HBT S-Parameter Neural Models for Different Transistor Sizes (From [6], © European Microwave Conference Management Committee 1998, Reprinted with Permission)

HBT Size	MLP (3-12-8)	MLP (3-15-8)	MLP (3-20-8)	MLP (3-25-8)
Size 1	1.98%	1.95%	1.82%	1.86%
Size 2	1.40%	1.34%	1.16%	1.23%
Size 3	1.30%	1.18%	1.08%	1.15%
Size 4	1.29%	1.12%	1.06%	1.11%
Size 5	1.23%	1.07%	1.04%	1.07%

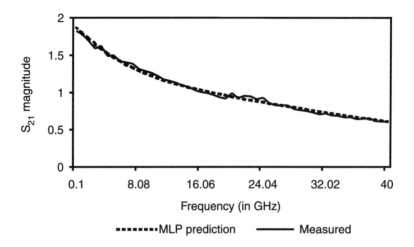

Figure 7.5 Comparison of neural model (MLP 3-20-8) prediction with measured S_{21} magnitude (from [6], © European Microwave Conference Management Committee 1998, reprinted with permission).

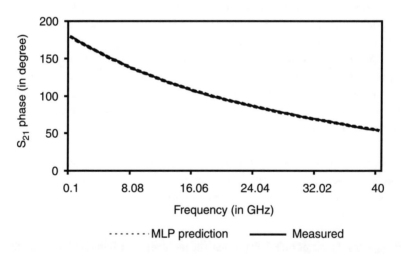

Figure 7.6 Comparison of neural model (MLP 3-20-8) prediction with measured S_{21} phase (from [6], © European Microwave Conference Management Committee 1998, reprinted with permission).

model is an analytical and continuous model, with the ability to link the nonlinear relationship between transistor S-parameters and transistor bias. The neural network approach is more effective in the cases where the device under consideration cannot be modeled accurately by existing equivalent circuit models.

Table 7.2
Average Test Errors of HBT Y-Parameter Neural Models for Different Transistor Sizes (From [6], © European Microwave Conference Management Committee 1998, Reprinted with Permission)

HBT Size	MLP (3-8-8)	MLP (3-12-8)	MLP (3-15-8)	MLP (3-20-8)
Size 1	2.95%	2.77%	2.81%	2.85%
Size 2	2.70%	2.27%	2.37%	2.81%
Size 3	2.31%	1.90%	1.98%	1.86%
Size 4	2.38%	2.22%	2.34%	2.38%
Size 5	2.78%	2.25%	2.26%	2.50%

7.2.3 Large-Signal Models

A neural network formulation of large-signal models to describe terminal currents and charges of transistors as nonlinear functions of the device parameters and terminal voltages was proposed in [7, 8]. The model in this form is very convenient for a harmonic balance simulator. In the case of a MESFET, the model outputs are gate, drain, and source currents (i_{gc}, i_{dc}, i_{sc}) and corresponding charges (q_g, q_d, q_s) on the gate, drain, and source terminals. The output vector is given by

$$y = [i_{gc} \quad i_{dc} \quad i_{sc} \quad q_g \quad q_d \quad q_s]^T \tag{7.1}$$

In order to connect the neural model to a simulator, we first need a circuit representation of the model consistent with the device under consideration. A circuit representation of the six-output neural network model is shown in Figure 7.7.

Incorporation of a Large Signal MESFET Neural Model into Circuit Simulation

In this case, we consider the use of the harmonic balance method (HBM) as introduced in Chapter 2 for steady state analysis of nonlinear periodic circuits. The technique can be applied to transient analysis as well. In HBM, the circuit under consideration is divided into linear and nonlinear subnetworks. It is thus appropriate to include the neural network model as an additional nonlinear subnetwork, as shown in Figure 7.8. In other words, when the neural model is used in place of an active or passive component, the overall circuit harmonic balance equation can be written as

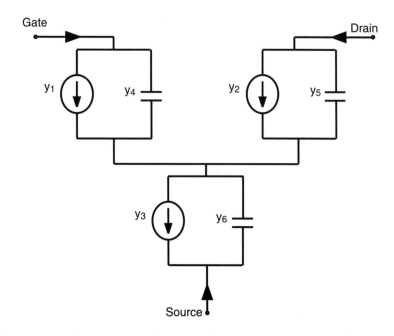

Figure 7.7 Circuit representation of the six-output neural network model (from [7], © 1995 IEEE, reprinted with permission).

$$I(V) + j\Omega Q(V) + I_{ANN}(V) + j\Omega Q_{ANN}(V) + YV + I_{ss} = 0 \quad (7.2)$$

where Y is the nodal admittance matrix of the linear subnetwork, V represents the voltages in the circuit, $I(V)$ and $Q(V)$ represent the currents and charges of the nonlinear subnetwork, I_{ss} represents the sources, and Ω is the angular frequency matrix. The vectors $I_{ANN}(V)$ and $Q_{ANN}(V)$ represent the Fourier coefficients of the time-domain conduction currents and charges i_{ANN} and q_{ANN} entering the circuit from the neural network model. For example, in the case of a MESFET neural model, $I_{ANN}(V)$ and $Q_{ANN}(V)$ are computed from the Fourier transform of the time-domain currents (i_{gc}, i_{dc}, i_{sc}) and charges (q_g, q_d, q_s) which are provided by the neural network outputs shown in Equation (7.1). It may be noted that solving the harmonic balance problem by incorporating neural models does not require the time-consuming repeated simulation of the original device physics equations in the standard approaches of [3, 4]. Another type of analysis is the Monte Carlo analysis, in which the circuit is repeatedly simulated with randomly generated device parameters. Again in this case the neural network approach speeds up analysis by avoiding repeated solving of the device physics equations.

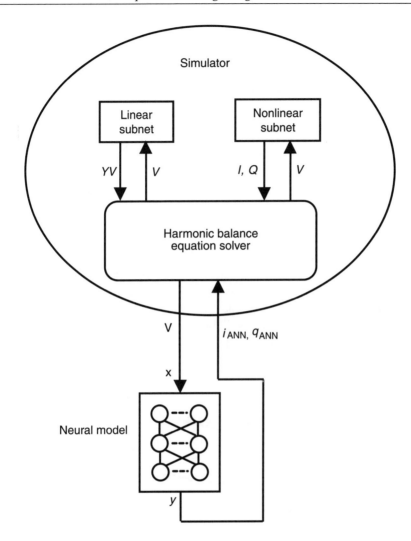

Figure 7.8 Incorporating the large signal neural network models of a transistor into harmonic balance circuit simulator (from [7], © 1995 IEEE, reprinted with permission).

Example of a Physics-Based Large-Signal MESFET Neural Model

In this example, a neural model to represent the large-signal behavior of a transistor is developed [7]. A three-layer perceptron neural network is used. The neural model has six inputs, gate length (L), gate width (W), channel thickness (a), doping density (N_d), gate-source voltage (V_{gs}), and drain-source voltage (V_{ds}) as shown in Figure 7.9. The neural model has six outputs, i_{gc}, i_{dc}, i_{sc}, q_g, q_d, and q_s.

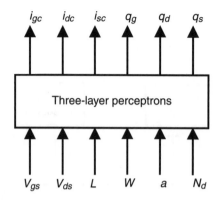

Figure 7.9 Large-signal neural model of a MESFET.

The training data was obtained by OSA90 simulation [5] with the Khatibzadeh and Trew model of the MESFET [4]. An independent set of data was used to test the quality of the final model. The DC characteristics as predicted by the neural model are shown in Figure 7.10. The small-signal S-parameters were computed by the circuit simulator by incorporating the

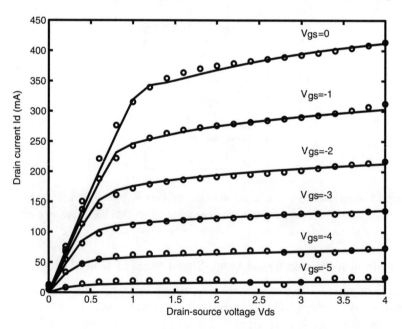

Figure 7.10 Comparison of the DC characteristics from the large-signal neural model (o) with those of the Khatibzadeh and Trew model (-) (from [7], © 1995 IEEE, reprinted with permission).

large-signal neural model. In Figure 7.11, the predictions of the S-parameters by the circuit simulator are compared with the original Khatibzadeh and Trew model.

The large-signal neural model has been incorporated into the harmonic balance simulator of OSA90 and HP-ADS for simulation and yield optimization of a three-stage MMIC amplifier circuit. The significant feature is that the same neural model (trained only once) is reused many times in place of all three transistors present in the circuit for different values of transistor physical/process/bias parameters during the Monte Carlo analysis and yield optimization. The amplifier optimization will be described in detail in Chapter 8.

7.2.4 Time-Varying Volterra Kernel-Based Model

The Volterra series approach has re-emerged through the use of time-varying Volterra kernels [9]. In the case of short-memory devices, the first-order kernel dominates the device behavior, allowing us to ignore higher-order kernels. The higher-order kernels have been partly responsible for the complexities and difficulties in the analysis and measurement using the classical Volterra series approach. It was suggested in [9] that the short memory condition is likely to be satisfied by many electronic devices. As such, the time-varying Volterra

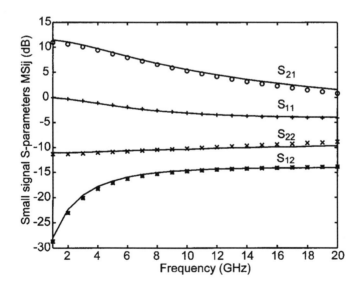

Figure 7.11 Comparison of the magnitude of the small-signal S-parameters from the large-signal neural model (o, +, *, and x) with those of the Khatibzadeh and Trew model (-) (from [7], © 1995 IEEE, reprinted with permission).

kernel approach is suitable for device modeling. A Volterra series with a first-order kernel for a FET can be expressed as

$$i(t) = I_{DC}(v(t)) + \sum_{p=-K}^{K} Y(v(t), \omega_p) V_p \exp(j\omega_p t) \qquad (7.3)$$

where ω_p, $|p| \leq K$, are the frequency points spanning the entire frequency range, $i(t)$ is a vector of large-signal gate and drain currents, I_{DC} is a vector of DC gate and drain currents, $v(t)$ and V_p are vectors of gate and drain voltages in time and frequency domains, respectively, and Y is the time-varying kernel (which is a 2-by-2 complex matrix).

The conventional way to represent the functions $I_{DC}(v(t))$ and $Y(v(t), \omega)$ is by either table-lookup or interpolation methods based on extensive bias-dependent DC and AC measurements. In the neural network approach, these functions are represented by the corresponding neural models [9]. The neural models have good learning and generalization capabilities, which can provide efficient data compression.

Example of Transistor Neural Models for the Volterra Kernel Approach

In [9], a time-varying Volterra kernel-based model was used to model a transistor for amplifier design. The inputs to the neural networks are gate-source voltage (V_{gs}), drain-source voltage (V_{ds}), and frequency (f). The objective is to develop neural models to represent each of the Volterra kernel parameters, drain-source current (I_{ds}), gate-source current (I_{gs}), and the real and imaginary parts of all Y_{ij} over the entire region of operation ($V_{ds} \in [0, 9V]$, $V_{gs} \in [-4, 0.8V]$, and $f \in [1, 86GHz]$). As shown in Figure 7.12, the overall neural network architecture is composed of 10 neural networks (two for modeling input/output currents, four for modeling real parts of the Y matrix, and four for modeling imaginary parts of the Y matrix). Each of the neural networks modeling the input/output currents (DC) is composed of two input neurons, six sigmoid neurons in the hidden layer, and one output neuron. Each Y-parameter neural network has three input neurons, 15 sigmoid neurons in the hidden layer, and one output neuron. The DC neural networks are trained using 350 measurement samples each, and the Y-parameter neural networks are trained using 7,000 measurement samples. A classical lumped equivalent electrical circuit is deduced from the same measurements. Figure 7.13 compares a good agreement of the neural model with the measurements for both DC and AC cases. The transistor neural models have been incorporated into a circuit simulator through the Volterra kernel formulation, and subsequently used in the design of a 13–14 GHz amplifier.

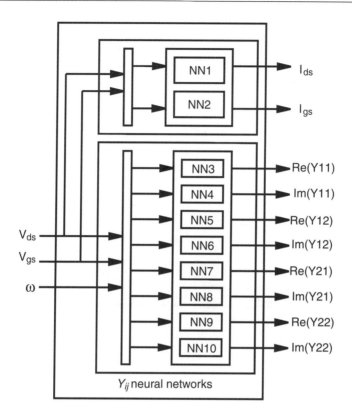

Figure 7.12 Transistor neural models to be subsequently used in the Volterra kernel approach (from [9], Harkouss, Y., et al, "The Use of Artificial Neural Networks in Nonlinear Microwave Devices and Circuits Modeling: An Application to Telecommunication System Design," *Int. J. RF and Microwave CAE,* pp. 198–215, © 1999, John Wiley and Sons. Reprinted with permission from John Wiley and Sons, Inc.).

7.3 Indirect Modeling Approach Through a Known Equivalent Circuit Model

The commonly used modeling approach for active devices is the lumped equivalent circuit. Developing such models requires experience and involves a trial-and-error process to find appropriate circuit topology and the values of the circuit elements. Moreover, an equivalent circuit model may not have direct links with the physical/process parameters of the device. Empirical formulas for such links may exist, but the accuracy cannot be promised when applied to different devices. Neural networks have the capability of learning highly nonlinear relationships through training samples. By combining the two

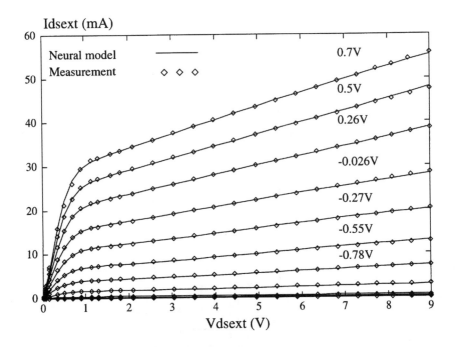

Figure 7.13 Comparison of a neural model with measurements. (a) DC model; (b) Y_{11} as an example of a Y-parameter model (from [9], Harkouss, Y., et al, "The Use of Artificial Neural Networks in Nonlinear Microwave Devices and Circuits Modeling: An Application to Telecommunication System Design," *Int. J. RF and Microwave CAE,* pp. 198–215, © 1999, John Wiley and Sons. Reprinted with permission from John Wiley and Sons, Inc.).[1]

modeling techniques, namely, the equivalent circuit representation (known a priori) and neural modeling, an efficient and flexible model can be developed. The concept is shown in Figure 7.14.

As an illustration, a small signal FET model based on the indirect modeling approach is shown in Figure 7.15. At a lower level, there is a neural network that models the nonlinear relationship between the device physical/process/bias parameters and the equivalent circuit element values (e.g., C_1, C_2, C_3, r, g_m), through a training process. The training data is obtained through the parameter extraction process of the known equivalent circuit model, where a set of C_1, C_2, C_3, r and g_m values are calculated per bias from small-signal S-parameter measurements. At a higher level, there is the equivalent circuit model that takes the element values and frequency as inputs and gives the

---

1. The original reproducible of this figure was kindly supplied by Dr. J. Rousset of IRCOM, University of Limoges, Limoges, France.

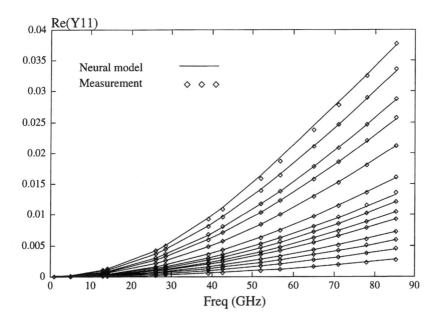

Figure 7.13 (continued).

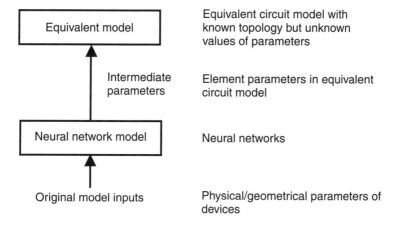

Figure 7.14 Indirect modeling approach by neural networks through a known equivalent circuit model.

circuit response as the output. The advantage of using the neural model lies in the fact that the circuit response can be obtained for bias parameters never seen during training, in the region of validity of the neural model. A further description of this approach is presented in Chapter 9.

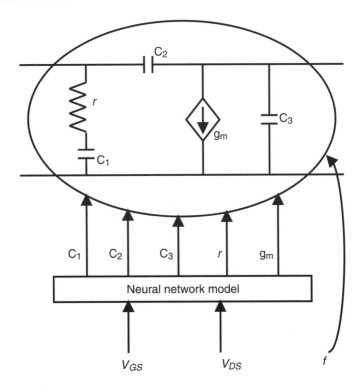

Figure 7.15 Illustrative example of a small signal FET model based on the indirect modeling approach.

Example of Indirect Neural Modeling for a HEMT

Large-signal models for active devices such as high electron-mobility transistors (HEMT) are essential for accurately designing microwave circuits. Recently, a neural network approach for modeling large-signal behavior of an HEMT has been proposed [10]. A multilayered neural network along with a known equivalent circuit model is used. In order to plug the HEMT model into a standard harmonic-balance simulator, the large-signal behavior is characterized by a small-signal equivalent circuit. The bias-dependent intrinsic elements (C_{gs}, R_i, C_{gd}, g_m, τ, g_{ds}, C_{ds}) of the small-signal equivalent circuit are realized by a neural network whose inputs are V_{gs} and V_{ds}.

The intrinsic element values at various bias settings V_{gs} and V_{ds} are obtained from the S-parameter measurements. The inputs and outputs of the neural model are defined as

$$\boldsymbol{x} = [V_{gs} \ V_{ds}]^T \tag{7.4}$$

$$\boldsymbol{y} = [C_{gs} \ R_i \ C_{gd} \ g_m \ \tau \ g_{ds} \ C_{ds}]^T \tag{7.5}$$

Through experiments, a six-layered MLP with 21 hidden neurons was chosen to represent all the large-signal parameters simultaneously, as shown in Figure 7.16. In [11], individual HEMT parameters such as C_{gs} were modeled using three-, four-, and five-layered perceptrons through a neural network structural optimization scheme using a genetic algorithm.

Example of Neural Networks for Global Modeling Applications

Global modeling of monolithic microwave integrated circuits is important for analyzing the electromagnetic coupling, the device-EM wave interaction, and

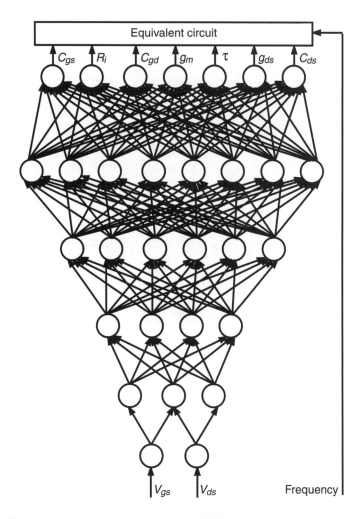

Figure 7.16 An indirect large-signal model of an HEMT using neural networks to feed a known equivalent circuit model, after [10].

the EM radiation effects of closely spaced active and passive components [12]. The CAD process can include numerically solving Maxwell's equations in passive structures and semiconductor device regions and solving physics-based equations of semiconductor devices such as the hydrodynamic model [13]. Maxwell's equations and semiconductor equations are coupled, and the overall computation can be very expensive. Recently, neural network models have been used to learn the solutions of full hydrodynamic models [13]. The neural model is subsequently used to interact with the Maxwell equations, thereby speeding up the overall global modeling process. This example illustrates the use of neural models for the global modeling of a microwave amplifier in [13].

A three-layer perceptron neural network is used. The neural model has two inputs, namely, the gate-source voltage (V_{gs}) and the drain-source voltage (V_{ds}). There are five outputs: drain current (I_{ds}), gate-drain capacitance (C_{gd}), gate-source capacitance (C_{gs}), drain-source capacitance (C_{ds}), and drain-source resistance (R_{ds}). The overall device model consists of an equivalent circuit whose parameters are the outputs of the neural model, as shown in Figure 7.17. To accurately represent the high-frequency effects in the amplifier, full hydrodynamic simulation of the transistor is used to generate training and test data. The neural network was trained to a good accuracy as compared with the original hydrodynamic model. A benchmark comparison in [13] indicates

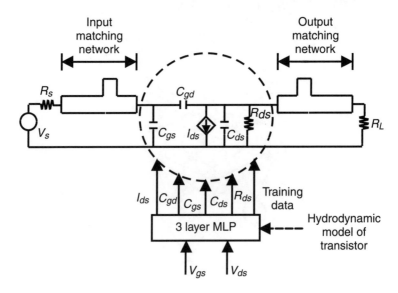

Figure 7.17 The neural-network–based global modeling approach for an amplifier with a transistor and input-output matching circuits. The transistor is modeled by a neural network through a known equivalent circuit (after [13]).

that the solution from the neural model for 21 samples of V_{gs} and V_{ds} is instantaneous, while the hydrodynamic model requires 45 minutes.

The neural model together with the known equivalent circuit model is subsequently used for the global modeling of the amplifier circuit [13]. The transistor's equivalent circuit parameters (at different bias conditions) as computed from the neural model are supplied to the FDTD marching time algorithm. The EM effects of the matching circuit, the semiconductor device, and the couplings between them were simulated by the FDTD method, as illustrated in Figure 7.18. Had the original hydrodynamic model of the semiconductor device been directly used with FDTD, the global modeling process would have been prohibitively slow.

7.4 Discussion

In general, neural network models of active components take any type of parameters as inputs, such as physical/geometrical/bias parameters, as long as training data can be generated. As such, design/adjustment of components' physical and bias parameters is feasible in the computer-aided design and tolerance analysis. In this chapter, two types of active device modeling approaches using neural networks have been described—the direct modeling approach and the indirect modeling approach through a known equivalent circuit model.

In the direct modeling approach, neural networks model external behaviors of the devices. The overall model could include all the practical effects, nonideal effects, or new semiconductor effects that are not yet included in

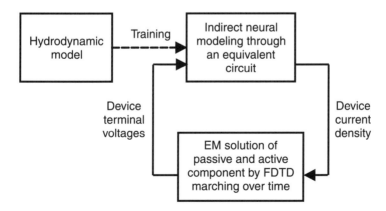

Figure 7.18 The iterative computation process of the global modeling approach with neural models interacting with the FDTD computation.

available commercial models because of neural network learning directly from measurements. The models can be developed even when theory/experience/ knowledge of the component is not yet available, again because of neural network learning from measurement. Since the neural network directly learns the external behavior of the device, the models involve either no or fewer assumptions than equivalent circuit models.

The indirect approach involves neural network modeling through known equivalent circuit models. Resulting models are easily compatible with the circuit simulators, including both time- and frequency-domain simulators. It is also possible to provide time-domain dynamic behavior. However, the models developed in this approach could be limited by the assumptions of the equivalent circuit models.

References

[1] Materka, A., and T. Kacprzak, "Computer Calculation of Large-Signal GaAs FET Amplifier Characteristics," *IEEE Trans. Microwave Theory and Techniques*, Vol. 33, 1985, pp 129–135.

[2] Wang, F., and Q. J. Zhang, "Knowledge Based Neural Models for Microwave Design," *IEEE Trans. Microwave Theory and Technniques*, Vol. 45, 1997, pp. 2333–2343.

[3] Bandler, J. W., et al., "Integrated Physics-Oriented Statistical Modeling, Simulation and Optimization," *IEEE Trans. Microwave Theory and Techniques*, Vol. 40, 1992, pp. 1374–1400.

[4] Khatibzadeh, M. A., and R. J. Trew, "A Large-Signal, Analytical Model for the GaAs MESFET," *IEEE Trans. Microwave Theory and Techniques*, Vol. 36, No. 2, 1988, pp. 231–238.

[5] *OSA90 3.0*, Optimization Systems Associates, P.O. Box 8083, Dundas, Canada, L9H 5E7, now HP EEsof (Agilent Technologies), 1400 Fountaingrove Parkway, Santa Rosa, CA 95403.

[6] Devabhaktuni, V. K., C. Xi, and Q. J. Zhang, "A Neural Network Approach to the Modeling of Heterojunction Bipolar Transistors from S-parameter Data," *Proc. European Microwave Conf.*, Vol. 1, Amsterdam, Netherlands, October 1998, pp. 306–311.

[7] Zaabab, A. H., Q. J. Zhang, and M. S. Nakhla, "Neural Network Modeling Approach to Circuit Optimization and Statistical Design," *IEEE Trans. Microwave Theory and Techniques*, Vol. 43, 1995, pp. 1349–1358.

[8] Zaabab, A., Q. J. Zhang, and M. S. Nakhla, "Device and Circuit Level Modeling using Neural Networks with Faster Training based on Network Sparsity," *IEEE Trans. Microwave Theory and Techniques*, Vol. 45, 1997, pp. 1696–1704.

[9] Harkouss, Y., et al., "The Use of Artificial Neural Networks in Nonlinear Microwave Devices and Circuits Modeling: An Application to Telecommunication System Design," *Int. Journal of RF and Microwave CAE*, Special Issue on Applications of ANN to RF and Microwave Design, Vol. 9, 1999, pp. 198–215.

[10] Shirakawa, K., et al., "A Large Signal Characterization of an HEMT using a Multilayered Neural Network," *IEEE Trans. Microwave Theory and Techniques*, Vol. 45, 1997, pp. 1630–1633.

[11] Shirakawa, K., et al., "Structural Determination of Multilayered Large-Signal Neural Network HEMT Model," *IEEE Trans. Microwave Theory and Techniques*, Vol. 46, 1998, pp. 1367–1375.

[12] Imtiaz, S. M. S., and S. M. El-Ghazaly, "Global Modeling of Millimeter-Wave Circuits: Electromagnetic Simulation of Amplifiers," *IEEE Trans. Microwave Theory and Techniques*, Vol. 45, 1997, pp. 2208–2216.

[13] Goasguen, S., S. M. Hammadi, and S. M. El-Ghazaly, "A Global Modeling Approach using Artificial Neural Network," *IEEE MTT-S International Microwave Symp. Digest*, Anaheim, CA, June 1999, pp. 153–156.

8

Design Analysis and Optimization

This chapter presents examples of ANN-model–based analysis and optimization of RF and microwave designs. Examples of component optimization, CPW circuits (a power divider and a folded double-stub filter), optimization of a CPW patch antenna, optimization of multilayer band pass filters, and yield optimization of a three-stage MMIC MESFET amplifier are presented.

8.1 Design and Optimization Using ANN Models

When ANN models for various RF and microwave components have been developed, they can be used in the design and optimization of circuits, antennas, integrated circuit-antenna modules, and other subsystems and systems. The role of models in the design process was discussed in Chapter 2, "Modeling and Optimization for Design."

For RF and microwave circuit design, ANN models are linked to the circuit simulator (such as HP-EEsof's MDS, ADS). Integration of ANN models with circuit simulators was discussed in Section 5.1. Once linked with the simulator, these models are used just as any other model available within the network simulator. The models can be used with other ANN models or combined with many other models available in the simulator to form a complete circuit.

Circuit and/or component optimization, where the gradient of a function is needed, can also be performed using the routines available within the circuit simulator. Two-layer networks with sigmoidal hidden units have the ability to simultaneously approximate both a function and its derivative [1]. This is due to having continuous hidden layer and output layer neuron activation functions (sigmoids), which are differentiable everywhere.

Several examples of the use of ANN models for component, circuit, and antenna design are available in literature. Some of these are reviewed in this chapter.

8.2 Optimization of Component Structure

ANN models can be used to find the optimal physical structure of a component for a given application. This can be accomplished by using standard techniques such as random and gradient optimization. The two examples demonstrate component optimization using ANN models.

Example 1: Stripline-to-Stripline Multilayer Interconnect

Development of the ANN model for a stripline-to-stripline multilayer interconnect was discussed in Section 5.2.4. For this EM-ANN model, two variable physical parameters are the diameter of the via and the diameter of the ground access opening. This structure has been considered in the literature earlier [2], and it was found that a good performance was obtained as long as the diameter of the via was large and the ratio of the diameter of the ground access to that of the via was near 4.2:1. In fact, this ratio is the same as that of the inner and outer conductors of a 50 Ω coaxial line with ϵ_r = 2.94. For optimizing this structure, one would expect that the ratio for the stripline-to-stripline interconnect might be less than that for a coaxial line. The resulting increase in the capacitance of the structure will compensate for only having a partial outer conductor.

This structure has been optimized [3] using the ANN model. Initial values for the physical parameters were set at 0.3 for D_{via}/W_1 and 6.0 for D_{gnd}/D_{via}. These are clearly not optimal values. Optimization was completed by maximizing the transmission coefficient yielding D_{via}/W_1 = 0.77 and D_{gnd}/D_{via} = 3.4. These results agree well with expectations mentioned previously.

Example 2: A Chamfered CPW 90° Bend

Optimization of chamfer in a 90° CPW bend was discussed in Section 5.3.2. In this example the use of ANN modeling resulted not only in the value of the optimum chamfer given by Equation (5.1), but also in a novel compensated bend design shown in Figure 5.8(b).

In principle, this component optimization could be carried out by using EM simulation directly without the use of ANN models. However, that process is computationally much more intensive and hence not very efficient.

8.3 Circuit Optimization Using ANN Models

A couple of interesting examples of circuit optimization using ANN models make use of the CPW components' models discussed in Chapter 5. These are reviewed in this section.

8.3.1 CPW Folded Double-Stub Filter

The first CPW circuit design example is a CPW folded double-stub filter, as shown in Figure 8.1. For this design, W = 70 μm and G = 60 μm, yielding $Z_o \approx 50$ Ω. CPW EM-ANN models used are CPW transmission line, 90° compensated bends, short circuit stubs, and symmetric T-junctions. The filter was designed for a center frequency of 26 GHz. Ideally, the length of each short circuit stub and the section of line between the stubs should have a

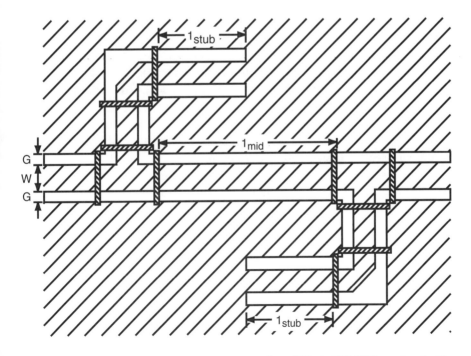

Figure 8.1 CPW folded double-stub filter geometry (from [20], © 1997 IEEE, reprinted with permission).

length of $\lambda/4$ at 26 GHz. However, due to the presence of discontinuities, these lengths need to be adjusted. Design and optimization were accomplished using an *HP-MDS* [4] network simulator and EM-ANN models for various components.

Parameters to be optimized are l_{stub} and l_{mid}. Initial values for these lengths were determined and the structure simulated, showing a less-than-ideal response. The circuit was then optimized, using gradient descent, to provide the desired circuit response. The effect of optimization was reductions in the two line lengths. Results are shown in Figure 8.2 for the original design, optimized design, and full-wave EM simulation of the entire optimized circuit. Good agreement was obtained between the optimized EM-ANN circuit design and the full-wave EM simulation over the 1 GHz to 50 GHz frequency range. This demonstrates applications of EM-ANN models in CPW circuit design and optimization.

8.3.2 CPW Power Divider

The second example of CPW circuit design using ANN models is a 50 Ω, 3 dB power divider, shown in Figure 8.3. For this design, EM-ANN models for CPW transmission lines, 90° compensated bends, T-junctions, and step-in-width transitions are used. The power divider was designed for a 3 dB power split between the output ports at 26 GHz. Ideally, $\lambda/4$, 70.7 Ω line sections are used to transform the 50 Ω output impedance at ports 2 and 3 into 100 Ω loads at the T-junction. Due to the presence of discontinuities, the lengths of the transformers need to be adjusted. Input and output lines are 50 Ω (W = 70 μm, G = 60 μm) and transformer lines are approximately 71 Ω (W = 20 μm, G = 60 μm). Again, *HP-MDS* [4] and ANN models have been used for design and optimization.

The optimizable parameter for this design is the length of the transformer line section, l_{trans}. An initial value for l_{trans} was determined. After the initial design failed to meet the prescribed design criteria, optimization was performed, resulting in a shorter line length than initially determined. Results are shown in Figure 8.4 for the original design, optimized design, and full-wave EM simulation of the entire optimized circuit. As with the double-stub filter design, excellent agreement has been obtained between the optimized EM-ANN circuit design and EM simulation results over the entire 1 GHz to 50 GHz range.

In this example, the use of EM-ANN models for CPW components allowed accurate and efficient design and optimization of the CPW filter. Thus we note that the use of EM-ANN models provides an efficient EM-simulation–based optimization procedure. Optimization time for the EM-ANN circuit was only three minutes and required seven circuit analyses.

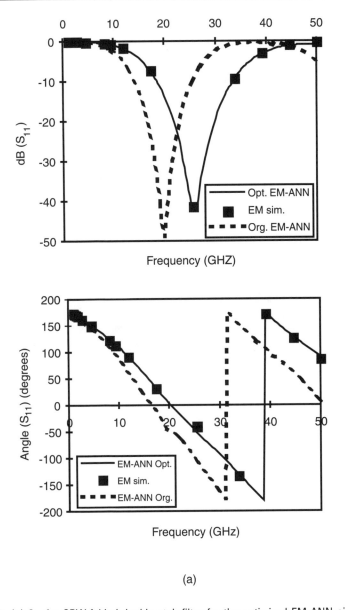

(a)

Figure 8.2 (a) S_{11} for CPW folded double-stub filter for the optimized EM-ANN circuit (EM-ANN Opt.), the original EM-ANN circuit (EM-ANN Org.), and EM simulation (EM sim.). (b) S_{21} for CPW folded double-stub filter for the optimized EM-ANN circuit (EM-ANN Opt.), the original EM-ANN circuit (EM-ANN Org.), and EM simulation (EM sim.) (from [20], © 1997 IEEE, reprinted with permission).

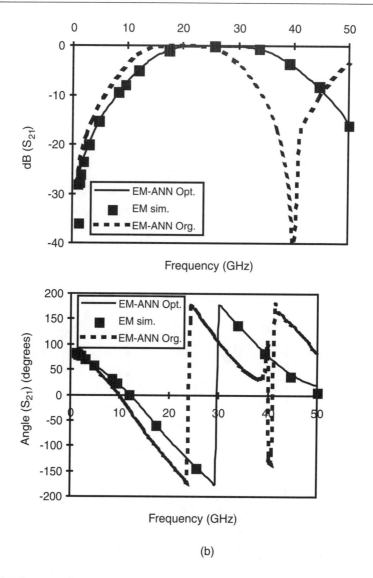

Figure 8.2 (continued).

The amount of time required to provide EM simulation results for 17 frequency points for the entire filter circuit was approximately 14 hours on the same HP 700 workstation. Optimization time for the power divider circuit was only two minutes and required six circuit analyses. EM simulation time for the entire power divider, at 15 frequency points, totaled almost 11 hours on this HP 700 workstation. This confirms that substantial savings in time are achievable by using EM-ANN component models, especially when optimization is desired,

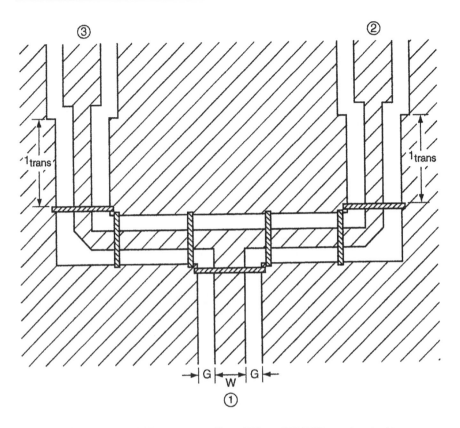

Figure 8.3 CPW power divider geometry (from [20], © 1997 IEEE, reprinted with permission).

requiring numerous circuit solutions, and when these components are to be used over and over in different circuit designs.

It should be mentioned that even larger and more complex circuits could be designed using the developed EM-ANN models. EM simulation of large, complex circuits is limited by the computer resources available and in many cases is not practical. These difficulties are overcome with EM-ANN component modeling.

8.4 Multilayer Circuit Design and Optimization Using ANN Models

Development of ANN models for analysis and synthesis of multilayer multiconductor transmission lines was discussed in Chapter 5. An example of application

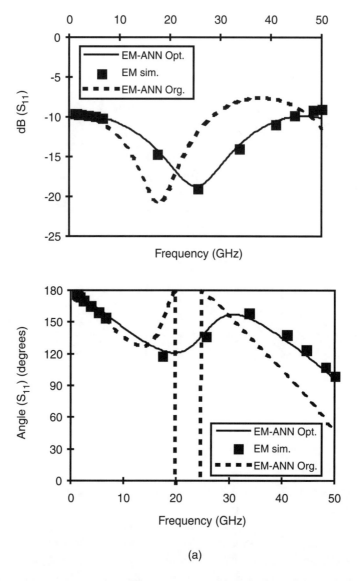

(a)

Figure 8.4 (a) S_{11} for CPW power divider for the optimized EM-ANN circuit (EM-ANN Opt.), the original EM-ANN circuit (EM-ANN Org.), and EM simulation (EM sim.). (b) S_{21} for CPW power divider for the optimized EM-ANN circuit (EM-ANN Opt.), the original EM-ANN circuit (EM-ANN Org.), and EM simulation (EM sim.) (from [20], © 1997 IEEE, reprinted with permission).

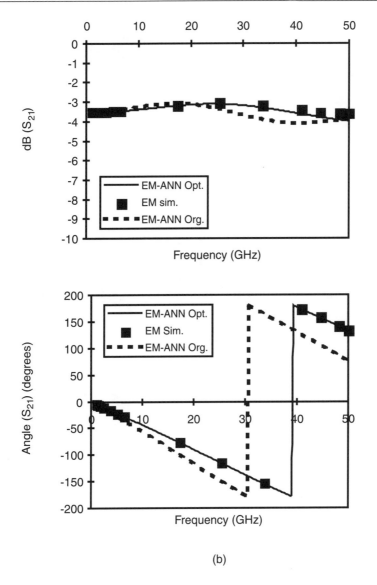

Figure 8.4 (continued).

of these models for multilayer circuit design and optimization is discussed in this section.

ANN models for multilayered lines have been used for the design of multilayer coupled line filters. The geometry of the type of filter under consideration is shown in Figure 8.5. Specifically, it is a two-layer asymmetric coupled line filter in an inhomogeneous medium consisting of three coupled line

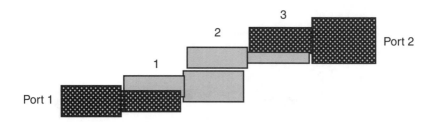

Figure 8.5 Top view of a two-layer coupled line filter consisting of three coupled line sections. Sections 1 and 3 couple from layer 1 to layer 2 and section 2 couples from layer 1 to layer 1. Input and output ports are on the top of layer 2.

sections. There are two types of open-ended coupling sections, each of $\lambda/4$ length at the desired center frequency of the filter. Sections 1 and 3 couple layer 1 to layer 2, while for section 2 both the coupled lines are on layer 1. The height and dielectric constant of each layer are $h_1 = 31$ mils, $h_2 = 10$ mils, and $\epsilon_{r1} = \epsilon_{r2} = 2.2$ (refer to Figure 5.19).

Design Example: Two-Layer Coupled Line Filter

One starts with the conventional filter design procedure. Beginning with filter specifications, J-parameters (admittance inverter parameters) are derived for each coupled line section. For selected values of Z_{o1}, Z_{o2} (impedances at two ends of a coupled line section), and another design parameter, 'a', related to coupling, normal mode parameters (NMPs) for coupled line sections are derived. These NMPs are the different voltage ratios (R_c and R_π), mode impedances (Z_{c1}, $Z_{\pi 1}$, Z_{c2}, and $Z_{\pi 2}$), and phase velocities for the two normal modes, known as c- and π-modes, used to characterize asymmetrical coupled line sections [5–7]. In the conventional design procedure, four of the NMPs (R_c, R_π, Z_{c1} and $Z_{\pi 1}$) are utilized to obtain the physical dimensions (for selected substrate, h, and ϵ_r values) of each coupled line section by using an optimization routine. This optimization process compares the NMPs obtained from specifications with those calculated from the physical geometry by evaluating capacitance and inductance matrices [5–7] as an intermediate step. The capacitance and inductance matrices for a specific coupled line geometry were determined using Segmentation and Boundary Element Method (*SBEM*) analysis [8]. This process (contained inside the dotted block A in Figure 8.6) continues in an iterative manner until the desired results are obtained.

ANN models were used to design a two-layer coupled line filter having the specifications given in Table 8.1. Physical dimensions, obtained from the ANN models, for each coupled line section are given in Table 8.2. To determine

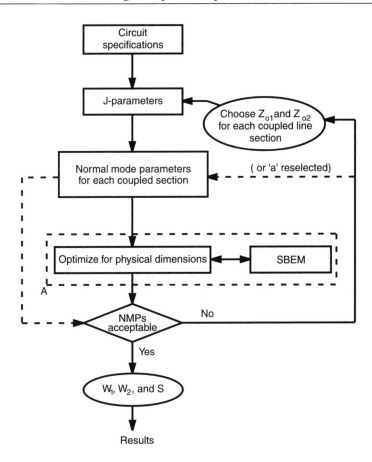

Figure 8.6 Procedure for the design of asymmetric multilayer coupled line sections using optimization [9] (from [21], Watson, P. M., C. Cho, and K. C. Gupta, "EM-ANN Model for Synthesis of Physical Dimensions for Multilayer Asymmetric Coupled Transmission Line Structures," *Int. J. RF and Microwave CAE*, pp. 175–186, © 1999, John Wiley and Sons. Reprinted with permission from John Wiley and Sons, Inc.).

the accuracy of the ANN model outputs, comparisons were made between the modeled 4-port S-parameters of each coupled line section with those obtained from specifications at the desired center frequency. These 4-port S-parameters are for general coupled lines and have been used to determine whether the output of the synthesis model is valid for the given input NMPs. Error bounds for an acceptable solution were set at 0.01 for magnitude and 5° for angle. Angle error was not considered when the magnitude of a given S-parameter was below 0.01. The 4-port S-parameters can then be transformed into the

Table 8.1
Filter Specifications Used for the Design of a Two-Layer Asymmetric Coupled Line Filter

Center frequency	2 GHz
Ripple Bandwidth	7.5%
Ripple Level	0.5 dB
Number of coupled sections	3
$\epsilon_{r1} = \epsilon_{r2}$	2.2
h_1	31 mil
h_2	10 mil

Table 8.2
Physical Dimensions Obtained From ANN Models for the Two-Layer Filter Example. Specifications for This Filter Are Given in Table 8.1 (From [21], Watson, P. M., C. Cho, and K. C. Gupta, "EM-ANN Model for Synthesis of Physical Dimensions for Multilayer Asymmetric Coupled Transmission Line Structures," *Int. J. RF and Microwave CAE*, pp. 175–186, © 1999, John Wiley and Sons. Reprinted with permission from John Wiley and Sons, Inc.)

Section #	1	2	3
W_1 (mm)	1.50803	1.71022	2.95975
W_2 (mm)	2.93213	3.06325	2.79627
S (mm)	0.05418	0.79071	−0.08822
W_i, W_o (mm)		3.2259	
Z_{o1} (Ω)	60	60	40
Z_{o2} (Ω)	50	40	50

appropriate 2-port S-parameters, which characterize the open-ended, $\lambda/4$ coupling sections used for filter design.

The modeled filter response is shown in Figure 8.7 along with the response obtained from SBEM analysis of the coupled line sections as given in Table 8.2. Also, center frequency, ripple bandwidth, ripple level, and 3 dB bandwidth are given in Table 8.3. Analysis ANN models for each coupled line section have been linked to a commercial microwave circuit simulator (HP-MDS [4]) to obtain the filter response. Good agreement is obtained confirming the accuracy of the ANN coupled line models.

If the 0.7 dB ripple level is not acceptable, the filter response may be improved to obtain (say) 0.5 dB ripple using the ANN analysis models. The optimized filter dimensions are given in Table 8.4. Filter response is given in Table 8.5 and shown in Figure 8.8. Note that only small changes in physical parameters were necessary to achieve the desired response. These changes are on the order of the errors for the ANN models.

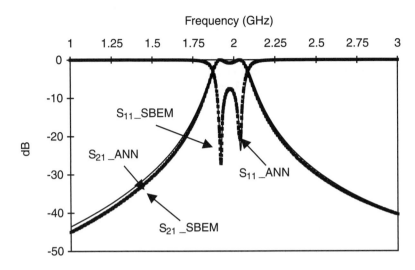

Figure 8.7 Two-layer filter response. Both ANN modeled (light solid lines) and SBEM (dark dashed lines) results are shown (from [21], Watson, P. M., C. Cho, and K. C. Gupta, "EM-ANN Model for Synthesis of Physical Dimensions for Multilayer Asymmetric Coupled Transmission Line Structures," *Int. J. RF and Microwave CAE*, pp. 175–186, © 1999, John Wiley and Sons. Reprinted with permission from John Wiley and Sons, Inc.).

Table 8.3
Comparison of Two-Layer Filter Responses (From [21], Watson, P. M., C. Cho, and K. C. Gupta, "EM-ANN Model for Synthesis of Physical Dimensions for Multilayer Asymmetric Coupled Transmission Line Structures," *Int. J. RF and Microwave CAE*, pp. 175–186, © 1999, John Wiley and Sons. Reprinted with permission from John Wiley and Sons, Inc.)

	Center Frequency (GHz)	3 dB Bandwidth (%)	Ripple Level (dB)	Ripple Bandwidth (%)
Specifications	2	—	0.5	7.5
ANN	1.98	11.11	0.702	7.87
SBEM	1.98	10.86	0.748	7.83

Comparison of Two-Layer Filter Design Using ANN Method and the Conventional Method

A filter with the same specifications given in Table 8.1 was also designed using the conventional method (without using ANN models). Table 8.6 records the physical dimensions, while Table 8.7 gives the center frequency and bandwidth

Table 8.4
Optimized Physical Dimensions Obtained From ANN Models for the Two-Layer Filter Example. Filter Specifications Are Given in Table 8.1 (From [21], Watson, P. M., C. Cho, and K. C. Gupta, "EM-ANN Model for Synthesis of Physical Dimensions for Multilayer Asymmetric Coupled Transmission Line Structures," *Int. J. RF and Microwave CAE*, pp. 175–186, © 1999, John Wiley and Sons. Reprinted with permission from John Wiley and Sons, Inc.)

Section #	1	2	3
W_1 (mm)	1.52068	1.70905	3.00076
W_2 (mm)	2.94101	3.06157	2.69997
S (mm)	−0.00732	0.75217	−0.10564
W_i, W_o (mm)		3.2259	
Z_{o1} (Ω)	60	60	40
Z_{o2} (Ω)	50	40	50

Table 8.5
Comparison of Optimized Two-Layer Filter Responses (From [21], Watson, P. M., C. Cho, and K. C. Gupta, "EM-ANN Model for Synthesis of Physical Dimensions for Multilayer Asymmetric Coupled Transmission Line Structures," *Int. J. RF and Microwave CAE*, pp. 175–186, © 1999, John Wiley and Sons. Reprinted with permission from John Wiley and Sons, Inc.)

	Center Frequency (GHz)	3 dB Bandwidth (%)	Ripple Level (dB)	Ripple Bandwidth (%)
Specifications	2	—	0.5	7.5
ANN (opt.)	1.98	11.61	0.456	8.08
SBEM (opt.)	1.98	11.36	0.501	8.02

parameters of the designed filter. The filter performance parameters obtained by using ANN modeling are also repeated here for comparison.

The filter designs using the conventional method and ANN modeling are comparable. The ripple level for the ANN design is slightly larger than desired. However, optimizing the physical geometry slightly, using the analysis ANN models, resulted in the correct response. Note that no analysis models are available when using the conventional method.

The advantage of using ANN models for the filter design is a large savings in required CPU time. Table 8.8 gives the CPU time on an HP700 workstation and the number of design iterations (changes in physical geometry) required to arrive at the final filter dimensions. The amount of time required for

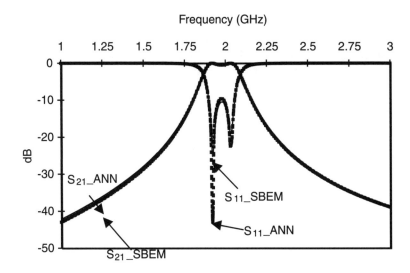

Figure 8.8 Two-layer filter response. Both ANN modeled (light solid lines) and SBEM (dark dashed lines) results are shown (from [21], Watson, P. M., C. Cho, and K. C. Gupta, "EM-ANN Model for Synthesis of Physical Dimensions for Multilayer Asymmetric Coupled Transmission Line Structures," *Int. J. RF and Microwave CAE,* pp. 175–186, © 1999, John Wiley and Sons. Reprinted with permission from John Wiley and Sons, Inc.).

Table 8.6
Physical Dimensions Obtained From Conventional Method for the Two-Layer Filter Example. Filter Specifications Are Given in Table 8.1 (From [21], Watson, P. M., C. Cho, and K. C. Gupta, "EM-ANN Model for Synthesis of Physical Dimensions for Multilayer Asymmetric Coupled Transmission Line Structures," *Int. J. RF and Microwave CAE,* pp. 175–186, © 1999, John Wiley and Sons. Reprinted with permission from John Wiley and Sons, Inc.)

Section #	1	2	3
W_1 (mm)	1.4928	1.7078	2.9706
W_2 (mm)	2.9259	3.1520	2.7987
S (mm)	0.0535	0.7880	−0.0904
W_i, W_o (mm)		3.2259	
Z_{o1} (Ω)	60	60	40
Z_{o2} (Ω)	50	40	50

Table 8.7
Center Frequency and Bandwidth Parameters for the Two-Layer Filter Designed Using the Conventional Method. Also, the Response of the Filter Using ANN Modeling Is Repeated Here for Comparison Purposes (From [21], Watson, P. M., C. Cho, and K. C. Gupta, "EM-ANN Model for Synthesis of Physical Dimensions for Multilayer Asymmetric Coupled Transmission Line Structures," *Int. J. RF and Microwave CAE*, pp. 175–186, © 1999, John Wiley and Sons. Reprinted with permission from John Wiley and Sons, Inc.)

	Center Frequency (GHz)	3 dB Bandwidth (%)	Ripple Level (dB)	Ripple Bandwidth (%)
Specifications	2	—	0.5	7.5
Con.	1.98	10.86	0.539	7.68
ANN	1.98	11.11	0.702	7.87

Table 8.8
Two-Layer Filter Design Times and Required Iterations for ANN Modeling and the Conventional Method of [9] (From [21], Watson, P. M., C. Cho, and K. C. Gupta, "EM-ANN Model for Synthesis of Physical Dimensions for Multilayer Asymmetric Coupled Transmission Line Structures," *Int. J. RF and Microwave CAE*, pp. 175–186, © 1999, John Wiley and Sons. Reprinted with permission from John Wiley and Sons, Inc.)

Section	No. Iterations	CPU Time
ANN: 1	4	0.0379 sec.
2	8	0.0758 sec.
3	3	0.0284 sec.
ANN optimization using HP-MDS:	2	0.76 sec.
Conventional: 1	4,212	64 min. 11 sec.
2	1,634	125 min. 4 sec.
3	1,285	19 min. 1 sec.

optimization of the filter response using ANN analysis models linked to HP-MDS [4] is also included in this table. It is evident that using ANN models for the filter design results in a much more efficient process than using the conventional method. Both designs were carried out on an HP700 workstation. For the design comparison, only 'a' was allowed to change. Z_{o1} and Z_{o2} were held constant for each section.

The above example illustrates that ANN modeling offers an accurate and efficient alternative to the conventional methods for the design of multilayer coupled line circuit components. Circuit elements can be designed in a small fraction of the time using ANN models.

Another advantage of the ANN modeling approach for multilayer circuit design is the availability of analysis models, which can be linked to commercial microwave simulators. These analysis models can be used in conjunction with other component models and optimization routines, available within commercial microwave circuit simulators, for the design and optimization of larger circuits. Analysis models are not available when using the conventional method for multilayer coupled line circuit design.

8.5 CPW Patch Antenna Design and Optimization

This section describes the use of ANN modeling in the design and optimization of CPW patch antennas. CPW patches are radiating elements using open-ended coplanar waveguide (CPW) resonators to provide an alternative for printed slot antennas and microstrip patch antennas. Motivation for exploring these antenna structures arises from the ease of connecting CPW resonators to CPW lines, which have received much attention due to several advantages of CPW lines over microstrip lines. For developing simple network models for these antennas (like the transmission line model for microstrip patch antennas), we need models for wide strip CPW lines or open-end discontinuities and their radiating properties. EM-ANN models have been developed for wide strip CPW lines, CPW open-end discontinuities (including radiation conductance modeling), and feed discontinuities used for connecting CPW resonators to CPW lines. These models are then used within commercial microwave circuit simulators for CPW patch antenna design.

8.5.1 Transmission Line Model for CPW Antennas

Transmission line models for microstrip patch antennas have been studied extensively [10]. In the same manner, a transmission line model has been developed for CPW antenna. Figure 8.9 shows the antenna structure without the feedline and its equivalent transmission line model. The transmission line model consists of a length of CPW transmission line open-circuited at the two ends. Thus, the patch can be represented by a uniform length of transmission line of characteristic impedance Z_o and phase velocity v_p (or phase constant, $\beta = \omega/v_p$). For this model, we consider that the line is operating in the TEM CPW mode and is lossless. The fringing fields associated with the open ends are represented by equivalent lumped admittances, consisting of a radiation conductance, G_r, and an edge capacitance, C.

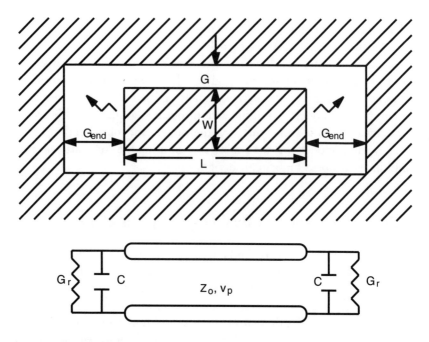

Figure 8.9 Ideal CPW patch antenna geometry and corresponding transmission line equivalent circuit model.

EM-ANN Model for CPW Transmission Line

An EM-ANN model for wide conductor CPW transmission lines on Duroid 5880 (ϵ_r = 2.2 and H_{sub} = 31.25 mil) was developed for design of C band antenna (4 GHz to 7 GHz). Model input variables and corresponding ranges are given in Table 8.9. Outputs of this ANN model are Z_o and β for the CPW line. Since the training data was relatively inexpensive to obtain for this component, EM simulations were performed, using HP Momentum [11], for a uniform distribution over the input variables' ranges. HP-Momentum is able to provide values for Re(Z_o) and β for CPW lines.

Table 8.9
Input Variables and Ranges for Model CPW Transmission Line on Duroid 5880 (ϵ_r = 2.2 and H_{sub} = 31.25 mil) (From [22], © 1998 IEEE, Reprinted with Permission)

Input Parameter	Minimum Value	Maximum Value
Frequency	4 GHz	7 GHz
W	0.5 cm	2.0 cm
G	0.05 cm	0.2 cm

EM-ANN Model for CPW Open-End Effects

An EM-ANN model for CPW open-end effects, including radiation conductance, was developed for this antenna design. Variable input parameters and their corresponding ranges are given in Table 8.9. For this model, the open-end spacing, G_{end}, was set equal to $0.5(W + 2G)$. EM-ANN model outputs are the magnitude and the phase of S_{11}. The reflection coefficient, S_{11}, for the open-end was obtained by simulating a $\lambda/2$ long CPW line open at one end and finding input impedance at the other end.

A DOE central composite design was used to select the parameters' values for obtaining the EM simulation data for neural network training. The parameters for the central composite design were frequency, W, and G. Initially, 15 samples were used for training, 14 for testing, and 10 for verification. However, errors were not as low as desired. Therefore, the 14 test samples were added to the training dataset resulting in a total of 29 examples. The final model contained 10 hidden layer neurons (62 weights).

Radiation Conductance and Capacitance Models

The radiation conductance and capacitance of the open end were obtained from the EM-ANN model by noting that S_{11} is the reflection coefficient of the open-end. When the line length used to characterize the open end is $\lambda/2$ long and the line is lossless, the load (that is, the open-end) impedance may be found as

$$Z_L = Z_0 \frac{1 + S_{11}}{1 - S_{11}} \quad (8.1)$$

Taking the inverse of Z_L to obtain Y_L yields the radiation conductance as

$$G_r = Re(Y_L) \quad (8.2)$$

and the end capacitance as

$$C = \frac{Im(Y_L)}{\omega} \quad (8.3)$$

As pointed out above, this calculation of the load impedance is valid only when the $\lambda/2$ line is lossless. Therefore, when radiation from this line increases, the values of G_r and C are not only for the open-end, but also include the effect of radiation from the line. However, the EM-ANN model with S_{11} as the output still gives useful results, but G_r and C for the open end alone cannot be extracted using the above equations.

ANN Model for CPW Patch Feed Line

The configuration of the CPW patch antenna designed is shown in Figure 8.10. The equivalent transmission line model is also shown in this figure. In addition to the model for the CPW line and open end, we need a model for the input feedline as well.

The central strip width in the feed region needs to be kept narrow in order to minimize interference with radiation from the radiating edge, but has to be wide enough to be realizable using available fabrication facilities. Therefore, the incoming CPW line was chosen to have W = 0.1 cm, G = 0.05 cm, and $Z_o \approx 96\ \Omega$. The output parameters for this model are the magnitudes and phases of S_{11}, S_{21}, and S_{22}. EM simulations have been performed on 27 structures over the 4 GHz to 7 GHz frequency range providing 111 training/

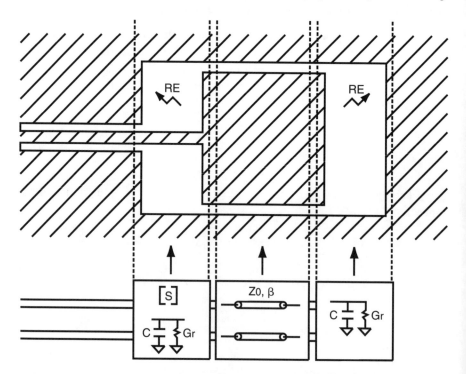

Figure 8.10 CPW patch fed on the radiating edge and its corresponding transmission line model (from [22], © 1998 IEEE, reprinted with permission).

test samples and 42 verification samples. The optimal ANN structure contained 15 hidden layer neurons (156 weights). As with the other models, excellent accuracy is obtained.

8.5.2 CPW Patch Antenna Design Using EM-ANN Models

CPW patch antennas can now be designed using the EM-ANN models developed for the CPW open-end, CPW line, and feed discontinuities. As an example, a CPW patch antenna was designed at 5 GHz with W = 1.5 cm and G = 0.1 cm. The layout of the antenna is shown in Figure 8.11. The length of the CPW section 'a' was selected by optimization using ANN models to yield a real value of impedance at its input. Impedance at the input of this section was found to be 821 Ω. Three matching sections have been included to transform the input impedance from 821 Ω to 50 Ω. The option of stub matching was not preferred, as that would need incorporation of air-bridges (to suppress the slot line mode). Table 8.10 and Figure 8.12 compare return loss for the EM-ANN model (EM-ANN), EM simulation using HP's Momentum [11] (EMsim), and measurement results (meas). Good accuracy was achieved by the EM-ANN model design when compared with EM simulation and measurements.

8.5.3 CPW Patch Antenna Design Optimization Using EM-ANN Models

The usefulness of the EM-ANN modeling approach has been demonstrated by investigating the effects of changes in certain design parameters on the

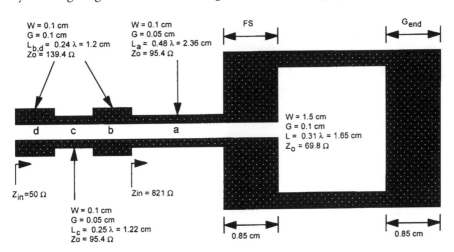

Figure 8.11 Layout of CPW patch antenna design. Electrical parameters referenced to the resonant frequency of 4.99 GHz.

Table 8.10
Comparison of Resonant Frequency and Bandwidth for EM-ANN Modeling, EM Simulation, and Measurements

	EM-ANN	**EM Simulation**	**Measured**
Center Frequency	4.99 GHz	5.01 GHz	5.09 GHz
% Bandwidth	2.6%	2.6%	2.7%

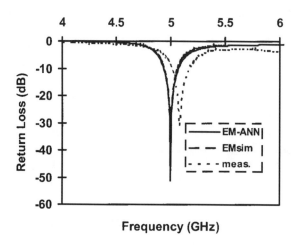

Figure 8.12 Comparison of return loss for the CPW patch antenna shown in Figure 8.11 with EM simulation (EMsim) and measured results.

response of the CPW patch antenna. Antenna performance was improved by changing the patch length using EM-ANN models.

It was noted that for several values of the patch design parameters, a loop forms in the S_{11} plot on the Smith Chart as shown in Figure 8.13. The patch with no feed structure is labeled as A1. As the feed section (FS in Figures 8.11 and 8.15), which is effectively a high impedance CPW line, is added, a loop forms showing another resonance created by the feed section length. The response after the feed section (FS) is added is labeled B1. Next, adding the incoming feedline, which has an impedance of 96 Ω, the response (C1) is rotated around the Smith Chart to the real axis for matching purposes. Note that the loop has effectively been pulled out.

By this kind of on-computer experimentation using ANN models, it was found that increasing the patch length increases the size of the loop as shown in Figure 8.14. If the antenna can be designed at a frequency where the loop

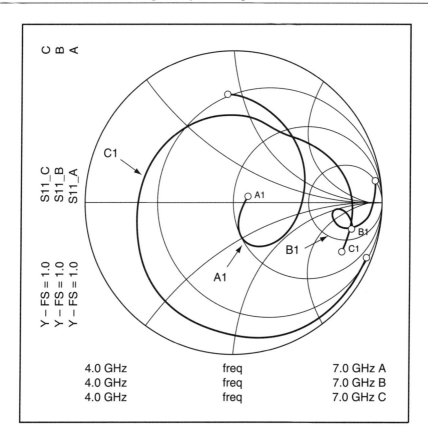

Figure 8.13 Effects of the radiating edge feed section and the input line on S_{11} response of the CPW patch antenna (W = 1.5 cm, G = 0.1 cm, and G_{end} = 0.85 cm). (A1) Ideal patch with no feed section (L_{patch} = 1.6465 cm); (B1) Addition of feed section; and (C1) Addition of incoming feedline.

appears, a wider bandwidth can be achieved. For this purpose, the length of the incoming feedline can be adjusted, rotating the loop to lie on the real line.

A CPW patch antenna with this modified geometry (with a longer-than-ideal patch length) was designed using ANN models, fabricated, and measured. The layout of this modified antenna is shown in Figure 8.16. Note that no additional impedance transforming sections was required since a match to 50 Ω was possible by selecting the length L_a. CPW length 'a' acts as a matching section between patch input point 'I' where Z_{in} = 169.7 Ω + j41.14 Ω at 5.56 GHz. The characteristic impedance of section 'a' is 96 Ω.

As expected, a wide bandwidth was achieved for this modified antenna as seen in Table 8.11 and in Figure 8.16. The return loss for this antenna (as given by EM simulation) is about −20 dB at resonance frequency. For a perfect

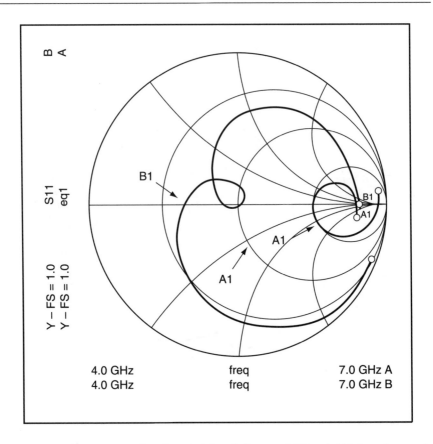

Figure 8.14 Effects of increasing the patch length (L_{patch} = 1.937 cm). (A1: L_a = 0 cm and B1: L_a = 3.4 cm.; L_a is the length of the section 'a'.)

match (according to EM simulation), the length of section 'a' (L_a) needs to be reduced by 0.025λ (0.123 cm). However, L_a was kept at the value given by the ANN model results (3.4 cm) for fabrication and measurements. As shown in Table 8.11 and Figure 8.15, EM-ANN modeling, EM simulation, and measurements agree well.

This example illustrates the usefulness of the neural network model in design improvement in a situation where such an improvement would otherwise be very difficult.

8.6 Yield Optimization of a Three-Stage MMIC Amplifier

This section demonstrates neural network-based optimization of an active circuit, using statistical design methodology. Manufacturing tolerances on

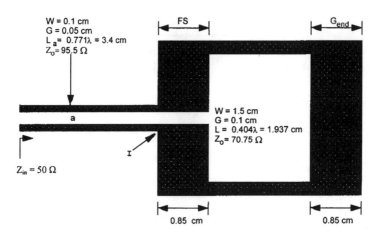

Figure 8.15 Layout for CPW patch antenna with longer-than-ideal patch length. All dimensions are in cm. Electrical parameters referenced to the resonant frequency of 5.56 GHz (from [22], © 1998 IEEE, reprinted with permission).

Table 8.11
Comparison of Resonant Frequency and Bandwidth for EM-ANN Modeling, EM Simulation, and Measurements for an Improved Bandwidth CPW Patch Antenna (From [22], © 1998 IEEE, Reprinted with Permission)

	EM-ANN	EM Simulation	Measured
Center Frequency	5.56 GHz	5.54 GHz	5.49 GHz
% Bandwidth	14.03%	11.69%	14.5%

geometrical parameters of circuit components and semiconductor process variations are important factors in designing active RF and microwave circuits such as amplifiers, mixers, and so on. Performance-driven optimization of these circuits is not adequate, and a yield-driven statistical design taking into account parameter variations becomes important. For semiconductor devices, the statistical variations are best described by spreads of physical/geometrical and process parameters. This leads to the need for physics-based device models [12]. However, repetitively evaluating device physics equations with statistically varying physical/geometrical/process parameters is extremely expensive. In this example, we demonstrate the use of physics-based neural network models for transistors in yield optimization of an active circuit.

The incorporation of neural models for active devices into a circuit simulator is done with two different commercial simulators, one through an

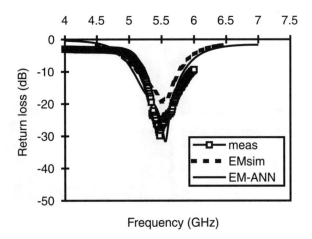

Figure 8.16 Comparison of return loss values obtained from ANN model design (EM-ANN), EM simulation (Emsim), and measured results (meas.) for the modified CPW patch antenna (from [22], © 1998 IEEE, reprinted with permission).

interprocess pipe connection to the OSA90/Hope [13] CAD system, another through an HP-ADS plug-in module called neuroADS [14] connecting neural models of active devices to the HP-ADS simulator. The solutions in this section are based on the OSA90/Hope yield optimization results published in [15].

We consider a three-stage, small-signal X-band cascadable MMIC (monolithic microwave integrated circuits) amplifier [12] as shown in Figure 8.17. The design is based on the circuit topology described in [16]. The amplifier contains three MESFETs. The matching circuits are composed of inductors and capacitors arranged in band-pass topology.

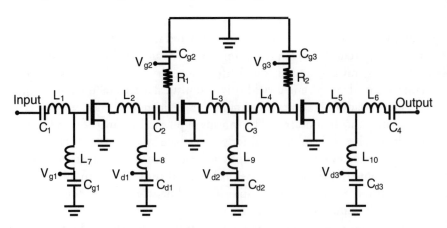

Figure 8.17 Circuit diagram of an X-band amplifier, after [16].

Physics-based models are used for both the MESFETs and passive elements of the amplifier. In this way, all the passive components, as well as the active devices, can be simulated and optimized in terms of physical parameters. Since all devices are made from the same material and on the same wafer, they share common parameters. All three MESFETs have the same values for the critical electric field, saturation velocity, relative permittivity, built-in potential, low-field mobility, and high-field diffusion coefficient [12]. Thus, the same neural network model, developed in Section 7.2.3, is used for all three MESFETs. All the MIM (metal-insulator-metal) capacitors have the same dielectric film, and all the bulk resistors have the same sheet resistance. The geometrical parameters, on the other hand, can have different values for different devices, including the gate length, gate width, channel thickness, and doping density of the MESFETs, the metal plate area of the MIM capacitors, and the number of turns of the spiral inductors. The MESFET neural model described in Section 7.2.3 is used during amplifier optimization. The input parameters to the neural models include gate length, gate width, channel thickness, doping density, and gate and drain voltages (L, W, a, N_d, V_{gs}, and V_{ds}). The neural network model is trained only once using physics-based data. In each evaluation of the circuit, the neural model is called three times, each time with a different set of input parameters L^i, W^i, a^i, N_d^i, V_{gs}^i, and V_{ds}^i, corresponding to MESFET^i, i = 1, 2, 3. The specifications for the amplifier include

- Passband (8–12 GHz): 12.4 dB $\leq$ gain $\leq$ 15.6 dB, input VSWR $\leq$ 2.8;
- Stopband (below 6 GHz or above 15 GHz): gain $\leq$ 2 dB.

There are 14 design variables, the area $S_{C1}, \ldots, S_{C4}$ of the metal plates of the MIM capacitors $C_1, \ldots, C_4$ and the number of turns $n_{L1}, \ldots, n_{L10}$ of the spiral inductors $L_1, \ldots, L_{10}$. As a first step, a nominal design optimization using the neural network was carried out reducing the value of minimax objective function for optimization from 6.7 to −0.15, all specifications being satisfied. Table 8.12 lists the 14 design variables before and after the minimax optimization. In Figure 8.18, the gain and input VSWR of the amplifier are compared before and after optimization using neural network models. To verify the optimization solution, the parameters in Table 8.12 were also used to simulate the amplifier with the Khatibzadeh and Trew model [17], which is a much more complex model. We found all specifications being satisfied as illustrated in Figure 8.19.

In the second step, yield optimization using l_1 centering algorithm [18] is performed with the minimax nominal design as a starting point. There are

Table 8.12
Variables for Nominal Design (From [15], © 1995 IEEE, Reprinted with Permission).

Design Variable	Before Optimization	After Optimization	Design Variable	Before Optimization	After Optimization
$S_{C1}(\mu m^2)$	353.1	326.8	n_{L4}	3.68	3.49
$S_{C2}(\mu m^2)$	2014.4	2001.5	n_{L5}	2.13	2.31
$S_{C3}(\mu m^2)$	212.3	224.6	n_{L6}	2.61	2.47
$S_{C4}(\mu m^2)$	354.2	343.8	n_{L7}	2.42	2.74
n_{L1}	3.06	3.50	n_{L8}	2.45	2.47
n_{L2}	3.56	3.76	n_{L9}	2.88	2.71
n_{L3}	2.84	2.91	n_{L10}	3.09	2.98

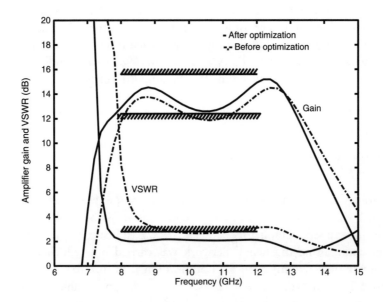

Figure 8.18 Gain and input VSWR of the X-band amplifier with neural network models before (dashed line) and after (solid line) nominal design optimization (from [15], © 1995 IEEE, reprinted with permission).

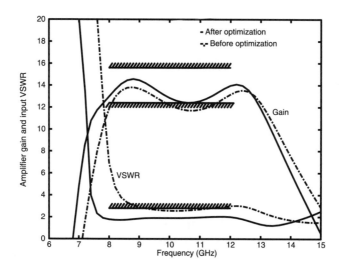

Figure 8.19 Gain and input VSWR of the X-band amplifier with Khatibzadeh and Trew model before (dashed line) and after (solid line) nominal design optimization (from [15], © 1995 IEEE, reprinted with permission).

37 statistical variables, including the neural network inputs gate length, gate width, channel thickness, and doping density of the MESFETs, as well as the geometrical parameters of the passive elements, namely, the conductor width W_L and spacing S_L of the 10 spiral inductors $L_1, L_2, \ldots, L_{10}$, the thickness d of the dielectric film for all MIM capacitors, and the area $S_{C1}, \ldots, S_{C4}$ of the metal plates of the MIM capacitors $C_1, \ldots, C_4$. The distributions for these 37 statistical variables are listed in Table 8.13. The conductor width of the 10 spiral inductors have the same distribution but different random values in the Monte Carlo analysis. Thus, in Table 8.13, the distribution for W_L actually represents the distributions for 10 random variables. Similarly, distribution for conductor spacing S_L means that for 10 variables, and distribution for transistor parameters (N_d, L, a, W) means that for 12 random variables (three transistors). The statistical variations between the three sets of transistor parameters should be correlated because they are manufactured in the same process. The correlation matrix between the three sets of MESFET parameters used in our optimization is shown in Table 8.14 following [12]. The yield

Table 8.13
Distributions for Statistical Variables (From [12], © 1992 IEEE, Reprinted with Permission)

Variable	Mean	Deviation (%)	Variable	Mean	Deviation (%)
$N_d(1/m^3)$	10^{23}	7.0	$d(\mu m)$	0.1	4.0
$L(\mu m)$	1.0	3.5	$S_{C1}(\mu m^2)$	326.8	3.5
$a(\mu m)$	0.3	3.5	$S_{C2}(\mu m^2)$	2022.4	3.5
$W(\mu m)$	300	2.0	$S_{C3}(\mu m^2)$	218.2	3.5
$W_L(\mu m)$	20	3.0	$S_{C4}(\mu m^2)$	352.2	3.5
$S_L(\mu m)$	10	3.0			

after minimax nominal design optimization was 26% with the neural network model and 32% with the Khatibzadeh and Trew model. The CPU time used for the Monte Carlo sweeps was 1 h and 30 min for the neural network approach and 40 h 34 min for the Khatibzadeh and Trew model, i.e., the neural-model-based approach is about 30 times faster. At the result of yield optimization using the neural network, the yield was improved to 58%. To verify this solution, we performed Monte Carlo analysis using the Khatibzadeh and Trew model. The yield was found to be 59%. This confirms the validity of the neural network approach. The variables for the yield optimization are given in Table 8.15. The Monte Carlo sweeps before and after yield optimization are shown in Figure 8.20. Yield optimization with 50 outcomes using the neural network model took 50 min CPU time per iteration on a Sun SPARCstation 2. The corresponding CPU time using the Khatibzadeh and Trew model with quadratic model [19] is 4 h and 14 min. Table 8.16 summarizes the CPU speed-up achievement.

8.7 Remarks

In this chapter we have discussed several examples in which ANN models have been used for the design and optimization of RF components, circuits, and antennas. These examples demonstrate effectiveness of the ANN modeling approach discussed in this book in diverse application areas.

Table 8.14
Correlation Coefficients Between the Three MESFET Parameters (From [12], © 1992 IEEE, Reprinted with Permission)

	a_{F1}	L_{F1}	W_{F1}	N_{dF1}	a_{F2}	L_{F2}	W_{F2}	N_{dF2}	a_{F3}	L_{F3}	W_{F3}	N_{dF3}
a_{F1}	1.0	0.0	0.0	−0.25	0.8	0.0	0.0	−0.2	0.78	0.0	0.0	−0.1
L_{F1}	0.0	1.0	0.0	−0.1	0.0	0.8	0.0	−0.05	0.0	0.78	0.0	−0.05
W_{F1}	0.0	0.0	1.0	0.0	0.0	0.0	0.8	0.0	0.0	0.0	0.78	0.0
N_{dF1}	−0.25	−0.1	0.0	1.0	−0.2	−0.05	0.0	0.8	−0.15	−0.05	0.0	0.78
a_{F2}	0.8	0.0	0.0	−0.2	1.0	0.0	0.0	−0.25	0.8	0.0	0.0	−0.2
L_{F2}	0.0	0.8	0.0	−0.05	0.0	1.0	0.0	−0.1	0.0	0.8	0.0	−0.1
W_{F2}	0.0	0.0	0.8	0.0	0.0	0.0	1.0	0.0	0.0	0.0	0.8	0.0
N_{dF2}	−0.2	−0.05	0.0	0.8	−0.25	−0.1	0.0	1.0	−0.2	−0.05	0.0	0.8
a_{F3}	0.78	0.0	0.0	−0.15	0.8	0.0	0.0	−0.2	1.0	0.0	0.0	−0.25
L_{F3}	0.0	0.78	0.0	−0.05	0.0	0.8	0.0	−0.05	0.0	1.0	0.0	−0.1
W_{F3}	0.0	0.0	0.78	0.0	0.0	0.0	0.8	0.0	0.0	0.0	1.0	0.0
N_{dF3}	−0.1	−0.05	0.0	0.78	−0.2	−0.1	0.0	0.8	−0.25	−0.1	0.0	1.0

Table 8.15
Design Variables for Yield Optimization (From [15], © 1995 IEEE, Reprinted with Permission)

Design Variable	Before Optimization	After Optimization	Design Variable	Before Optimization	After Optimization
$S_{C1}(\mu m^2)$	272.8	232.2	n_{L4}	3.49	3.58
$S_{C2}(\mu m^2)$	2001.5	2006.9	n_{L5}	2.31	2.38
$S_{C3}(\mu m^2)$	244.4	277.8	n_{L6}	2.47	2.49
$S_{C4}(\mu m^2)$	343.8	346.1	n_{L7}	2.74	2.72
n_{L1}	3.50	3.55	n_{L8}	2.47	2.49
n_{L2}	3.76	3.73	n_{L9}	2.71	2.73
n_{L3}	2.91	2.99	n_{L10}	2.98	3.00

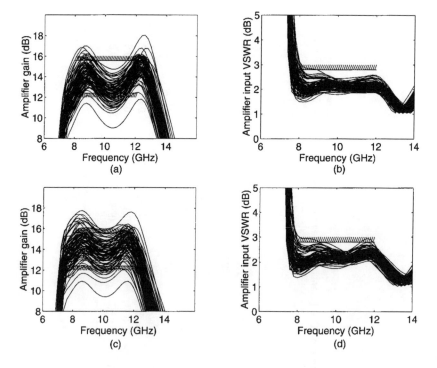

Figure 8.20 Monte Carlo sweep of gain and input VSWR versus frequency (GHz) of the X-band amplifier using neural network models before optimization (a) gain (b) input VSWR, and after optimization (c) gain (d) input VSWR (from [15], © 1995 IEEE, reprinted with permission).

Table 8.16
Summary of CPU Comparison (From [15], © 1995 IEEE, Reprinted with Permission)

Application	Khatibzadeh & Trew Model	Neural Network Model	Speed-Up Ratio
Optimization	7 min 8 sec	1 min 6 sec	6
Monte Carlo	40 hours 34 min	1 hour 30 min	30
Yield Optimization	4 hours 14 min	50 min	5

References

[1] Hornik, K., M. Stinchcombe, and H. White, "Universal Approximation of an Unknown Mapping and Its Derivatives using Multilayer Feedforward Networks," *Neural Networks*, Vol. 3, No. 5, May 1990, pp. 551–560.

[2] Daigle, R. C., G. W. Bull, and D. J. Doyle, "Multilayer Microwave Boards: Manufacturing and Design," *Microwave Journal*, April 1993, pp. 87–97.

[3] Watson, P. M., "Artificial Neural Network Modeling for Computer-Aided Design of Microwave and Millimeter-Wave Circuits," Ph.D. Thesis (Advisor: Prof. K. C. Gupta), University of Colorado at Boulder, 1998.

[4] *HP-MDS*, ver. 6.02.00, Hewlett-Packard Co. (Agilent Technologies, Inc.), Santa Rosa, CA, 1995.

[5] Tripathi, V. K., "Asymmetric Coupled Transmission Lines in an Inhomogeneous Medium," *IEEE Transactions on Microwave Theory and Techniques*, Vol. 23, No. 9, Sept. 1975, pp. 734–739.

[6] Tripathi, V. K., "Equivalent Circuits and Characteristics of Inhomogeneous NonSymmetrical Coupled-Line Two-Port Circuits," *IEEE Transactions on Microwave Theory and Techniques*, Vol. 25, No. 2, Feb. 1977, pp. 140–142.

[7] Dworsky, L. N., *Modern Transmission Line Theory and Applications*, Robert E. Krieger Publishing Company, Malabar, FL, 1988, pp. 109–136.

[8] Tsai, C. M., *Field Analysis, Network Modeling, and Circuit Applications of Inhomogeneous Multi-Conductor Transmission Lines*, Ph.D. Dissertation (Advisor: Prof. K. C. Gupta), University of Colorado at Boulder, Dept. of Electrical and Computer Eng., 1993.

[9] Cho, C., and K. C. Gupta, "Design Methodology for Multilayer Coupled Line Filters," *IEEE Int. Microwave Symp. Digest*, June 1997, pp. 4603–4606.

[10] Van de Capelle, A., "Transmission Line Model for Rectangular Microstrip Antennas," in *Handbook of Microstrip Antennas*, J. R. James and P. S. Hall, eds., London, UK: Peter Peregrinus Ltd., Vol. 1, 1989, pp. 527–578.

[11] HP-*Momentum*, ver. A.02.51, Hewlett-Packard Co. (now Agilent Technology Inc.), Santa Rosa, CA, 1996.

[12] Bandler, J. W., et al., "Integrated Physics-Oriented Statistical Modeling, Simulation and Optimization," *IEEE Trans. Microwave Theory and Techniques*, Vol. 40, 1992, pp. 1374–1400.

[13] *OSA90/HOPE V 2.0*, Formerly by Optimization Systems Associates Inc., Dundas, Ont. Canada, now HP-EESof (Agilent Technology Inc.), 1400 Fountaingrove Parkway, Santa Rosa, California 95403-1799.

[14] Zhang, Q. J., et al., *NeuroADS*, Carleton University, Ottawa, Canada K1S 5B6, 1999.

[15] Zaabab, H., Q. J. Zhang, and M. S. Nakhla, "A Neural Network Modeling Approach to Circuit Optimization and Statistical Design," *IEEE Trans. Microwave Theory and Techniques*, Vol. 43, 1995, pp. 1349–1358.

[16] Kermarrec, C., and C. Rumelhard, "Microwave Monolithic Integrated Circuits," In *GaAs MESFET Circuit Design*, R. Soares, ed., Boston, MA: Artech House, 1988.

[17] Khatibzadeh, M. A., and R. J. Trew, "A Large-Signal, Analytical Model for the GaAs MESFET," *IEEE Trans. Microwave Theory and Techniques*, Vol. 36, No. 2, 1988, pp. 231–238.

[18] Bandler, J. W., and S. H. Chen, "Circuit Optimization: The-State-of-the-Art," *IEEE Trans. Microwave Theory and Techniques*, Vol. 36, 1988, pp. 424–443.

[19] Biernacki, R. M., et al., "Efficient Quadratic Approximation for Statistical Design," *IEEE Trans. Circuit Systems*, Vol. 36, No. 11, 1989, pp. 1449–1454.

[20] Watson, P. M., and K. C. Gupta, "Design and Optimization of CPW Circuits using EM-ANN Models for CPW Components," *IEEE Trans. Microwave Theory and Techniques*, Vol. 45, Dec. 1997, pp. 2515–2523.

[21] Watson, P. M., C. Cho, and K. C. Gupta, "EM-ANN Model for Synthesis of Physical Dimensions for Multilayer Asymmetric Coupled Transmission Line Structures," *Int. J. of RF and Microwave CAE*, Vol. 9, 1999, pp. 175–186.

[22] Watson, P. M., and K. C. Gupta, "EM-ANN Models for Design of CPW Patch Antennas," IEEE APS International Symp. Antennas and Propagation, June 1998, Atlanta, GA, Symp. Digest, pp. 648–651.

9

Knowledge-Based ANN Models

This chapter[1] describes some of the recent advances in the neural network area where existing RF/microwave knowledge is combined with neural networks. The chapter starts with a brief description of knowledge issues in conventional neural network applications (e.g., pattern classification) and in microwave design. We then study various effects of adding knowledge on the performance of the neural models, such as generalization ability, extrapolation ability, and model reliability, versus different sizes of training data through a knowledge-based neural network (KBNN) technique and demonstrate examples of comparisons with conventional MLP (without any knowledge base). We also describe several ways of combining existing circuit models with neural networks, including the source difference method, the prior knowledge input method, and the space-mapped neural models. Finally, an advanced hierarchical neural network structure for the task of neural model library development is discussed.

9.1 Introduction

9.1.1 Motivation

With continuing developments in applications of neural networks to microwave design, there is a growing need for reduction in the cost of model development and improvement in model reliability. The commonly used MLP model belongs to the type of black-box models structurally embedding no problem-dependent information/knowledge. Therefore, it derives the entire information about the

---

1. Sections 9.2 and 9.6 are based on F. Wang's Ph.D. thesis [1] at Carleton University.

RF/microwave behaviors from the training data. Consequently, a large amount of training data is needed to ensure model accuracy. In microwave applications, training data is obtained either by simulation of original EM/device-physics problems, or by measurements. Generating a large amount of training data could be very expensive, because simulation/measurement may have to be performed at many points in the model input parameter space (e.g., for various combinations of geometrical/process/bias parameters). Without sufficient training data, the resulting neural models may not be reliable. In addition, even with a sufficient amount of training data, the reliability of MLP, when used for extrapolation, is not guaranteed and in most cases is very poor.

This chapter describes some of the recent advances in the neural network area where existing RF/microwave knowledge is combined with neural networks. Such knowledge provides additional information of the original problem that may not be adequately represented by limited training data. Instead of relying on a pure neural network to represent a microwave behavior, existing microwave empirical models are used to enhance generalization and extrapolation capability of neural models. At the same time, neural networks can help to bridge the gap between empirical models and EM solutions.

9.1.2 Rule-Based Knowledge Networks

Initially, knowledge-based neural networks were developed in the neural network community for pattern-classification applications. The specific knowledge of these applications is in the form of propositional symbolic rules [2]. These rules are also known as hand-built classifiers because the rules are manually extracted from the domain theory. The use of knowledge-based neural networks here is to insert a set of hand-constructed symbolic rules into a neural network. Networks created in this way make the same classification as the rules upon which they are based. The performance of the network is then refined using standard neural network training algorithms with a set of training samples. After training, the resulting neural model can function as a highly accurate classifier. Such knowledge-based neural networks are hybrid learning systems that use both a learning-by-being-told approach of hand-built classifiers and a knowledge-free empirical learning approach. Two types of methods have been used. The first method is to use symbolic knowledge in the form of rules to establish the structure and weights of the neural network [2–3]. The weights, for example, the certainty factor associated with rules [4], or both topology and weights of the network [5], can be revised during training. The second method is to build prior information into neural networks, for example, restricting the network architecture through the use of local connections and constraining the choice of weights by the use of weight sharing [6].

Besides these propositional symbolic rules, there are other kinds of rules known as fuzzy rules in Fuzzy Systems. Neurofuzzy networks [7, 8] have been developed to utilize existing fuzzy rule sets in the applications. A neurofuzzy network structure encodes if-then fuzzy rules that are in linguistic form. There is a direct link between a neuron activation function of RBF network (e.g., Gaussian function) and the membership function of fuzzy rules. As such, the domain knowledge embedded in the fuzzy rule set can be inserted into the neural network structure through a mapping that realizes such a link [9].

9.1.3 Microwave-Oriented Knowledge Structures

In the microwave modeling area, the design functions are often continuous, multidimensional, and nonlinear relationships instead of symbolic or discrete relationships in pattern classification. Knowledge about the RF and microwave problems is often in the form of empirical or equivalent circuit models, Maxwell's equations, Kirchoff's equations, and so forth. Detailed models based on EM/physics equations are accurate, but can be computationally intensive. Simple empirical and equivalent models exist for passive and active components, but such models may not be as accurate as desired, and a single model may fail to represent the component behavior in the entire input space (x-space). However, the simple empirical and equivalent circuit knowledge can be used to add to the learning ability of neural networks in learning the EM or physics behavior of the original problem. The use of such knowledge can help to reduce the need for a large amount of training data and to improve model reliability. The resulting knowledge-based models would have the speed of empirical or neural models and the accuracy of EM/physics models. With this point of view, the comparison of various modeling techniques previously described in Table 3.1 can be expanded to include knowledge-based models as presented in Table 9.1.

9.2 Knowledge-Based Neural Networks (KBNN)

This section serves two purposes. Firstly, through a knowledge-based neural network (KBNN) approach, we illustrate how microwave knowledge can be integrated into neural network structures. Secondly, the advantages of adding knowledge in terms of the amount of training data required and the generalization capability of the resulting neural models are demonstrated through examples.

The knowledge-based neural network (KBNN) structure was proposed in [1, 10]. In this method microwave knowledge in the form of empirical

Table 9.1
A Comparison of Various Modeling Techniques for RF/Microwave Applications

Basis of Comparison	EM/Physics Models	Empirical and Circuit Equivalent Models	Pure Neural Network Models	Knowledge-Based Neural Models
Speed	Slow	Fast	Fast	Same as empirical models
Accuracy	High	Limited	Could be close to EM/physics models	Could be close to EM/physics models
Number of Training Data	0	A few	Sufficient training data is required, which could be large for high dimensional problems	Small to medium
Circuit/EM Theory of the Problem	Maxwell, or semiconductor physics equations	Partially involved	Not involved	Partially involved

functions or analytical approximations is embedded into the neural network internal structure. Switching boundary and region neurons are introduced in the model structure to reflect RF/microwave designs, in which different equations or formulas with different parameters can be interchangeably used in different regions of the input parameter space. By inserting the microwave empirical formulas into the neural network structure (neuron activation functions), the empirical formulas can be refined/adjusted as part of the overall neural network training process. This technique enhances neural model accuracy, especially for the data not seen during training (generalization capability), and reduces the need for a large amount of training data.

9.2.1 Model Structure

The basic idea of this method is illustrated in Figure 9.1. The existing microwave knowledge is embedded as part of the overall neural network structure. Figure

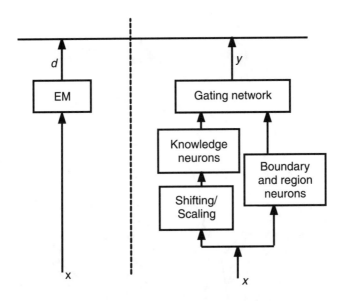

Figure 9.1 Basic idea behind knowledge-based neural network model development.

9.2 shows the details of the arrangement of neurons in the KBNN structure [10]. There are six layers in the structure, namely the input layer x, the knowledge layer z, the boundary layer b, the region layer r, the normalized region layer r', and the output layer y. The input layer x accepts the external inputs to the model. The knowledge layer z is the place where microwave knowledge resides in the form of single or multidimensional functions $\psi(\cdot)$. For knowledge neuron i in the z layer

$$z_i = \psi_i(x, w_i), \; i = 1, 2, \ldots, N_z \qquad (9.1)$$

where x is a vector including neural network inputs $x_i (i = 1, 2, \ldots, n)$, N_z is the number of knowledge neurons, and w_i is a vector of parameters in the knowledge formula. The knowledge function $\psi_i(x, w_i)$ is usually in the form of empirical or semi-analytical functions. For example, the drain current of a FET is a function of its gate length, gate width, channel thickness, doping density, gate voltage, and drain voltage. The boundary layer b can incorporate knowledge in the form of problem-dependent boundary functions $b(\cdot)$, or in the absence of such boundary knowledge, it can represent just linear boundaries. Neuron i in this layer is calculated as

$$b_i = b_i(x, v_i), \; i = 1, 2, \ldots, N_b \qquad (9.2)$$

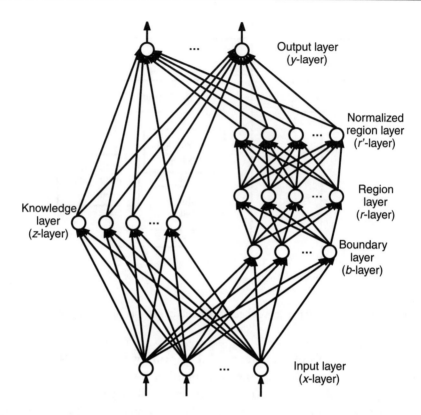

Figure 9.2 Structure of a knowledge-based neural network (from [9], © 1997 IEEE, reprinted with permission).

where v_i is a vector of parameters in b_i defining an open or closed boundary in the input space x, and N_b is the number of boundary neurons.

Let $\sigma(\cdot)$ be a sigmoid function. The region layer (r) contains neurons to construct regions from boundary neurons,

$$r_i = \prod_{j=1}^{N_b} \sigma(\alpha_{ij} b_j + \theta_{ij}), \ i = 1, 2, \ldots, N_r \qquad (9.3)$$

where α_{ij} and θ_{ij} are the scaling and bias parameters, respectively, and N_r is the number of region neurons. The normalized region layer r' contains rational function-based neurons to normalize the outputs of region layer,

$$r'_i = \frac{r_i}{\left(\sum_{j=1}^{N_r} r_j\right)}, \ i = 1, 2, \ldots, N_{r'} \text{ where } N_{r'} = N_r \qquad (9.4)$$

The output layer y contains second-order neurons combining knowledge neurons and normalized region neurons as

$$y_j = \sum_{i=1}^{N_z} \beta_{ji} z_i \left(\sum_{k=1}^{N_{r'}} \rho_{jik} r'_k \right) + \beta_{j0}, \quad j = 1, 2, \ldots, m \quad (9.5)$$

where β_{ji} reflects the contribution of the ith knowledge neuron to output neuron y_j and β_{j0} is the bias parameter. If $\rho_{jik} = 1$, it indicates that the region r'_k is the effective region of the ith knowledge neuron contributing to the jth output. A total of $N_{r'}$ regions are shared by all the output neurons. As a special case, if we assume that each normalized region neuron selects a unique knowledge neuron for each output j, the function for output neurons can be simplified as

$$y_j = \sum_{i=1}^{N_z} \beta_{ji} z_i r'_i + \beta_{j0}, \quad j = 1, 2, \ldots, m \quad (9.6)$$

9.2.2 Neural Network Training

The neural model can directly learn from the training data (x, d) obtained from EM simulation and/or measurements. A gradient-based l_2 optimization technique is employed in the training of KBNN, which requires the derivative of the training error for individual training samples with respect to each weight parameter in KBNN. The derivatives form a matrix called the Jacobian Matrix (J). The network is not a regularly layered MLP structure, and also microwave empirical functions are used instead of standard activation functions. As such, a conventional back propagation algorithm is not applicable. A new scheme extending the idea of error back propogation is derived to obtain the Jacobian matrix in [9].

Using the Jacobian matrix, various training techniques described in Chapter 4 can be applied. During the training process, shift and scale parameters of the empirical functions, other parameters in the empirical functions, boundary locations (defined by boundary neurons), and shape of regions for the empirical functions (defined by region neurons) are all automatically adjusted such that the overall KBNN model matches the training data.

9.2.3 Finished Model for User and Discussion

The final model provides the overall input-output relationship through knowledge formulas and other layers in the network. This model can be used together with the existing models in a circuit simulator. An example of such implementation is *NeuroADS* [11], which plugs neural network models such as KBNN into the ADS [12] simulator.

Several forms of functions $\psi(\cdot)$ and/or $b(\cdot)$ can co-exist in the network. The constant coefficients in the original empirical functions can be replaced by trainable weight parameters, and additional bias/scale parameters can be added to provide extra freedom. If some input parameters are not present in the original empirical models, they can be added to the knowledge function $\psi(\cdot)$ in the weighted sum form. The KBNN structure was inspired from the fact that practical empirical functions are usually valid only in a certain region of the parameter space. To build a neural model for the entire space, several empirical formulas and a mechanism to switch among them are needed. The KBNN model retains the essence of neural networks in the sense that the exact location of each switching boundary and the scale and position of each knowledge function are initialized randomly and then refined during training. When the model is used after training, the KBNN output usually contains the contributions from one or a few of the knowledge neuron responses according to the location of the input parameters in the model input space.

9.2.4 KBNN Examples

Example 1: Circuit Waveform

This is a simple illustration example first presented in [10] showing the effect of the incorporation of electrical knowledge into the neural network. A circuit shown in Figure 9.3, with three transmission lines representing signal integrity analysis of high-speed VLSI interconnects and terminations [13], is selected. The excitation of the circuit is a step-voltage source with 0.1ns rise-time at node 1. Neural networks (both KBNN and MLP) were used to model the output waveform at nodes 3 and 4. The inputs to the neural models are resistor value R and time t, and the outputs are voltages v_3 and v_4. From circuit knowledge, the output waveform of this circuit can be estimated by a two-pole approximation with exponentially decaying sinusoids [14]. This two-pole approximation knowledge, plus an additional resistor variable R as part of the model input, was incorporated into KBNN by providing the knowledge function $\psi(\cdot)$ at layer z as

$$\begin{aligned}z_i &= \psi_i(R, t, \boldsymbol{w}_i) \\ &= e^{-(w_{i1}*R+w_{i2})t} \sin((w_{i3}*R+w_{i4})t + w_{i5}*R + w_{i6}), \\ i &= 1, 2, \ldots, N_z\end{aligned} \qquad (9.7)$$

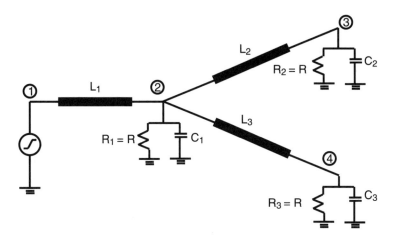

Figure 9.3 A circuit with three transmission lines representing high-speed VLSI interconnects.

With this two-pole approximation, the boundary in the parameter space is independent of time t. This is the knowledge embedded in layer b realized as a linear function of resistance only. The output layer is constructed following the summation in (9.6). The size of KBNN is represented by the number of neurons in b and z layers, for example, $b1z2$ representing a KBNN with one boundary layer neuron and two knowledge layer neurons. For the traditional MLP, the size of the model is represented by the number of hidden neurons in the network, for example, 7 representing a three-layer MLP with seven hidden neurons. The training and testing data were obtained by simulating the original circuit using HSPICE in the time interval from 0 ns to 7 ns. The training data was created with five resistor values of 25, 50, 75, 100, and 125 Ω and with only 10 time points per resistor value and is used to train the neural models (both KBNN and MLP). A KBNN of size $b1z4$ and four MLP networks of sizes 12, 20, 28, and 36 were trained. Testing data includes 69 time points in the waveform on a different set of resistor values (R = 35, 55, 70, 90, and 115 Ω).

Table 9.2 shows the testing results of both models. For each model, the average accuracy from three trainings with different random starting points was used. The accuracy of the model is represented by the error and correlation coefficient between neural model output and testing data. A value of a correlation coefficient closer to 1 indicates a good accuracy of the neural model. As seen in Table 9.2, the training data with 10 time points per resistor value were insufficient for MLP to model the waveforms. Figure 9.4 shows the waveforms from the original HSPICE simulation, the KBNN model, and the MLP model,

Table 9.2
Model Accuracy Comparison Between Standard MLP and Knowledge-Based Neural
Network (KBNN) for Circuit Waveform Modeling Example
(From [10], © 1997 IEEE, Reprinted with Permission)

Neural Net Type	Model Size	No. of Weights	Average Test Error (%)	Worst-case Test Error (%)	Correlation Coefficient
Standard (MLP)	12	62	2.40	18.00	0.9449
	20	102	2.06	13.52	0.9641
	28	142	1.65	9.13	0.9784
	36	182	2.04	12.98	0.9664
Knowledge-based (KBNN)	$b1z4$	36	1.00	8.71	0.9894

for a resistor value of 115 Ω. As can be seen from the figure, the MLP model does not match well with the original waveforms. With the same set of training data, KBNN achieved very good accuracy. This illustrates that the prior knowledge provides additional information that is not adequately represented by the original training data.

The boundary neuron divides the input space into two parts, namely, the low resistance and high resistance parts. The two normalized region layer neurons correspond to these two parts respectively and select the appropriate knowledge neurons. The final output of the network is the sum of products between normalized region layer and knowledge layer neurons. All the parameters in waveform functions and boundary layer and region layer neurons were determined during training through the information encoded in the limited training data. Another interesting point is that the empirical knowledge, that is, two-pole approximation with decay sinusoids, is not an adequate model by itself, since it cannot represent the waveform change with respect to resistor values. This example illustrates that through knowledge-based neural networks a simple empirical function can be used in a large input parameter space, by using different sets of function parameters in different regions of the space. The smooth switching between the regions is realized in the KBNN network by region layer neurons.

Example 2: MESFET

This example demonstrates the use of knowledge-based neural networks to model the physics-based MESFET [10]. Device physical/process parameters (channel length L, channel width W, doping density N_d, channel thickness a) and terminal voltages (gate-source voltage V_{GS}, drain-source voltage V_{DS}) are neural network input parameters, and drain current (I_D) is the neural

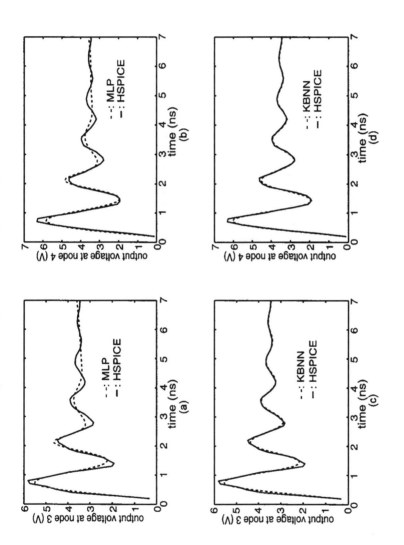

Figure 9.4 Model accuracy comparison with original HSPICE simulation for circuit waveform modeling example; (a) MLP; (b) MLP; (c) KBNN at node 3; (d) KBNN at node 4. All models were trained with only 10 data points per waveform (from [10], © 1997 IEEE, reprinted with permission).

network output [15]. The original problem is physics based [16] and requires a slow numerical simulation procedure.

Empirical formulas for MESFET modeling exist relating FET response to physical/geometrical parameters [17]. The KBNN is developed incorporating empirical formulas in [17] in knowledge layer z. Training samples were obtained by simulating original Khatibzadeh and Trew models [16] using *OSA90* [18] at randomly selected points. The training data range is shown in Table 9.3. Three sets of training data with 100, 300, and 500 samples were used. Neural networks were trained on SPARC station 5.

The ability to extrapolate beyond the boundaries of the training data is a challenging and important aspect of a model. Two sets of testing data were simulated. The first set of test data was in the same range as the training data in input parameter space. The second set of test data was out of the range by 12.5% on either side (i.e., extrapolation region as shown in Table 9.4). These two sets were used to test the neural network models of various sizes, and the results are tabulated in Table 9.5 and Table 9.6. In both cases, KBNN outper-

Table 9.3
Training Data Ranges of Neural Model Input Parameters for MESFET Modeling Example (From [10], © 1997 IEEE, Reprinted with Permission)

Parameters	Notation	Range
Gate length	L	0.336–0.504 μm
Gate width	W	0.8–1.2 mm
Channel thickness	a	0.28–0.42 μm
Doping density	N_d	1.68×10^{23}–2.52×10^{23} $1/m^3$
Gate voltage	V_{GS}	–5–0 V
Drain voltage	V_{DS}	0–4 V

Table 9.4
Extrapolation Data Ranges of Neural Model Input Parameters for MESFET Modeling Example (From [10], © 1997 IEEE, Reprinted with Permission)

Parameters	Notation	Range
Gate length	L	0.315–0.525 μm
Gate width	W	0.75–1.25 mm
Channel thickness	a	0.263–0.438 μm
Doping density	N_d	1.58×10^{23}–2.63×10^{23} $1/m^3$
Gate voltage	V_{GS}	–5–0 V
Drain voltage	V_{DS}	0–4 V

Table 9.5
Model Accuracy Comparison Between Standard MLP and Knowledge-Based Neural Network (KBNN) for MESFET Modeling Example with Testing Data from the Same Region as Training Data. The Results Shown Are the Average of Three Different Trainings for Each Model (From [10], © 1997 IEEE, Reprinted with Permission)

Training Sample Size	Neural Net Type	Model Size	Average Test Error (%)	Worst-case Test Error (%)
100	Standard (MLP)	7	2.14	45.20
		10	2.91	30.80
		14	3.38	52.53
		18	2.75	35.05
		25	3.78	39.71
	Knowledge-based (KBNN)	$b5z6$	1.12	8.99
		$b6z8$	1.03	8.49
300	Standard (MLP)	7	0.90	8.74
		10	0.90	12.89
		14	0.96	14.16
		18	0.89	11.09
		25	1.11	13.66
	Knowledge-based (KBNN)	$b5z6$	0.74	5.68
		$b6z8$	0.72	5.72
500	Standard (MLP)	7	0.74	6.15
		10	0.70	5.39
		14	0.68	6.59
		18	0.69	7.51
		25	0.78	10.08
	Knowledge-based (KBNN)	$b5z6$	0.61	5.37
		$b6z8$	0.61	5.47

forms MLP in all the accuracy measures. The superiority of KBNN is even more significant when fewer training data is available. The overall tendency is that a KBNN trained with 100 samples achieves an accuracy that is similar to an MLP trained with 300 samples. KBNN trained by 300 samples is as accurate as MLP trained by 500 samples. The KBNN and MLP models are compared in terms of average and worst-case test errors in Figure 9.5 and Figure 9.6, respectively.

Moving to the extrapolation region, the accuracy of KBNN deteriorates much more slowly than that of MLPs. This is because the built-in knowledge

Table 9.6
Model Accuracy Comparison Between Standard MLP and Knowledge-Based Neural Network (KBNN) for MESFET Modeling Example with Testing Data in the Extrapolation Region. The Results Shown Are the Average of Three Different Trainings for Each Model (From [10], © 1997 IEEE, Reprinted with Permission)

Training Sample Size	Neural Net Type	Model Size	Average Test Error (%)	Worst-case Test Error (%)
100	Standard (MLP)	7	2.88	51.52
		10	3.86	53.38
		14	4.22	96.31
		18	3.82	71.04
		25	5.04	88.70
	Knowledge-based (KBNN)	$b5z6$	1.56	21.69
		$b6z8$	1.43	11.31
300	Standard (MLP)	7	1.24	12.67
		10	1.43	28.31
		14	1.47	25.91
		18	1.38	39.71
		25	1.52	31.83
	Knowledge-based (KBNN)	$b5z6$	1.04	6.96
		$b6z8$	1.02	7.92
500	Standard (MLP)	7	1.10	12.67
		10	1.00	7.64
		14	0.96	9.35
		18	0.99	10.86
		25	1.29	15.53
	Knowledge-based (KBNN)	$b5z6$	0.83	6.64
		$b6z8$	0.83	9.40

in the KBNN gives it more information that is not seen in the training data. Figure 9.7 shows an example of I-V curves from the best performing MLP (with seven hidden neurons) and KBNN($b5z6$) models, both trained with an insufficient training data of 100 samples. The performance of KBNN is visibly better than MLP. Figure 9.8 shows an example of I-V curves from both models when trained by 300 samples, and both models give good accuracy.

Example 3: Transmission Line

This example demonstrates the knowledge-based neural networks in modeling transmission lines for analysis of high-speed VLSI interconnects [10]. Electro-

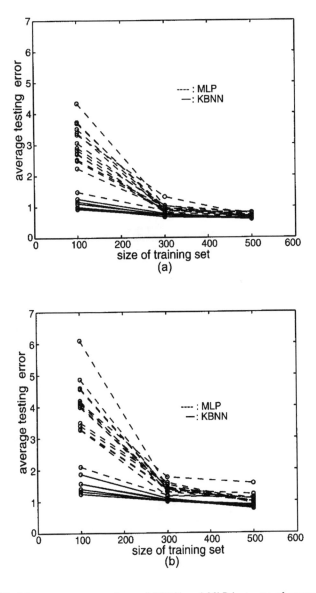

Figure 9.5 Model accuracy comparison of KBNN and MLP in terms of average testing error for the MESFET example; (a) Testing data sampled within the same range as training data; (b) Testing data sampled outside the boundary of training data as shown in Table 9.4. The curves are from models of various sizes and trainings with different initial weights. The advantage of KBNN over MLP is more significant when less training data is available. KBNN is more reliable than MLP in the extrapolation region, i.e., case (b) (from [10], © 1997 IEEE, reprinted with permission).

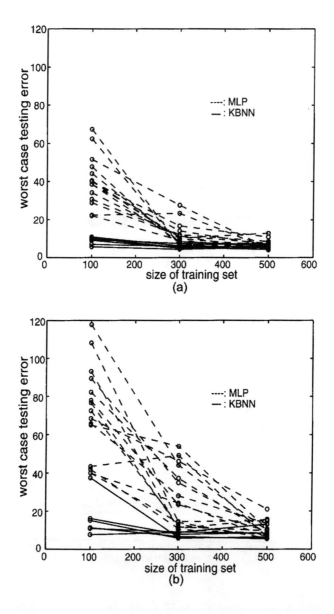

Figure 9.6 Model accuracy comparison of KBNN and MLP in terms of the worst-case testing error for the MESFET example; (a) Testing data sampled within the same range as training data; (b) Testing data sampled outside the boundary of training data (from [10], © 1997 IEEE, reprinted with permission).

Knowledge-Based ANN Models

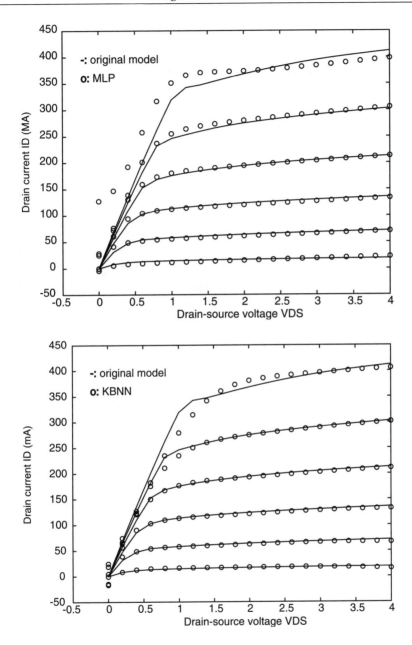

Figure 9.7 An example of *I-V* curves from (a) MLP and (b) KBNN for the MESFET modeling example. Both models were trained with insufficient training data of 100 samples generated in the six-dimensional space of L, W, a, N_d, V_{GS}, and V_{DS} (from [10], © 1997 IEEE, reprinted with permission).

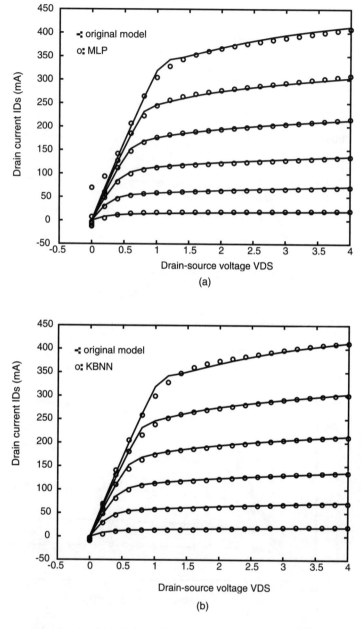

Figure 9.8 An example of *I-V* curves from (a) MLP and (b) KBNN for the MESFET modeling example. Both models were trained with a reasonable size of training data, that is, 300 samples (from [10], © 1997 IEEE, reprinted with permission).

magnetic (EM) simulation of transmission lines is slow, especially if there is a need for repeated evaluations. In this example, MLP and KBNN were used to model the cross sectional per unit length mutual inductance, l_{12}, between two conductors of a transmission line. The inputs to the model are width of conductor (x_1), thickness of conductor (x_2), separation between two conductors (x_3), height of substrate (x_4), relative dielectric constant (x_5), and frequency (x_6). There are empirical formulas for mutual inductance, for example, [19],

$$l_{12} = \frac{\mu_r \mu_0}{4\pi} \ln\left[1 + \frac{(2x_4)^2}{(x_1 + x_3)^2}\right] \quad (9.8)$$

This equation becomes the knowledge to be incorporated into the knowledge layer neurons as,

$$z_i = \psi_i(\boldsymbol{x}, \boldsymbol{w}_i)$$
$$= \ln\left[1 + e^{w_{i1}} \frac{(x_4 - w_{i2})^2}{(x_1 + x_3 - w_{i3})^2}\right] \quad (9.9)$$
$$+ w_{i4}x_2 + w_{i5}x_5 + w_{i6}x_6 + w_{i7}, \; i = 1, 2, \ldots, N_z$$

Linear activation functions were used for boundary layer neurons. Two KBNNs of size ($b2z3$ and $b4z6$) were trained and compared with three MLPs (6-7-1, 6-15-1 and 6-20-1). Five sets of data were generated by EM simulation. The first three sets with 100, 300, and 500 samples were generated within the parameter range shown in Table 9.7 and used for training. A set of 500 test samples was generated in the same range as the training data. These test data were never used in training, and the test results are shown in Table 9.8. Another set of test data with 4,096 samples was simulated around/beyond the boundary

Table 9.7
Ranges of Training Data for Neural Model Input Parameters for the Transmission Line Modeling Example (From [10], © 1997 IEEE, Reprinted with Permission)

Parameters	Notation	Range
Conductor width	x_1	0.10–0.25 mm
Conductor thickness	x_2	17–71 μm
Conductor separations	x_3	0.10–0.76 mm
Substrate height	x_4	0.10–0.31 mm
Relative dielectric constant	x_5	3.7–4.8
Frequency	x_6	0.5–2 GHz

Table 9.8
Model Accuracy Comparison Between MLP and KBNN for Transmission Line Modeling Example with Testing Data in the Same Region as Training Data. The Results Shown Are the Average of Three Different Trainings for Each Model
(From [10], © 1997 IEEE, Reprinted with Permission)

Training Sample Size	Neural Net Type	Model Size	Average Test Error (%)	Worst-case Test Error (%)
100	Standard (MLP)	7	0.95	9.30
		15	1.18	10.07
		20	1.33	10.04
	Knowledge-based (KBNN)	$b2z3$	0.51	4.18
		$b4z6$	0.64	4.16
300	Standard (MLP)	7	0.58	3.12
		15	0.56	3.29
		20	0.58	3.39
	Knowledge-based (KBNN)	$b2z3$	0.44	2.59
		$b4z6$	0.41	2.64
500	Standard (MLP)	7	0.51	3.38
		15	0.54	3.30
		20	0.56	3.28
	Knowledge-based (KBNN)	$b2z3$	0.41	2.02
		$b4z6$	0.38	2.19

of training range, in order to compare extrapolation capability of KBNN and MLP as shown in Table 9.9.

A significantly superior performance of KBNN showed up in the case of smaller training dataset, e.g., 100 samples. Furthermore, the overall tendency suggests that KBNN trained by small training dataset is comparable to MLP trained by a larger training data set. Figure 9.9 and Figure 9.10 compare the performance of KBNN and MLP in terms of average and worst-case testing errors, respectively. A much more stable performance of KBNN is observed over MLP, illustrated by a much smaller error jump in KBNN as opposed to MLP as test data moves to the extrapolation region. Figure 9.11 shows the scattering plots of predicted mutual inductance from neural models versus simulated mutual inductance for 500 test samples within the training data

Table 9.9
Model Accuracy Comparison Between MLP and KBNN for Transmission Line Modeling Example with Testing Data in the Extrapolation Region. The Results Shown Are the Average of Three Different Trainings for Each Model
(From [10], © 1997 IEEE, Reprinted with Permission)

Training Sample Size	Neural Net Type	Model Size	Average Test Error (%)	Worst-case Test Error (%)
100	Standard (MLP)	7	2.38	12.09
		15	2.66	13.57
		20	3.09	16.0
	Knowledge-based (KBNN)	$b2z3$	1.04	5.43
		$b4z6$	1.05	5.01
300	Standard (MLP)	7	1.01	3.61
		15	0.91	3.66
		20	0.96	5.11
	Knowledge-based (KBNN)	$b2z3$	0.69	2.95
		$b4z6$	0.98	3.47
500	Standard (MLP)	7	0.85	3.05
		15	0.89	3.65
		20	0.91	4.87
	Knowledge-based (KBNN)	$b2z3$	0.77	2.70
		$b4z6$	0.87	2.66

boundary. In an ideal case (correlation coefficient is equal to 1), the scattering plot should appear as one diagonal line. The comparison of scattering plots shows that the prediction from KBNN is closer to the diagonal line. Figure 9.12 shows the histograms of errors of MLP and KBNN trained with 100 samples, for the 4,096 test samples. The errors from KBNN are mostly concentrated near zero, whereas MLP has a higher error distribution.

9.3 Source Difference Method

The idea of combining RF/microwave empirical and equivalent circuit information together with neural network learning is an active research topic. In this section, we describe the source difference method [20], which is one of the earlier methods in this direction. The idea here is to exploit the existing

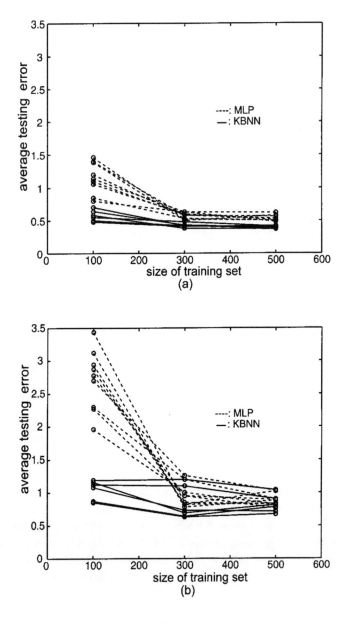

Figure 9.9 Model accuracy comparison of KBNN and MLP in terms of average testing error for the transmission line example; (a) Testing data sampled within the same range as training data; (b) Testing data sampled around/beyond the boundary of training data. The curves are from models of various sizes and trainings with different initial weights. The advantage of KBNN over MLP is more significant when less training data is available. KBNN is more reliable than MLP in the extrapolation region, i.e., case (b) (from [10], © 1997 IEEE, reprinted with permission).

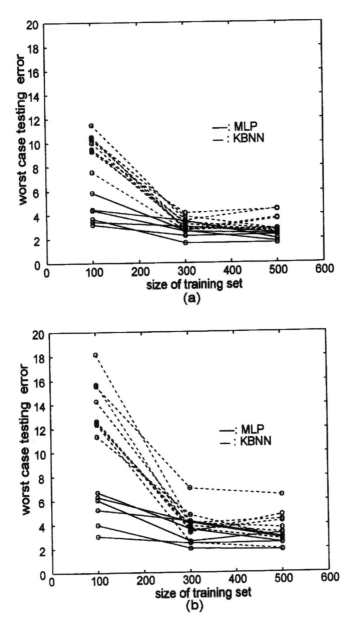

Figure 9.10 Model accuracy comparison of KBNN and MLP in terms of worst-case testing error for the transmission line example; (a) Testing data sampled within the same range as training data; (b) Testing data sampled around/ beyond the boundary of training data. The curves are from models of various sizes and trainings with different initial weights. The advantage of KBNN over MLP is more significant when less training data is available. KBNN is more reliable than MLP in the extrapolation region, i.e., case (b) (from [10], © 1997 IEEE, reprinted with permission).

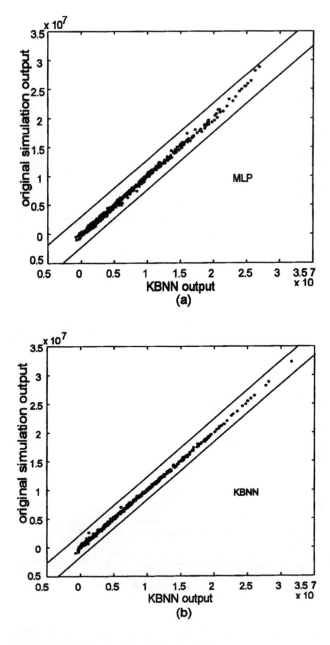

Figure 9.11 Scattering plot of mutual inductance I_{12} for the transmission line modeling example; (a) MLP; and (b) KBNN for 500 testing samples. Both models were trained with insufficient training data of only 100 samples in the six-dimensional space of L, W, a, N_d, V_{GS}, and V_{DS} (from [9], © 1997 IEEE, reprinted with permission).

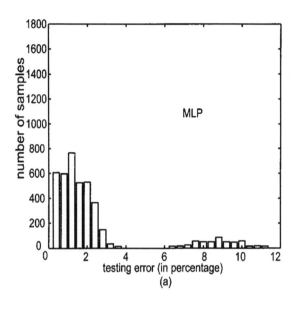

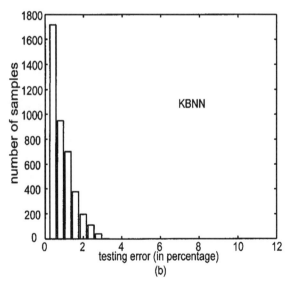

Figure 9.12 Histograms of testing error for the transmission line modeling example; (a) MLP; and (b) KBNN for 4,069 testing samples around/beyond training data boundary. Both models were trained with only 100 training samples (from [10], © 1997 IEEE, reprinted with permission).

information in the form of empirical or equivalent circuit models together with the neural models to develop fast and accurate hybrid EM-ANN models. In other words, one can use the known information of the component to simplify the input-output relationship to be modeled by a neural network. The microwave behaviors of the component (e.g., S-parameters) are generated using EM simulation in the region of interest (i.e., model utilization range in input space). The behaviors of the component are also computed from the existing approximate model. A fast neural model is then developed to learn the difference between the EM simulation results and the approximate model. Some of the application examples in Chapter 5 were developed using this method.

9.3.1 Model Structure

The model structure in the source difference approach is shown in Figure 9.13. The overall structure consists of an empirical or equivalent circuit model (approximate model) to represent the available knowledge, and a neural network that represents the difference between EM simulation data and approximate model.

9.3.2 Preprocessing

For each input sample x_k, the corresponding output d_k is computed from EM simulation. As a next step, for each input sample x_k, the corresponding output $\bar{d}_k$ from the empirical/approximate model is computed. In the preprocessing stage, the difference between EM simulation results and empirical model outputs, that is $\Delta d_k = d_k - \bar{d}_k$, is computed for all the samples. The sample pairs $(x_k, \Delta d_k)$ are used to train the neural network.

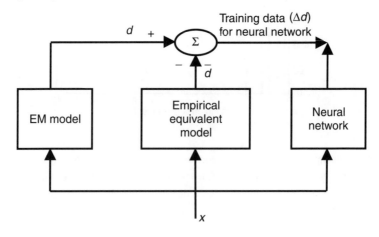

Figure 9.13 Model structure in source difference method.

9.3.3 Neural Network Training

The neural network structure used here can be an MLP with one hidden layer. The training data for the neural network are the sample pairs $(x_k, \Delta d_k)$. Using Δd_k as the desired outputs of the neural network results in a smaller range of the outputs Δy as compared to the original outputs y and a simpler input-output relationship. The neural network training requires less EM simulation points to capture important data trends. The neural network is trained to a respectable accuracy. After training, given input x, the neural model can predict the difference between EM simulation results and the empirical model outputs, that is, Δy.

9.3.4 Finished Model for User

The finished model for the user has two components, that is, the empirical model and the trained neural model. This hybrid EM-ANN model, shown in Figure 9.14, will be used during online microwave design. The empirical model computes the approximate outputs, and the trained neural model predicts the difference. In this way, the neural model can help to correct for the differences in the outputs. The EM-ANN model offers the accuracy of EM simulation, but at a speed much faster than EM simulation. Using the source difference method, fast and accurate microwave component models can be developed with less training data. The time required for model development is also shorter.

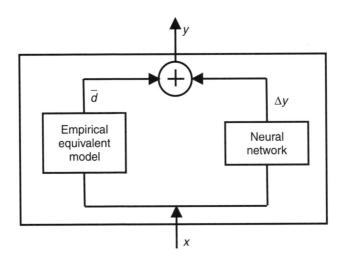

Figure 9.14 Finished model for user from the source difference method.

9.4 Prior Knowledge Input Method (PKI)

The prior knowledge input method was proposed in [21]. In general, the existing empirical models do not give the required accuracy over the desired range of operation. In this method, the empirical model outputs are used as inputs to the neural network model, in addition to the original problem inputs. In this case, the input-output mapping to be learned by the neural network is that between the outputs of the existing approximate model and the original problem. For the case in which the target outputs are the same as the approximate model outputs, the learning problem is reduced to a one-to-one mapping. Some of the application examples in Chapter 5 were developed using this method.

9.4.1 Model Structure

The model structure in the PKI method is shown in Figure 9.15. The overall structure consists of an empirical or equivalent circuit model (approximate model) to represent the available knowledge. It also has a neural network that represents the mapping between the outputs of the approximate model and the original problem. The quality of this mapping is enhanced by including the original problem inputs as additional inputs to the neural network.

9.4.2 Preprocessing

The training data obtained from the EM simulation needs to be preprocessed. For each x_k in the training data, a corresponding vector $\bar{y}_k$ is computed using the empirical or equivalent circuit model. The neural network will then learn

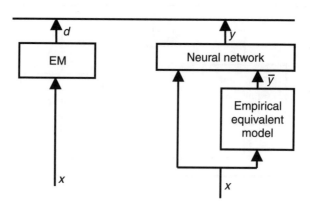

Figure 9.15 Model structure in PKI method.

the mapping from original problem inputs and empirical model outputs (x_k, $\bar{y}_k$), to the EM simulation results d_k.

9.4.3 Neural Network Training

The training data for the neural network in this approach is $\left(\begin{bmatrix} x_k \\ \bar{y}_k \end{bmatrix}, d_k\right)$, resulting in a simpler input-output relationship as compared to the original (x_k, d_k) problem. This simpler input-output relationship requires less EM simulation points to capture important data trends. Standard neural network structures such as MLP can be used to learn this relationship. After training, given an input vector x, $\bar{y}$ is first computed using an approximate model, and then for the input $\begin{bmatrix} x \\ \bar{y} \end{bmatrix}$, the overall output is predicted by the neural network.

9.4.4 Finished Model for User

The trained neural model will be used together with the empirical model during online microwave design. Figure 9.16 shows the finished model for the users. The evaluation of the neural network starts from the outputs of the approximate model, and the result of the evaluation will be the overall model responses, matching the accuracy of EM simulation. The PKI model retains the speed of the approximate models and the neural models.

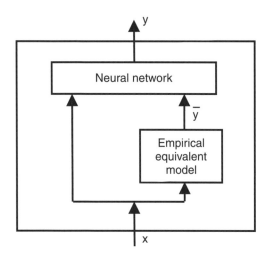

Figure 9.16 Finished model for user from the PKI method.

9.5 Space-Mapped Neural Networks

The space-mapped neural network approach was proposed in [22]. It is based on the space-mapping (SM) concept proposed in [23] for microwave optimization. The space-mapping technique combines the computational efficiency of coarse models with the accuracy of fine models. The coarse models are typically empirical functions or equivalent circuit models, which are computationally very efficient. However, such models are often valid only in a limited range of input-space, beyond which the model predictions become inaccurate. On the other hand, detailed or "fine" models can be provided by an electromagnetic (EM) simulator, or even by direct measurements. The detailed models are very accurate, but can be expensive (e.g., CPU-intensive simulations). The SM technique establishes a mathematical link between the coarse and the fine models and directs the bulk of the CPU-intensive computations to the coarse model, while preserving the accuracy offered by the fine model.

9.5.1 Model Structure

The space-mapped neural model structure is shown in Figure 9.17. The neural network module maps the original problem input-space denoted by vector x (i.e., fine model input-space) into a coarse model input-space denoted by vector x_c. The coarse model then produces the overall output y, which should match the EM data (i.e., fine model output).

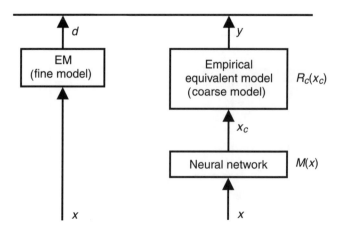

Figure 9.17 Space-mapped neural network structure.

9.5.2 Space-Mapping Concept

Let $f(x)$ and $R_c(x_c)$ represent the corresponding fine and coarse model responses. Let M represent a mapping from the fine model input-space x to the coarse model input-space x_c, that is,

$$x_c = M(x) \qquad (9.10)$$

As shown in Figure 9.18, the objective of SM is to find an appropriate M such that,

$$R_c(M(x)) \approx f(x) \qquad (9.11)$$

Once such mapping is found, the coarse model can be used for fast and accurate simulations.

9.5.3 Space-Mapped Neuromodeling

In the space-mapped neuromodeling (SMN) method, the fine model corresponds to EM simulation, and the coarse model is the empirical or equivalent circuit model. The mapping $M(x)$ is implemented using an artificial neural network. The mapping can be found by solving the optimization problem

$$\min_w \left\| [e_1^T \; e_2^T \; \ldots \; e_P^T]^T \right\| \qquad (9.12)$$

where w represents the weight parameters of the neural network (optimization variables), P is the total number of training samples, and e_k is the error vector given by

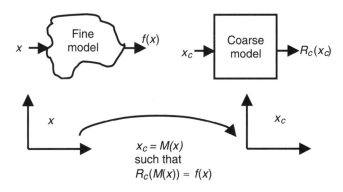

Figure 9.18 Illustration of the objective of space-mapping approach (from [22], © 1999 IEEE, reprinted with permission).

$$e_k = f(x_k) - R_c(M(x_k, w)), \quad k = 1, 2, \ldots, P \quad (9.13)$$

Figure 9.19 illustrates the space-mapped neuromodeling concept. Once the mapping is found, that is, once the neural network is trained, a space-mapped neuromodel for fast and accurate evaluations is readily available.

9.5.4 Frequency in Neuromapping

A typical limitation of many available empirical models is that they only yield good accuracy over a limited low range of frequencies. To establish an overall model applicable at a wider frequency range, we use a frequency-sensitive mapping from the fine model input-space to the coarse model input-space. This is realized by considering frequency as an extra input variable to the neural network that implements the mapping. Two techniques, namely, the frequency dependent space-mapped neuromodeling (FDSMN) and the frequency space-mapped neuromodeling (FSMN), have been developed [22].

In the FDSMN approach, both coarse and fine models are simulated at the same frequency. The mapping from the fine to coarse model input-space is dependent on the frequency, effectively making the coarse model itself frequency dependent as shown in Figure 9.20. With a more comprehensive domain, the FSMN technique establishes a mapping not only for the design variables, but also for the frequency variable. This is realized by adding frequency as an extra output to the neural network as shown in Figure 9.21. The coarse model is then simulated at the mapped frequency f_c to match the fine model response. The FDSMN and FSMN schemes can lead to enhanced model accuracy over an extended frequency range compared to original coarse models.

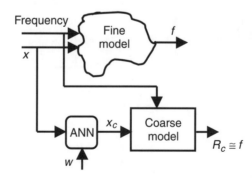

Figure 9.19 The space-mapped neuromodeling concept (from [22], © 1999 IEEE, reprinted with permission).

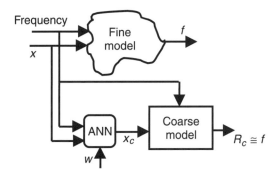

Figure 9.20 The frequency dependent space-mapped neuromodeling (FDSMN) approach (from [22], © 1999 IEEE, reprinted with permission).

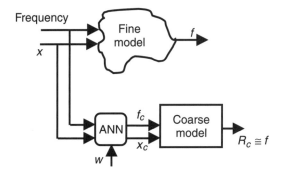

Figure 9.21 The frequency-space-mapped neuromodeling (FSMN) approach (from [22], © 1999 IEEE, reprinted with permission).

9.6 Hierarchical Neural Networks and Neural Model Library Development

In this section, we describe an advanced neural network structure incorporating both structural and functional knowledge of the problem. We also introduce a new task for neural model development, namely, the development of libraries of neural models for passive and active components, a task with potential significance to many microwave design tools. Developing libraries of neural models is very expensive due to massive data generation and repeated neural network training. This problem was addressed by the recently proposed hierarchical neural network approach [1, 24]. This section is based on the work of [1, 24]. The models in the library are developed through a set of base neural models that capture the basic characteristics common to the entire library, and high-level neural modules that map the outputs of the base models to the library model outputs. The hierarchical method substantially reduces the cost of library development through the reduced need for data collection and the shorter time of training.

9.6.1 Development of a Library of Models: Problem Statement

The purpose here is to develop libraries of passive and active microwave component models. Suppose the library consists of N_C component models. For each model, say, the qth model in the library, the input and output parameters are represented by vectors $\boldsymbol{x}^q$ and $\boldsymbol{y}^q$, respectively. The objective of library development is to create models to represent the multidimensional nonlinear relationships

$$\boldsymbol{y}^q = \boldsymbol{y}^q(\boldsymbol{x}^q) \qquad (9.14)$$

for each value of q, $q = 1, 2, \ldots, N_C$. The spaces spanned by $\boldsymbol{x}^q$ and $\boldsymbol{y}^q$ are known as the $\boldsymbol{x}^q$-space and the $\boldsymbol{y}^q$-space.

For illustration, let us consider modeling a multi-conductor transmission line for subsequent use in designing high-speed VLSI interconnects. In this case, the output vector $\boldsymbol{y}^q$ could represent self and mutual inductances of the coupled conductors. The input vector $\boldsymbol{x}^q$ could represent the physical/ geometrical parameters of the transmission line such as conductor width, separation between coupled conductors, substrate height, and dielectric constant. Many such neural models would be needed to cover a variety of transmission lines in interconnect design, for example, single-conductor line, dual strip lines, three-conductor coupled lines, and so forth, leading to the need for a library of transmission line models. Using the standard neural modeling approach, in which an MLP is used to represent each model, costly data collection and extensive model training have to be performed for each model in the library. As such, the overall library development could be expensive.

9.6.2 Hierarchical Neural Network Structure

In the neural network research community, an advanced concept called combining neural networks that addresses issues of neural network accuracy and training efficiency is being exploited. There are two categories of approaches, namely, the ensemble-based approach and the modular approach. In the ensemble-based approach [25, 26], several neural networks are trained such that each network approximates the original problem behavior in its own way. The outputs from these networks are then combined to produce the final output. In the modular approach (e.g., [5, 27–29]), the overall model structure consists of several neural networks, each trained to represent a particular sub-task of the original problem. An integrating unit then selects or combines the outputs of the subnetworks to form the final output. By combining neural networks and incorporating problem knowledge into network structures, the complexity of the overall problem can be divided and conquered more effectively [30, 31].

Based on the concept of combining neural networks, a hierarchical neural network approach was developed [24], using existing microwave information/knowledge in the formulation of sub-modules (networks) and in defining the interactions between modules. In this approach, a distinctive set of base neural models is established. The basic microwave functional characteristics common to various models in a library are first extracted and incorporated into base neural models. A hierarchical neural network is then constructed for each model in the library with low-level modules realized by the base neural models. A high-level neural module is trained to map the outputs of the low-level module to the final outputs of the library model. Figure 9.22 shows the library development using the hierarchical neural network approach.

9.6.3 Base Models and Their Training

First, a set of base models is developed to capture the basic RF/microwave characteristics common to all the models in the library. For example, in a

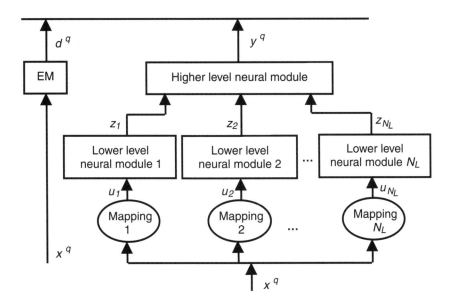

Figure 9.22 Library development using the hierarchical neural network approach.

library of multi-conductor transmission line models, the self-inductance of a conductor is one of the common characteristics needed for all the models in the library. Let x_B^j and y_B^j represent the inputs and outputs of the jth base model, $j = 1, 2, \ldots, N_B$, where N_B is the total number of base models in the library. Let the spaces spanned by x_B^j and y_B^j be called the x_B^j space and the y_B^j space, respectively. The jth base model realized by a neural network, relates x_B^j to y_B^j as

$$y_B^j = B_j(x_B^j, w_j) \tag{9.15}$$

where B_j represents the jth base model and w_j is a vector including all the neural network weights of the jth base model.

The definition of x_B^j (or y_B^j) space is done by choosing a form of space conversion between x_B^j and x^q (or, y_B^j and y^q) and/or examining the common characteristics in the library. Typical forms of space conversion include same-space mapping, subspace mapping, or linear transformation. The space conversions used here are similar to the concept of space-mapping [23]. The mapping between x_B^j and x^q (or, y_B^j and y^q) is called same-space mapping if they contain the same set of parameters with equal or unequal values. Subspace mapping means that x_B^j is a subspace of x^q. Linear transformation means x_B^j is obtained by linearly transforming x^q. The concept of space conversion is well explained in the examples of Section 9.6.8.

For microwave design, empirical formulas for the base relationships often exist. For example, an empirical approximation of self-inductance of a transmission line (y_B^j) as a function of physical/geometrical parameters (x_B^j) is known [19]. In such cases, knowledge-based neural networks (KBNN) [10] could be used to incorporate the empirical formulas into base neural models to further enhance their reliability.

Suppose $(x_B^{j,k}, d_B^{j,k})$ are pairs of training samples for the jth base model, $k = 1, 2, \ldots, M_B^j$, where M_B^j is the total number of training samples. The base neural models should be trained such that

$$\min_{w_j} \sum_{k=1}^{M_B^j} \| B_j(x_B^{j,k}, w_j) - d_B^{j,k} \|^2 \tag{9.16}$$

for $j = 1, 2, \ldots, N_B$. Each base model is trained to a high accuracy by using a sufficient number of training samples. Training of the base models can be considered as an overhead for the library development, because it is done only once in the beginning. The benefit of creating the base models is that they can be subsequently reused to develop all the component models in the library.

9.6.4 Hierarchical Neural Model and Its Training

For each model in the library, a hierarchical neural network structure is defined as shown in Figure 9.23. The purpose of this structure is to construct the overall model from several modules such that the base relationship of the library can be maximally reused for every model in the library. This structure consists of a high-level neural module denoted as H^q and several low-level neural modules denoted as L_i^q, $i = 1, 2, \ldots, N_L^q$ where N_L^q is the number of low-level modules for the qth model in the library. The low-level modules are realized by directly using the base models. The modules $U_i(\cdot)$, $i = 1, 2, \ldots, N_L$, called the knowledge hubs are the realizations of the mapping modules shown in Figure 9.22.

Let index function $j = \phi^q(i)$ be defined such that base model B_j is selected as the ith low-level neural module, that is

$$j = \phi^q(i) = \begin{cases} 1, & \text{for } 0 < i \leq N_{B_1}^q \\ 2, & \text{for } N_{B_1}^q < i \leq N_{B_1}^q + N_{B_2}^q \\ \vdots & \vdots \\ N_B, & \text{for } \sum_{k=1}^{N_B-1} N_{B_k}^q < i \leq \sum_{k=1}^{N_B} N_{B_k}^q \end{cases} \quad (9.17)$$

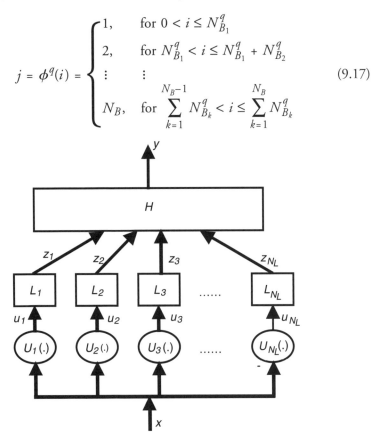

Figure 9.23 The hierarchical neural network structure (from [24], © 1998 IEEE, reprinted with permission).

where $N_{B_j}^q$ is the number of times B_j is reused in the low-level of the qth library model. Let u and z be vectors representing the inputs and outputs of low-level modules. Since the ith low-level module of the qth library model is realized by the $\phi^q(i)$th base model,

$$L_i^q = B_j, \; u_i^q = x_B^j, \; z_i^q = y_B^j, \text{ where } j = \phi^q(i) \qquad (9.18)$$

where u_i^q and z_i^q represent the input and output vectors of low-level module i in library model q. For each L_i^q, that is, the ith low-level module, a structural knowledge hub $U_i^q(\cdot)$ is defined, such that it extracts inputs only relevant to the $\phi^q(i)$th base model out of x^q based on the particular configuration of the qth library component, that is,

$$u_i^q = x_B^j = U_i^q(x^q), \text{ where } j = \phi^q(i) \qquad (9.19)$$

The low-level neural modules produce z^q by recalling the trained base models in the library,

$$z_i^q = L_i^q(u_i^q) = B_{\phi(i)}(u_i^q, w_{\phi(i)}), \; i = 1, 2, \ldots, N_L^q \qquad (9.20)$$

where $w_{\phi(i)}$ are the weights of the $\phi(i)$th base model, and $\phi(i) = \phi^q(i)$. Let vectors U^q and Z^q be defined by concatenating u_i^q and z_i^q for all i, $i = 1, 2, \ldots, N_L^q$, respectively, that is,

$$U^q = \begin{bmatrix} u_1^q \\ u_2^q \\ \vdots \\ u_{N_L}^q \end{bmatrix}, \; Z^q = \begin{bmatrix} z_1^q \\ z_2^q \\ \vdots \\ z_{N_L}^q \end{bmatrix} \qquad (9.21)$$

where $N_L = N_L^q$.

All the low-level modules together provide a map from the U^q space to the Z^q space. A high-level module H^q is defined mapping the Z^q space to the y^q space for each qth model in the library. The high-level module is realized by a neural network

$$y^q = H^q(Z^q, v^q) \qquad (9.22)$$

where v^q includes all neural network weights for module H^q. The relationship in (9.22) is much easier to model than the original $y^q = y^q(x^q)$ relationship since most of the information is already contained in the base models in the low-level. For example, even a linear two-layer perceptron for H^q might be sufficient to produce the final y^q. Consequently, the amount of data needed to train H^q is much less than that for training standard MLP to learn original $y^q = y^q(x^q)$.

Suppose $(x^{q,k}, d^{q,k})$ are pairs of training samples for the qth library model, where $k = 1, 2, \ldots, M^q$, and M^q is the total number of training samples. The $x^{q,k}$ data are mapped to the U^q space through knowledge hubs, and then feedforwarded through the low-level modules (i.e., various reuse of base neural models) into the Z^q space. Consequently, a new set of training samples, denoted by pairs of $(Z^{q,k}, d^{q,k})$ is obtained, where $Z^{q,k}$ is the vector constructed by concatenating $z_i^{q,k} = L_i^q(U_i^q(x^{q,k}))$, for all i, $i = 1, 2, \ldots, N_L^q$. The high-level neural module H^q should be trained such that

$$\min_{v^q} \sum_{k=1}^{M^q} \| H^q(Z^{q,k}, v^q) - d^{q,k} \|^2, \; q = 1, 2, \ldots, N_C \qquad (9.23)$$

With a linear two-layer perceptron neural network as the high-level module, this optimization is simply a quadratic programming problem. In this case, any training method will easily and quickly lead to a globally optimal training solution. This is in contrast to the standard MLP approach with the original nonlinear $y^q(x^q)$ relationship, where training usually takes a longer time and might end at a local optimal solution of the neural network, further prolonging the training process. In this method, the training of the high-level module is the only training needed for each model in the library. No training is needed for the low-level modules because all the low-level modules are obtained by reusing base models that were trained only once in the beginning of the library development.

9.6.5 Finished Model for User

The hierarchical neural model has both functional and structural microwave knowledge embedded inside the neural network structure. The structural knowledge is used to decompose the complicated component modeling into several small tasks, while the functional knowledge is incorporated in the form of pre-trained base models. The finished hierarchical neural model uses the high-level modules together with the set of pre-trained base models during online microwave design.

9.6.6 Algorithm for Overall Library Development

The overall library development is summarized in the following steps:

Step 1: Define the input and output spaces of base models, i.e., x_B^j and y_B^j for $j = 1, 2, \ldots, N_B$, and extract basic characteristics from library, using microwave empirical knowledge if available.

Step 2: Collect training data corresponding to each base model inputs and outputs, i.e, generate sample data $(x_B^{j,k}, d_B^{j,k})$ for the jth base model, where $k = 1, 2, \ldots, M_B^j$, and $j = 1, 2, \ldots, N_B$.

Remark: Training data for base models should be adequate in order to obtain reliable base models.

Step 3: Construct and train base neural models incorporating the knowledge from Step 1. Specifically, solve the optimization problem of (9.16) to find w_j such that $B_j(u, w_j)$ matches base model training data, for $j = 1, 2, \ldots, N_B$. Let $q = 1$.

Remark: Steps 1, 2, and 3 are done in the beginning of library development and are considered overhead effort for the library. The next several steps, i.e., Steps 4–8, are the incremental effort for each component model in the library.

Step 4: According to the base model input space definition in Step 1, set up the structural knowledge hubs $u_i^q = U_i^q(x^q)$, which map the model input space x^q into base model input space x_B^j, where $j = \phi^q(i)$, as defined in (9.17), and $i = 1, 2, \ldots, N_L^q$. This automatically sets up the low-level modules.

Step 5: Collect training data corresponding to the qth model in the library, i.e., generate sample data $(x^{q,k}, d^{q,k})$, where $k = 1, 2, \ldots, M^q$.

Remark: Only a small amount of training data is needed by the hierarchical technique.

Step 6: Map the $x^{q,k}$ data into the z^q space through knowledge hubs and low-level modules following Equations (9.19) and (9.20).

Step 7: Train the high-level neural module H^q, that is, solve the optimization problem of (9.23) to find v^q such that the outputs of high-level module match the training data.

Remark: This training step is very easy and fast since the module H^q is very simple and in most cases, a simple linear two-layer perceptron network. Therefore, only a small and incremental effort is needed to obtain each model in the library.

Step 8: If $q = N_C$, then stop; otherwise, proceed to train the next library model by setting $q = q + 1$ and go to Step 4.

This algorithm allows the hierarchical neural models to be developed systematically and helps the library development process to be maximally automated.

9.6.7 Discussion

The standard MLP approach to library development is an extreme case of the hierarchical neural network formulation. To illustrate this, consider each library model as a base model, that is, $N_B = N_C$. The base model input and output spaces are defined to be the same as the library model input and output spaces, i.e., $x_B^j = x^q$, $y_B^j = y^q$, $j = q$, $q = 1, 2, \ldots, N_C$. There is only one low-level module in each library model. The knowledge hub is simply a relay block passing x^q directly to the U^q space, that is, $u_1^q = U_1^q(x^q) = x^q$. The high-level module H^q will also perform a relay from Z^q space to the y^q space. Therefore, in the worst extreme case in which basic characteristics common to various models in the library are difficult to identify, the hierarchical technique falls back to the standard MLP approach. The hierarchical approach becomes very advantageous, when a few base models can capture the common characteristics of all the component models in the library.

9.6.8 Hierarchical Neural Model Examples

Example 1: Library of Stripline Models

As discussed earlier in Chapter 6, multi-conductor transmission line models are essential for delay and crosstalk analysis in high-speed VLSI interconnect design [13]. EM simulation of transmission line responses is slow, especially if it needs to be massively and highly repetitively evaluated. Neural models, trained off-line from EM data, can be used online during VLSI interconnect design providing instant solutions of the original EM problem. For practical VLSI interconnect design, libraries of 1-conductor, 2-conductor, ..., N-conductor transmission line models are needed. A brute force approach is to train each library model separately, requiring massive data generation and extensive training. Here, the hierarchical approach is used to develop a library of neural models for N-conductor striplines shown in Figure 9.24 [24]. In this example, the self and mutual inductances of coupled conductors are modeled. There are five models in the library, $q = 1, 2, 3, 4, 5$, as shown in Figure 9.25. For the qth model, the number of conductors $N = q$. Table 9.10 defines the notations for input and output parameters of stripline neural models and the effective ranges of their input parameters. A detailed list of input and output parameters of each model in the stripline library is shown in Table

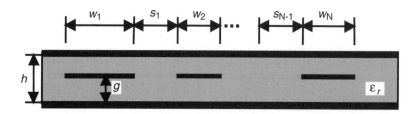

Figure 9.24 A typical N-conductor stripline component.

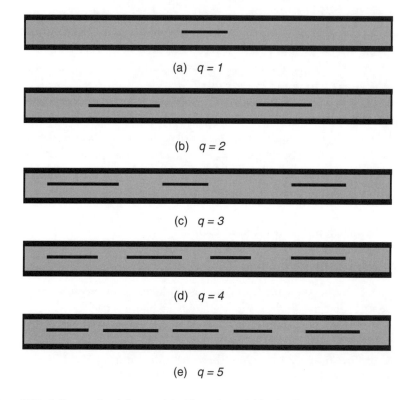

Figure 9.25 A library of stripline models. The qth model in the library represents a q-conductor coupled stripline component (from [24], © 1998 IEEE, reprinted with permission).

9.11. Training and test data were obtained using the LINPAR [32] simulator, which is based on the method of moments.

Base Model Selections

Two base models, B_1 for self inductance and B_2 for mutual inductance, are defined. The inputs to the base models include physical/geometrical parameters

Table 9.10
Notations for Input and Output Parameters of Stripline Neural Models and the Effective Range of Their Input Parameters (From [24], © 1998 IEEE, Reprinted with Permission)

Parameters	Notation	Range
The ith conductor width	w_i	0.05 mm–0.25 mm
The separation between the ith and $i + 1$th conductors	s_i	0.1 mm–0.82 mm
Conductor height above ground	g	0.08 mm–0.25 mm
Substrate height	h	0.16 mm–0.5 mm
Relative dielectric constant	ϵ_r	2–10.2
Self inductance of the ith conductor	L_{ii}	N/A
Mutual inductance between the ith and jth conductors	L_{ij}	N/A

N/A: Not Applicable

Table 9.11
Detailed List of Input and Output Parameters of Each Model in the Stripline Library (From [24], © 1998 IEEE, Reprinted with Permission)

Library Model Index q	Model Name	Neural Model Inputs x^q	Neural Model Outputs y^q	Number of Times Each Base Model (B_1, B_2) Is Used $N_{Bj}^q \times B_j$
$q = 1$	1 conductor stripline model	$w\ g\ h\ \epsilon_r$	L_{11}	$1 \times B_1$
$q = 2$	2 conductor stripline model	$w_1\ w_2\ s\ g\ h\ \epsilon_r$	L_{11}, L_{12}, L_{22}	$2 \times B_1, 1 \times B_2$
$q = 3$	3 conductor stripline model	$w_1\ w_2\ w_3\ s_1\ s_2\ g\ h\ \epsilon_r$	$L_{11}, L_{12}, L_{13},$ L_{22}, L_{23}, L_{33}	$3 \times B_1, 3 \times B_2$
$q = 4$	4 conductor stripline model	$w_1\ w_2\ w_3\ w_4\ s_1\ s_2\ s_3\ g\ h\ \epsilon_r$	$L_{11}, L_{12}, L_{13},$ $L_{14}, L_{22}, L_{23},$ $L_{24}, L_{33}, L_{34}, L_{44}$	$4 \times B_1, 6 \times B_2$
$q = 5$	5 conductor stripline model	$w_1\ w_2\ w_3\ w_4\ w_5$ $s_1\ s_2\ s_3\ s_4\ g\ h$ ϵ_r	$L_{11}, L_{12}, L_{13},$ $L_{14}, L_{15}, L_{22},$ $L_{23}, L_{24}, L_{25},$ $L_{33}, L_{34}, L_{35}, L_{44},$ L_{45}, L_{55}	$5 \times B_1, 10 \times B_2$

such as conductor width (w), distance of conductor from ground plane (g), substrate height (h), separation between conductors (s), and relative dielectric constant (ϵ_r). The outputs of $\boldsymbol{B}_1$ and $\boldsymbol{B}_2$ are self and mutual inductances, respectively. Since the relationship between the self inductance of a single conductor (and the mutual inductance between any two conductors) and the corresponding physical/geometrical parameters is always a useful partial solution to the modeling problem, these two base models do represent basic characteristics common to all five stripline models in the library. The stripline empirical formulas in [19] are adopted as functional knowledge incorporated into the KBNNs (knowledge-based neural networks) [10], which realize the base models $\boldsymbol{B}_1$ and $\boldsymbol{B}_2$. The KBNN structure is denoted by the number of boundary and knowledge neurons, e.g., $b2z3$ represents two boundary layer neurons and three knowledge layer neurons. The base models $\boldsymbol{B}_1$ and $\boldsymbol{B}_2$ are trained to an average testing accuracy of 0.39% and 0.16%, respectively, as shown in Table 9.12. Linear transformation is used as the form of space-mapping between $\boldsymbol{x}_B^j$ (of Table 9.12) and $\boldsymbol{x}^q$ (of Table 9.11), while subspace mapping is used between $\boldsymbol{y}_B^j$ and $\boldsymbol{y}^q$. The number of times base models $\boldsymbol{B}_1$ and $\boldsymbol{B}_2$ are reused in each library model, i.e., N_{Bj}^q, is shown in Table 9.11.

Example of Library Model $q = 1$

For $q = 1$, the library model represents a single conductor transmission line and is directly the base model $\boldsymbol{B}_1$. Therefore, $N_{B1}^q = 1$, $N_{B2}^q = 0$. The knowledge hub is $\boldsymbol{U}_1^q(\boldsymbol{x}^q) = \boldsymbol{x}^q$. The low-level module is $\boldsymbol{L}_1^q = \boldsymbol{B}_1$, and the high-level module is $\boldsymbol{H}^q(\boldsymbol{Z}^q) = \boldsymbol{Z}^q$.

Example of Library Model $q = 3$

For $q = 3$, the library model represents a 3-conductor coupled stripline component. The base models are reused as the lower level neural modules as shown

Table 9.12
Base Models for Stripline Library (From [24], © 1998 IEEE, Reprinted with Permission)

Base Model Index j	Base Model Symbol B_j	Base Model Inputs x_B^j	Base Model Outputs y_B^j	Base Model Structure (KBNN)	Model Accuracy (Average Error as %)
1	B_1	$w\ g\ h\ \epsilon_r$	Self inductance L	$b2z3$	0.39
2	B_2	$w_1\ w_2\ s\ g$ $h\ \epsilon_r$	Mutual inductance L_m	$b4z6$	0.16

in Figure 9.26. There are six low-level neural modules. The knowledge hubs for this library model are defined in Table 9.13. The six low-level neural modules make a preliminary prediction of self inductance for each of the three conductors (L_{11}, L_{22}, L_{33}), and mutual inductance between each pair of conductors, i.e., conductors 1 and 2, conductors 2 and 3, and conductors 1 and 3 (L_{12}, L_{23}, L_{13}). The high-level neural module H^3 is realized by a two-layer MLP with six inputs (i.e., the preliminary predictions of self and mutual inductances from the six low-level modules) and six outputs (i.e., the refined predictions of self and mutual inductances of the overall 3-conductor stripline model). The output neuron activation functions in the two-layer MLP are linear functions, which means a linear combination of low-level neural modules with no gating functions taking advantage of the modular neural network concept. Each low-level neural module provides a portion of the inductance prediction contributing to the overall inductance prediction at the high-level neural module. Only a small amount of training data (15 samples) is needed to train this high-level module of 3-conductor stripline model since the preliminary relationships of the model have already been captured in the base models.

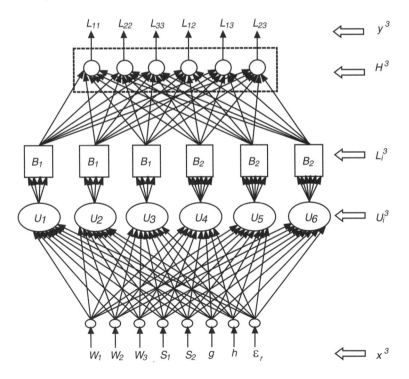

Figure 9.26 The hierarchical neural model for the third model (3 coupled striplines) in the stripline library, i.e., $q = 3$ (from [24], © 1998 IEEE, reprinted with permission).

Table 9.13
Low-Level Modules and Structural Knowledge Hubs for 3-Conductor Stripline, i.e., Library Model $q = 3$ (From [24], © 1998 IEEE, Reprinted with Permission)

Low-Level Modules L_l^q	Inputs to Module/Knowledge Hub $u_l^q = U_l^q(x^q)$	Index Function $\phi_q(i) = j$	Base Model Used $B_{\phi(i)}$
L_1^3	$u_1^3 = U_1^3(x^3) = [w_1\ g\ h\ \epsilon_r]$	$\phi^3(1) = 1$	B_1
L_2^3	$u_2^3 = U_2^3(x^3) = [w_2\ g\ h\ \epsilon_r]$	$\phi^3(2) = 1$	B_1
L_3^3	$u_3^3 = U_3^3(x^3) = [w_3\ g\ h\ \epsilon_r]$	$\phi^3(3) = 1$	B_1
L_4^3	$u_4^3 = U_4^3(x^3) = [w_1\ w_2\ s_1\ g\ h\ \epsilon_r]$	$\phi^3(4) = 2$	B_2
L_5^3	$u_5^3 = U_5^3(x^3) = [w_2\ w_3\ s_2\ g\ h\ \epsilon_r]$	$\phi^3(5) = 2$	B_2
L_6^3	$u_6^3 = U_6^3(x^3) = [w_1\ w_3\ s_1 + s_2\ g\ h\ \epsilon_r]$	$\phi^3(6) = 2$	B_2

However, with the conventional MLP neural model (8, 12, and 16 hidden neurons), 500 samples are needed to achieve similar model accuracy, as shown in Table 9.14. The variation of the library model accuracy versus the amount of training data is plotted in Figure 9.27. As training data becomes less and less available, the error of standard MLP grows quickly, but the hierarchical library model remains reasonable and reliable.

All Library Models

All library models, $q = 2, 3, 4, \ldots$, can be developed systematically in a similar way as model #3. It should be noted that efforts in developing those additional library models are small and incremental, since only a few training

Table 9.14
Model Accuracy Comparison (Average Error on Test Data Expressed as %) Between Standard MLP and the Model Developed by Hierarchical Approach for 3-Conductor Stripline (From [24], © 1998 IEEE, Reprinted with Permission)

Number of Training Samples	MLP (8-8-6)	MLP (8-12-6)	MLP (8-16-6)	Hierarchical Model
15	15.40	14.25	14.17	0.52
25	10.61	9.66	9.96	0.48
50	4.01	1.79	5.30	0.41
100	1.36	0.96	1.80	0.39
300	0.87	0.83	0.86	0.38
500	0.84	0.73	0.79	0.39

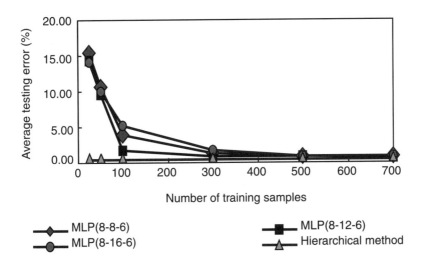

Figure 9.27 Model accuracy comparison (average error on test data) between standard MLP and the model developed by hierarchical approach for 3-conductor stripline (from [24], © 1998 IEEE, reprinted with permission).

data samples are needed, and only the high-level neural module H^q needs to be trained for each q.

Overall Library Accuracy and Development Cost—A Comparison

Using standard MLP for each library model, the total amount of training data needed for the library is 2,764 samples. On the other hand, the hierarchical approach needed only 649 samples (564 samples for base models and 85 samples for subsequent library models) as shown in Table 9.15. The total training time for all library models using standard MLP approach is two hours and 10 minutes on Sun Ultra 1 Workstation. Using the hierarchical approach, the total training time is only 12 minutes.

Example 2: Library of MESFET Models

The drive for first-pass-success in designing active microwave circuits leads to the need for physics-based transistor device models, which give more accurate predictions of device behavior than empirical or equivalent models. However, such physics-based models are too slow when used repetitively in circuit design. Fast neural models trained with physics-based transistor data can be used in place of physics-based models to speed up circuit design and optimization [15]. In this example, the hierarchical approach is used to develop a library of

Table 9.15
Comparison of Number of Training Samples Needed and Library Model Accuracy (expressed as %) for Stripline Library When Developed by Standard MLP and the Hierarchical Neural Network Structure, Respectively (From [24], © 1998 IEEE, Reprinted with Permission)

Library Model Index	Stripline Model Name	Number of Training Samples Needed, (and Corresponding Model Accuracy)	
		Standard MLP	Hierarchical Model
Overhead for base models		0	$264^1 + 300^2$
$q = 1$	1 conductor stripline model	264, (0.42)	0, (0.39)
$q = 2$	2 conductor stripline model	400, (0.75)	10, (0.56)
$q = 3$	3 conductor stripline model	500, (0.73)	15, (0.52)
$q = 4$	4 conductor stripline model	700, (0.78)	25, (0.74)
$q = 5$	5 conductor stripline model	900, (0.99)	35, (0.63)
	stripline library	Total = 2764	Total = 649

^{1} For Base Model B_1 Training and ^{2} for Base Model B_2 Training

MESFET models [24]. The library consists of bias-dependent S-parameter models for MESFETs with different gate-length values. A typical model in the library represents the intrinsic MESFET structure shown in Figure 9.28. There are 10 models, $q = 1, 2, \ldots, 10$ with gate length values of 0.35μm, 0.4μm, 0.45μm, 0.5μm, 0.55μm, 0.6μm, 0.65μm, 0.7μm, 0.75μm, 0.8μm, respectively. The model inputs x^q include frequency (f), channel thickness (a), gate bias voltage (V_{GS}), and drain bias voltage (V_{DS}). The model outputs

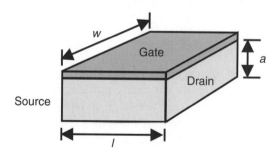

Figure 9.28 Physics-based intrinsic MESFET.

y^q include real and imaginary parts of the two-port S-parameters S_{11}, S_{12}, S_{21}, and S_{22}. Training and test data were obtained by using *OSA90* [18] with the Khatibzadeh and Trew models [16]. In this library, all the transistor models have equal gate-width of 1 *mm*. Other model input parameters and their ranges are shown in Table 9.3.

Base Model Selections

In this library, the relationships between the real and imaginary parts of the two-port S-parameters (S_{11}, S_{12}, S_{21}, and S_{22}) and model input parameters (f, a, V_{GS} and V_{DS}) are taken as the common characteristics required for all transistor models. To represent these common characteristics, eight base models, $B_1, B_2, \ldots, B_8$ are defined corresponding to four S-parameters of two MESFETs, one with smaller gate length ($l = 0.40 \mu m$) and another with larger gate length ($l = 0.80 \mu m$) as shown in Table 9.16. Each base model has two outputs, that is, real and imaginary parts of the corresponding S-parameter. Same-space mapping is used between x_B^j and x^q, and subspace mapping is used between y_B^j and y^q. In this example, MLP structures were used to construct the base models, with test accuracy shown in Table 9.16.

Example of Library Model $q = 2$

For $q = 2$, the library model is constructed from four base models, B_1, B_2, B_3, and B_4, directly, without any further training, i.e., $L_1^2 = B_1$, $L_2^2 = B_2$, $L_3^2 = B_3$, $L_4^2 = B_4$, $H^2(Z^2) = Z^2$.

Table 9.16
Base Models for the MESFET Library (From [24], © 1998 IEEE, Reprinted with Permission)

Base Model Index j	Base Model Symbol B_j	Base Model Inputs x_B^j	Base Model Outputs y_B^j	Base Model Structure (MLP)	Model Accuracy (Average Error as %)
1	B_1	a f V_{GS} V_{DS}	S_{11} of $l = 0.4 \mu m$	4-60-2	0.17
2	B_2	a f V_{GS} V_{DS}	S_{12} of $l = 0.4 \mu m$	4-60-2	0.20
3	B_3	a f V_{GS} V_{DS}	S_{21} of $l = 0.4 \mu m$	4-60-2	0.21
4	B_4	a f V_{GS} V_{DS}	S_{22} of $l = 0.4 \mu m$	4-80-2	0.18
5	B_5	a f V_{GS} V_{DS}	S_{11} of $l = 0.8 \mu m$	4-80-2	0.28
6	B_6	a f V_{GS} V_{DS}	S_{12} of $l = 0.8 \mu m$	4-80-2	0.38
7	B_7	a f V_{GS} V_{DS}	S_{21} of $l = 0.8 \mu m$	4-80-2	0.36
8	B_8	a f V_{GS} V_{DS}	S_{22} of $l = 0.8 \mu m$	4-80-2	0.29

Example of Library Model $q = 5$

For library model $q = 5$, the inputs and outputs of the model are the same as that of the base models. The only difference is that the gate-length is equal to 0.55 μm. The overall model structure is shown in Figure 9.29. There are eight low-level modules, i.e., $N_L^5 = 8$, and $L_1^5 = B_1$, $L_2^5 = B_2$, ..., $L_8^5 = B_8$. Base models are used in the low-level neural modules to predict the S-parameters for different model inputs. Since model input space is exactly the same as that of base models, knowledge hubs simply perform relay operations, i.e. $u_i^5 = U_i^5(x^5) = x^5$, $i = 1, 2, \ldots, 8$. The high-level neural module H^5 is a two-layer perceptron with 16 inputs and 8 outputs (real and imaginary parts of S_{11}, S_{22}, S_{12} and S_{21}). Out of the 16 inputs, 8 inputs correspond to the predictions from base models B_1, B_2, B_3 and B_4, while the other 8 inputs correspond to the predictions from base models B_5, B_6, B_7 and B_8. For the purpose of comparison, all the library models were also developed

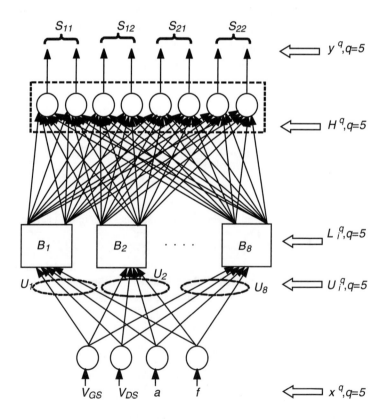

Figure 9.29 The hierarchical neural model for the fifth model in the MESFET library, i.e., $q = 5$ (from [24], © 1998 IEEE, reprinted with permission).

using standard MLP networks. Table 9.17 and Figure 9.30 show the comparison of model accuracy between MLP models (with 60, 80, and 100 hidden neurons) and the hierarchical model.

Overall Library Accuracy and Development Cost—A Comparison

Model $q = 10$ can be developed similarly as model $q = 2$. All other library models, $q = 1, 3, 4, 6, 7, 8,$ and 9, can be developed easily in a similar way as library model #5. Using the hierarchical approach, the training time and training data required are much less as compared to the standard MLP approach as shown in Table 9.18.

9.7 Summary

This chapter demonstrates an important trend far beyond the straightforward use of standard neural networks to solve a standard microwave example. A new concept featuring interdependent neural network and RF/microwave formulation is described, allowing the incorporation of the microwave designer's knowledge and understanding of the problem into neural model structures. The effects of using knowledge on model reliability, extrapolation capability, and amount of training data needed are demonstrated through examples. The knowledge neural network techniques introduced in this chapter include KBNN, source difference method, PKI, and space-mapped neural modeling. An advanced hierarchical neural network approach that simplifies the task of developing libraries of neural models is also presented. Various types of knowledge structures (models) covered in this chapter can be incorporated into circuit simulators, such as MDS [33] and ADS [12], to be used together with existing models for circuit simulation and optimization.

Table 9.17
Model Accuracy Comparison (Average Error on Test Data Expresses as Percentage) Between Standard MLP and the Hierarchical Model for Library Model, $q = 5$, of MESFET Library (From [24], © 1998 IEEE, Reprinted with Permission)

Number of Training Samples	MLP (4-60-8)	MLP (4-80-8)	MLP (4-100-8)	Hierarchical Model
25	13.97	15.15	14.78	2.14
50	4.51	4.30	4.97	0.99
100	2.29	2.25	2.57	0.87
150	1.71	1.57	1.62	0.82
200	1.46	1.35	1.37	0.79
300	0.96	0.86	0.94	0.74

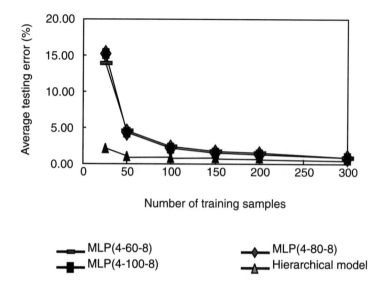

Figure 9.30 Model accuracy comparison (average error on test data) between standard MLP and the hierarchical model for the MESFET library model, $q = 5$, whose gate length equals 0.55 μm (from [24], © 1998 IEEE, reprinted with permission).

References

[1] Wang, F., *Knowledge Based Neural Networks for Microwave Modeling and Design*, Ph.D. Thesis (Supervisor: Q. J. Zhang), Department of Electronics, Carleton University, Ottawa, Canada, 1998.

[2] Towell, G. G., and J. W. Shivalik, "Knowledge-Based Artificial Neural Networks," *Artif. Intell.*, Vol. 70, 1994, pp. 119–165.

[3] Fu, L. M., "Integration of Neural Heuristics into Knowledge-Based Inference," *Connection Sci.*, Vol. 1, 1989, pp. 325–340.

[4] Lacher, R. C., S. I. Hruska, and D.C. Kuncicky, "Backpropagation Learning in Expert Networks," *IEEE Trans. Neural Networks*, Vol. 3, 1992, pp. 62–72.

[5] Mahoney, J. J., and R. J. Mooney, "Combining Connectionist and Symbolic Learning to Refine Certainty Factor Rule Bases," *Connection Sci.*, Vol. 5, 1993, pp. 339–364.

[6] Haykin, S., *Neural Networks: A Comprehensive Foundation*. New York, IEEE Press, 1994.

[7] Figueiredo, M., and F. Gomide, "Design of Fuzzy Systems Using Neurofuzzy Networks," *IEEE Trans. Neural Networks*, Vol. 10, 1999, pp. 815–827.

[8] Reyneri, L. M., "Unification of Neural and Wavelet Networks and Fuzzy Systems," *IEEE Trans. Neural Networks*, Vol. 10, 1999, pp. 801–814.

[9] Halgamuge, S. K., and M. Glesner, "Fuzzy Neural Networks: Between Functional Equivalence and Applicability," *International Journal of Neural Systems*, World Scientific Publishing, June 1995, Vol. 6, No. 2, pp. 185–196.

Table 9.18
Comparison of Number of Training Samples Needed and Training Time Used for the MESFET Library When Developed by Standard MLP and the Hierarchical Neural Network Structure, Respectively (From [24], © 1998 IEEE, Reprinted with Permission)

Library Model Index q	MESFET Model	Number of Training Samples Needed, (and Corresponding Model Accuracy in %, Training Time)	
		Standard MLP	Hierarchical Model
Overhead for base models		0	300^1, (0.23, 4.22hrs) 300^2, (0.33, 4.54hrs)
$q = 1$	($l = 0.35$ μm)	300, (0.81, 2.52 hrs)	50, (0.70, 1.4 min)
$q = 2$	($l = 0.4$ μm)	300, (0.88, 2.64 hrs)	0, (0.23, 0 min)
$q = 3$	($l = 0.45$ μm)	300, (0.86, 2.48 hrs)	50, (0.66, 1.4 min.)
$q = 4$	($l = 0.5$ μm)	300, (0.88, 2.32 hrs)	50, (0.71, 1.4 min.)
$q = 5$	($l = 0.55$ μm)	300, (0.86, 2.66 hrs)	50, (0.99, 1.4 min.)
$q = 6$	($l = 0.6$ μm)	300, (0.89, 2.18 hrs)	50, (1.22, 1.4 min)
$q = 7$	($l = 0.65$ μm)	300, (0.82, 2.45 hrs)	50, (1.08, 1.4 min)
$q = 8$	($l = 0.7$ μm)	300, (0.87, 2.58 hrs)	50, (1.05, 1.4 min)
$q = 9$	($l = 0.75$ μm)	300, (0.79, 2.50 hrs)	50, (0.77, 1.4 min)
$q = 10$	($l = 0.8$ μm)	300, (0.88, 2.78 hrs)	0, (0.33, 0 min)
	Library	Total = 3,000, (25.11 hrs)	Total = 1,000, (8.95 hrs)

[1] For training base models B_1, B_2, B_3, and B_4; [2] For training base models B_5, B_6, B_7, and B_8.

[10] Wang, F., and Q. J. Zhang, "Knowledge Based Neural Models for Microwave Design," *IEEE Trans. Microwave Theory and Techniques*, Vol. 45, 1997, pp. 2333–2343.

[11] Zhang, Q. J., et al., *NeuroADS*, Department of Electronics, Carleton University, 1125 Colonel By Drive, Ottawa, Canada, K1S 5B6.

[12] *ADS*, HP EEsof (now Agilent Technologies), 1400 Fountaingrove Parkway, Santa Rosa, CA, 95403.

[13] Zhang, Q. J., F. Wang, and M. S. Nakhla, "Optimization of High-Speed VLSI Interconnects: A Review," *Int. J. Microwave and Millimeter-Wave CAE*, Vol. 7, 1997, pp. 83–107.

[14] Zhou, D., et al., "A Simplified Synthesis of Transmission Lines with a Tree Structure," *Int. J. Analog Integrated Circuits and Signal Processing*, Vol. 5, 1994, pp. 19–30.

[15] Zaabab, A. H., Q. J. Zhang, and M. S. Nakhla, "A Neural Network Modeling Approach to Circuit Optimization and Statistical Design," *IEEE Trans. Microwave Theory and Techniques*, Vol. 43, 1995, pp. 1349–1358.

[16] Khatibzadeh, M. A., and R. J. Trew, "A Large-Signal, Analytical Model for the GaAs MESFET," *IEEE Trans. Microwave Theory and Techniques*, Vol. 36, 1988, pp. 231–238.

[17] Ladbrooke, P. H., *MMIC Design: GaAs FET's and HEMT's*. Boston, MA: Artech House, 1989.

[18] *OSA90 Version 3.0*, Optimization Systems Associates Inc., P.O. Box 8083, Dundas, Ontario, Canada L9H 5E7, now HP EEsof (Agilent Technologies), 1400 Fountaingrove Parkway, Santa Rosa, CA 95403.

[19] Walker, C. S., *Capacitance, Inductance and Crosstalk Analysis*, Boston, MA: Artech House, 1990.

[20] Watson, P. M., and K. C. Gupta, "EM-ANN Models for Microstrip Vias and Interconnects in Multilayer Circuits," *IEEE Trans. Microwave Theory and Techniques*, Vol. 44, 1996, pp. 2495–2503.

[21] P. M. Watson, K. C. Gupta, and R. L. Mahajan, "Development of Knowledge Based Artificial Neural Network Models for Microwave Components," in *IEEE Int. Microwave Symp. Digest*, Baltimore, MD, 1998, pp. 9–12.

[22] Bandler, J. W., et al., "Neuromodeling of Microwave Circuits Exploiting Space Mapping Technology," in *IEEE Int. Microwave Symp. Digest*, Anaheim, CA, 1999, pp. 149–152.

[23] Bandler, J. W., et al., "Space Mapping Technique for Electromagnetic Optimization," *IEEE Trans. Microwave Theory and Techniques*, Vol. 42, 1994, pp. 2536–2544.

[24] Wang, F., V. K. Devabhaktuni, and Q. J. Zhang, "A Hierarchical Neural Network Approach to the Development of a Library of Neural Models for Microwave Design," *IEEE Trans. Microwave Theory and Techniques*, Vol. 46, 1998, pp. 2391–2403.

[25] Hashem, S., and B. Schmeiser, "Improving Model Accuracy Using Optimal Linear Combinations of Trained Neural Networks," *IEEE Trans. Neural Networks*, Vol. 6, 1995, pp. 792–794.

[26] Hashem, S., "Algorithms for Optimal Linear Combinations of Neural Networks," *Proc. IEEE Intl. Conf. Neural Networks*, Houston, TX, June 1997, pp. 242–247.

[27] Sharkey, A. J., "On Combining Artificial Neural Nets," *Connection Science*, Vol. 8, 1996, pp. 299–314.

[28] Jansen, W. J., et al., "Assembling Engineering Knowledge in a Modular Multilayer Perceptron Neural Network," *Proc. IEEE Intl. Conf. Neural Networks*, Houston, TX, June 1997, pp. 232–237.

[29] Lendaris, G. G., A. Rest, and T. R. Misley, "Improving ANN Generalization using *A Priori* Knowledge to Pre-Structure ANNs," *Proc. IEEE Intl. Conf. Neural Networks*, Houston, TX, June 1997, pp. 248–253.

[30] Auda, G., M. Kamel, and H. Raafat, "A New Neural Network Structure with Cooperative Modules," *Proc. IEEE Intl. Conf. Neural Networks*, Orlando, FL, June 1994, pp. 1301–1306.

[31] Wang, L. C., N. M. Nasrabadi, and S. Der, "Asymptotical Analysis of a Modular Neural Network," *Proc. IEEE Intl. Conf. Neural Networks*, Houston, TX, June 1997, pp. 1019–1022.

[32] Djordjevic, A., et al., *Matrix Parameters for Multiconductor Transmission Lines: Software and User's Manual*, Boston, MA: Artech House, 1989.

[33] *MDS*, HP EEsof (now Agilent Technologies), 1400 Fountaingrove Parkway, Santa Rosa, CA, 95403.

10

Concluding Remarks and Emerging Trends

This is a short chapter in which some conclusions are drawn to complete the book. The current status of ANN applications to RF/microwave design is summarized from the earlier chapters. Some thoughts on the likely impact of neural network technology on RF and microwave design are articulated. Current trends and challenges in applications of neural network technology are summarized.

10.1 Summary of the Book

The objective of this book has been to introduce neural networks as an important tool for RF and microwave modeling and design. In Chapter 1, we started with a very brief introduction to RF and microwave design and to artificial neural networks. Also, an overview of the book was included in this chapter.

Chapter 2 presented an overview of modeling and optimization for RF and microwave design. An understanding of the overall generic design process brings out two major segments of any design:

1. Design-in-the-large where one starts from specifications identification and arrives at an initial design;
2. Design-in-the-small where the initial design is processed into the final optimized design.

Currently available RF and microwave design tools concentrate on the second segment whereas it is recognized that quite often many significant decisions affecting the design are made in the first segment of the design process. Conventional design and computer-aided design methodologies were presented. Concepts of knowledge-aided design were introduced. The importance of modeling and optimization in the design process was brought out. Techniques for RF and microwave circuits CAD and also for printed RF and microwave antennas were reviewed. It was pointed out that ANNs are likely to play very crucial roles in modeling RF and microwave components, circuits, antennas, and systems; in implementation of efficient optimization strategies; and in implementation of knowledge-aided design techniques.

Chapter 3 described various different types of neural network structures (or architectures) that have been used or could be used for RF and microwave design. Most commonly used are feedforward neural network configurations known as multilayer perceptrons (MLP). The back propagation method is used for training such ANNs. Other ANN structures discussed in this chapter include radial basis function (RBF) network, wavelet neural network, and self-organizing maps (SOM). Brief reviews of arbitrary structures for ANNs and recurrent neural networks are also included.

Training of neural networks is the theme of Chapter 4. Training algorithms are integral and the most important part of neural network development. The first part of this chapter documents key issues related to training, namely, data generation, range, and distribution of samples in the input parameter space, data scaling, initialization of weight parameters, and quality measures for the trained neural network. The actual training process—algorithms for finding values of weights associated with various neurons—can be viewed as an optimization process. Thus, various well-known optimization techniques can be used for training ANN structures. Methods discussed in this chapter include: the back propagation algorithm based on the steepest descent principle, the conjugate gradient algorithm, the Quasi-Newton algorithm, Levenberg-Marquardt and Gauss-Newton algorithms, use of decomposed optimization, the simplex method, genetic algorithms, and simulated annealing algorithms. A section on comparison of various training techniques is also included.

Chapter 5 includes several examples of the techniques described in earlier chapters for the development of CAD models for RF/microwave components. The general discussion on the modeling procedure includes: selection of model inputs and outputs, training data generation, error measures, and integration of EM-ANN models with RF and microwave circuit simulators. Component examples for which EM-ANN models have been developed are divided into three groups. The first group includes microstrip transmission lines, one-port and two-port microstrip vias, and vertical interconnects (stripline-to-stripline and microstrip-to-microstrip) used in multilayer circuits. The second group consists of components used in coplanar waveguide (CPW) circuit design.

These are CPW lines, CPW bends, and other discontinuities (opens, shorts, steps-in-width, T-junction) occurring in CPW circuits. The third group consists of various other passive components such as spiral inductors, multiconductor transmission lines, microstrip patch antennas, and waveguide filter components. These examples convey the message that neural network models could be developed for any component needed for design.

High-speed IC interconnects used in digital circuits constitute an important enabling technology item in high-speed computer circuits. ANN macromodeling techniques allow high-speed interconnect optimization to be carried out. Chapter 6 is a detailed description of this important application of neural networks. The neural network approach is compared with the traditional techniques for the high-speed interconnect modeling and signal integrity analysis. Applications and numerical examples presented include three parallel coupled interconnects, asymmetric two-line interconnect, an eight-bit digital bus configuration, a transmission line circuit with nonlinear termination, and optimization of coupled transmission lines. The chapter concludes with a comparison of run time, and a performance evaluation. It is pointed out that the ANN approach can be used for model reduction to smaller equivalent macromodels, for hierarchical signal integrity design, and for concurrent design optimization, including electrical, physical, thermal, EMC/EMI, and reliability parameters.

Chapter 7 introduces applications of neural networks to active component models. Active device modeling is an important and critical area of microwave CAD, and it is an area where ANNs can play a significant role. This chapter describes direct modeling of the device's external behavior (for DC, small signal, and nonlinear models), incorporation of large signal neural model into circuit simulations, and indirect modeling through a known equivalent circuit model of the device. Discussion also includes a time-varying Volterra kernel-based model for a transistor and circuit representation of the neural network models.

Use of neural network models for analysis and optimization of RF and microwave designs is the topic of Chapter 8. Examples discussed in this chapter include CPW circuits (a power divider and a folded double stub filter), optimization of a CPW patch antenna, optimization of multilayer band pass filters, and yield optimization of a three-stage MMIC amplifier.

Chapter 9 on knowledge-based ANN models addresses the important topic of embedding knowledge in ANN models as well as use of prior knowledge in reducing the training time of ANN structures. The discussion starts with a mention of rule-based knowledge networks and microwave-oriented model structure issues. An approach for knowledge-based neural networks (KBNN) and three examples of its implementation are included. Both the source-difference method and prior-knowledge method of using prior knowledge in the training process are described. Space-mapped neural networks that combine the space-mapping approach with the neural network approach is a recent development with interesting potential applications. ANN structures using a "knowledge layer" are described. Examples of these techniques are included. Another important topic in this chapter relates to hierarchical neural networks and neural model library development.

10.2 Impact of Neural Nets on RF and Microwave Design

The examples of ANN applications to RF and microwave design as presented in this book point out clearly the usefulness of this approach in improving the efficiency and accuracy of the design process. We can anticipate considerably increased use of this approach in several different ways.

10.2.1 Insertion in Design Tools

The first one is increased insertion of ANN models in commercially available RF and microwave design tools. The advantages resulting from this activity may be listed as follows:

- Increased accuracy of the models available in the software;
- Validity of the models over increased range of parameter values;
- Ease of updating the models in subsequent versions of the software;
- Savings in time and effort required in introducing models for newer components and devices.

All these changes will be made by RF and microwave software vendors and will be transparent to the end user of these design tools. As is the current practice, software users are told very little about what kind of modeling techniques have been used for various models inserted in these design tools.

10.2.2 ANN Models Linked to Design Software

The second and perhaps more widespread usage of ANN approach is likely to be by the design groups who use newer components and devices for which adequate models are not available in commercially available design software tools. Most of the examples reported in this book would fall into this category. Here the neural net approach presents a systematic and efficient method for development of accurate and efficient models that could be linked with commercially available RF and microwave network simulators. Most of the commercially available RF and microwave network simulators do have provisions for linking externally developed component models. This will extend the use of commercially available tools into design areas where they are not used currently. Currently some of these areas are multilayer circuit design, CPW circuit design, design of printed antennas, design of integrated circuit-antenna modules, design of active circuits, and incorporation of RF considerations in high-speed digital circuits.

10.2.3 Efficient Use of EM Simulators

Closely related to the second item above is the application of the neural net modeling to facilitate an efficient use of electromagnetic (EM) simulators for RF, microwave, and millimeter-wave design. As pointed out in Chapter 2 in relation to printed antennas, frequency domain and time domain EM simulators are considered as the last resort for accurate analysis of a component, circuit, or antenna design for which there is no simpler method of analysis available. More and more RF and microwave designers have started making use of these simulators for their design problems. However, intensive computer time and memory requirements associated with these EM simulators have restricted their widespread usage. The use of ANN models for storing the results of segments of the design that are used repeatedly opens up a new vista for more efficient usage of these EM simulators. Using the concept of design decomposition (also known as "segmentation" or "diakoptics") in association with electromagnetic simulations is recognized as one of the emerging trends in RF, microwave, and millimeter-wave designs [1]. Neural network modeling can play a critical role in efficient implementation of these design decomposition techniques. Once ANN models are developed for design segments (which may or may not be a physically identifiable part of the design) and for interconnections among these segments, these models can be used for design modifications and optimization. An example of this approach is the design of antenna arrays as discussed in [2]. In this case, multiport network models were developed for various elements of the array and also for mutual couplings among these elements. ANN models for antenna elements and for mutual coupling among these elements, based on the accurate electromagnetic simulation of these segments of the design, will make the implementation of this design decomposition approach more accurate and extendable to other kinds of antenna array systems.

10.2.4 Development of Efficient Optimization Strategies

The importance of the optimization process in design has been discussed in Chapter 2. As mentioned earlier, optimization is another important and time-consuming part of the CAD process. Repeated analysis of the design that is needed for the optimization process can be accelerated by using ANN macromodels of the design to be optimized. Chapters 6 and 8 have illustrated this approach. In Chapter 6 ANN models were used for yield optimization of high-speed IC interconnects. Chapter 8 provided examples of optimization of multilayer circuits, CPW circuits, CPW fed slot antennas, and a three-stage MMIC amplifier using an ANN modeling approach. The possibility of developing accurate ANN macromodels opens up a new dimension in implementing design optimization.

10.2.5 Implementation of Knowledge-Aided Design

The role of knowledge-based approaches to the initial stages of RF and microwave circuit and antenna design was mentioned in Chapter 2 as an area that needs to be explored. Suitably trained ANN macromodels can become a useful vehicle for incorporating knowledge in the design process. Discussion on knowledge-based ANN models included in Chapter 9 is a pointer in that direction. As pointed out in that chapter, prior knowledge about a component embedded in the ANN model helps reduce the training time needed in developing an accurate ANN model. This knowledge-based design approach can be extended to the overall design process, say for high efficiency RF amplifiers as follows. Based on the design rules developed for various classes of high-efficiency amplifiers, an ANN model can be trained to provide the designer with an appropriate configuration for a given set of specifications. An ANN macromodel for this configuration can be used to modify an embedded typical design into a design that meets the given specifications. We can look forward to increased research activity in neural network-based knowledge-aided design.

10.3 Trends and Challenges

The past several years represent the beginning of artificial neural network concepts for RF and microwave CAD. The area is continuously emerging and has a promising potential in helping shape the design automation methodology for high-frequency circuits in the 21st century. Further research will continue in ANN applications for both EM and active circuit design, ANN architectural and training methodologies, and knowledge-aided design.

The work in incorporating RF and microwave knowledge with neural networks continues to be an important area of research. The goal is to develop hybrid circuit/ANN model structures such that the combined model requires even fewer training data and can extrapolate even more robustly. Such models should be much faster than detailed EM or physics-based models, and much more accurate across a larger range of parameters than empirical and equivalent circuit models. Different ways of incorporating knowledge into neural networks, different types of knowledge such as time or frequency domains, and empirical or equivalent electrical information can be further exploited to achieve this goal.

Continued work in ANN architecture and training methodology will make the ANN approach effective to more microwave design problems. One trend is the drive for increased ability to handle a large number of model parameters, for example, large number of physical and geometrical parameters involved in modeling electromagnetic behavior of a three-dimensional high frequency component or nonlinear behavior of semiconductor devices. As the number of model input parameters increase, the amount of training data, size of neural network, and training time would all increase. Scalable neural model approaches need to be addressed such that the neural network structural and training complexity would be manageable [3]. This article discusses the general issue of performance and efficiency of neural nets. It is pointed out that a combination of several individually trained models may be able to improve upon the performance of a single model. Further research related to application of these ideas is needed.

Another trend to be addressed is the ability to handle a higher degree of nonlinearity in microwave applications. Increased nonlinearity in a model also requires increased training data, larger neural network size, and longer training time. More research in this area is very much needed. Design of modular neural network (MNN) architectures is another future possibility. In this approach an overall complicated modeling problem can be divided and handled by a set of submodules [4]. Modular neural networks have already demonstrated impressive success in systematic design and development in many application areas. It is expected [4] that there is a great future potential for MNNs as a convenient tool for coherently integrating information processing modules.

More automated training algorithms that integrate training and validation data generation, automatic neural network structural optimization, and weight training are on the way. This would simplify the human cost of microwave model development through computerized automated model generation. A long-term direction is to develop automated training methodology for the general knowledge-based structures. Training data generation by measurement or by simulation can be automatically driven on demand during training. Necessary criteria for data distribution and generation, error control, parameter extraction, knowledge selection, knowledge modification through training, and knowledge network structure selection will be allowed, and the overall model development process automated.

Modeling of large-signal nonlinear behavior of active devices and circuits continue to be a very active research topic. This is because new types of semiconductor devices continue to evolve, and efforts in creating new models continue to be an issue. In addition, nonlinear circuit level modeling is becoming important in order to facilitate higher-level simulation, such as complete transmitter and receiver module simulation. Large signal nonlinear behavior is best described in the time domain, and the time domain model requires temporal information in neural networks, and structures such as recurrent neural networks become necessary. The recently available large signal time domain measurements could provide very useful form of training data. An alternative approach is to use frequency domain models based on harmonic information. Continued work in embedding empirical and equivalent circuit knowledge into neural networks for both time domain and frequency domain approaches are needed. The overall objective is to achieve automated computer-based model development using knowledge and neural network learning as an alternative to the present human-based trial-and-error process in active component modeling. Potential applications of these approaches in modeling and design of nonlinear devices and circuits such as HBTs, FETs, power amplifiers, oscillators and mixers can be further exploited.

Because of their learning capability, neural networks are very effective in representing diverse range of problems. This will continue to motivate research in a variety of circuit design applications [5, 6]. For each circuit abstraction level—from device/component level circuit level to system level—there is a potential contribution with neural network modeling. ANN can fit well in multidisciplinary applications such as the electrical, thermal, and physical design of circuits and systems. Increased use in different design steps such as modeling, reverse modeling, simulation, and synthesis of microwave circuits is a possible trend. An important role of ANN that can be further exploited is its potential contribution to next generation optimization strategies for high-speed and high-frequency circuit design [7]. One of the frontiers that remain in optimization is the successful application of optimization procedures in problems where direct use of traditional optimization approaches is not practical. To address this need, a new direction of research has emerged in the optimization community. This approach makes use of surrogates in conjunction with true models for optimizing expensive engineering problems [8]. The development of artificial neural network approaches to microwave modeling, combined with space mapping, opens a new possibility in that direction. Further research in this direction can make extensive design optimization involving prohibitive simulations into a manageable process, efficiently helping to perform EM and physics-based optimization, hierarchical optimization from devices and circuits to systems, and large-scale, yield-driven optimization.

Another possible application of ANN is in increasing the efficiency of algorithms used in numerical methods like FDTD, MoM, FEM, TLM, SDA, and so forth, which are widely used in microwave CAD-oriented EM simulation. An example of this trend has been called the neurospectral technique [9]. In this technique the basic formulation of the problem is through the spectral domain approach (SDA). A neural network is then used to evaluate the integrals involving singular Green's functions. Finally, the characteristic equations are solved for nontrivial solutions using the reverse modeling approach in ANN. An example of this approach applied to determine the resonant frequency of a rectangular patch resonator is given in [9]. Another possible application is the combination of FDTD technique with ANN. The FDTD, being a spatial-temporal discretization method, requires huge computational storage and processing time. The computation time can possibly be improved by combining FIR-NN with it. Further research in applications of neural networks in accelerating parts of an algorithm for numerical computations would be of much broader interest beyond the domain of RF and microwave design.

As pointed out earlier, knowledge-aided design (KAD) could be a very exciting new direction that could transform the presently human intensive initial design into an automated design process [6]. The initial circuit and system design stage involves many fundamental decisions, such as circuit architecture decisions, process/material and component selections, and formulation of design specifications for subcircuits. These fundamental decisions could dictate the circuit cost and performance achievable in later design stages, and poor initial decisions could lead to a prolonged design cycle [10]. However, since there are no precise equations governing many of these issues, the knowledge-based approach becomes very essential. A significant role of ANNs that we propose to explore is the use of ANN for making the knowledge base available to designers of RF circuits for wireless systems. Within this scenario, ANN modules can be first trained for critical steps of design of representative RF circuits in a multilayer configuration (or in other design families). Such trained ANN models can be incorporated in knowledge-based CAD methods for initial design of RF circuits. A set of design modifier methods should be developed to allow designers to transform a representative circuit design to a custom design. Some decision making in an initial design, including a circuit structural design, requires judgment that does not always follow crisp, clear rules. A future possibility is to investigate fuzzy rules in the knowledge base. A long-term future potential that may be explored is the application of computational intelligence concepts [11] integrating artificial neural networks, fuzzy systems, and evolutionary algorithms [12] with overall circuit design process. An overview of the current trends in computational intelligence research is available in a special issue of the IEEE Proceedings [13]. Applications of some of these concepts would lead to much increased design automation throughout the entire RF and microwave design process, from conceptual design to final detailed design.

References

[1] Gupta, K. C., "Emerging Trends in Millimeter-Wave CAD," *IEEE Trans. Microwave Theory and Techniques*, Vol. MTT-46, June 1998, pp. 747–755.

[2] Gupta, K. C., A. Benalla, and R. Chew, "Computer-Aided Design of Microstrip Patch Arrays—A Multiport Network Modeling Approach," *Electromagnetics*, Vol. 11, 1991 (a special issue on "Microstrip Arrays," R. J. Mailloux, Ed.), pp. 89–106.

[3] Ma, S., and C. Ji, "Performance and Efficiency: Recent Advances in Supervised Learning," *Proc. IEEE*, Vol. 87, 1999, pp. 1519–1535.

[4] Caelli, T., L. Guan, and W. Wen, "Modularity in Neural Computing," *Proc. IEEE*, Vol. 87, 1999, pp. 1497–1518.

[5] Zhang, Q. J., and G. L. Creech, eds., *International Journal of RF and Microwave Computer-Aided Engineering*, Special Issue on Applications of Artificial Neural Networks to RF and Microwave Design, Vol. 9, New York: John Wiley and Sons, 1999.

[6] Gupta, K. C., "ANN and Knowledge-Based Approaches for Microwave Design," in *Directions for the Next Generation of MMIC Devices and Systems*, N. K. Das and H. L. Bertoni, Eds. New York: Plenum, 1996, pp. 389–396.

[7] Bandler, J. W., and Q. J. Zhang, "Next Generation Optimization Methodologies for Wireless and Microwave Circuit Design," (plenary session invited paper), *IEEE MTT-S Int. Topical Symp. on Technologies for Wireless Applications Digest*, Vancouver, BC, Feb. 1999, pp. 5–8.

[8] Booker, A. J., et al., "A Rigorous Framework for Optimization of Expensive Functions by Surrogates," *Structural Optimization*, Vol. 17, No. 1, 1999, pp. 1–13.

[9] Mishra, R. K., and A. Patnaik, "Neurospectral Computation for Complex Resonant Frequency of Microstrip Resonators," *IEEE Microwave and Guided Wave Letters*, Vol. 9, No. 9, September 1999, pp. 351–353.

[10] D. Montuno, D., et al. "Towards PCB Physical Design Automation: Architectural Analysis and Synthesis," *IEEE Electronic Components and Technology Conf*, (San Jose, CA), May 1997, pp. 991–995.

[11] Palaniswami, M., *Computational Intelligence*, New York: IEEE Press, 1995.

[12] Bonissone, P., et al., "Hybrid Soft Computing Systems: Industrial and Commercial Applications," *Proc. IEEE*, Vol. 87, 1999, pp. 1641–1667.

[13] Fogel, D. B., T. Fukuda, and L. Guan, eds., "Special Issue on Computational Intelligence," *Proc. IEEE*, Vol. 87, No. 9, Sept. 1999, pp. 1434–1700.

Appendix A

NeuroModeler Introductory Version

The CD-ROM attached to this book contains the introductory version of NeuroModeler. NeuroModeler program was developed to help RF and microwave engineers to develop neural network models for passive and active components/circuits for high-frequency circuit design. The objective is that the neural model development process will become easier for typical RF and microwave engineers, and that features important to RF and microwave design are implemented. The introductory version of NeuroModeler is a complimentary version for the purpose of evaluation and trial.

The software is developed by Professor Q. J. Zhang and his research team at the Department of Electronics, Carleton University. For information on advanced versions of NeuroModeler, please visit www.doe.carleton.ca/~qjz.

A.1 System Requirements

The introductory version of the software runs on Windows NT4.0, or Windows 95/98. The minimum system requirement is 200MHz CPU, 64MB RAM (NT) 32MB RAM (Windows 95). It occupies about 32MB of disk space.

A.2 How to Install the Software

Step 1: Put the NeuroModeler CD in your CD-ROM drive.
Step 2: Your computer will recognize the CD and run the Setup. If it does not, choose Run from the Start Menu. At the Open box, type *D:\setup* or *E:\setup* depending on which represents your CD-ROM drive.

Step 3: Follow the instructions that appear on the screen to complete the installation of the CD.

Step 4: If the installation is on Windows 95, please read the readme file when prompted at the end of installation for additional remarks.

A.3 Quick Start the Program Using an Example

This simple example will help you quickly get started with NeuroModeler. The example is the modeling of FET I-V characteristics.

The model has two inputs, that is $x = [V_{DS}\ V_{GS}]^T$ and one output, that is, $y = [I_D]$. The problem is to model the original $y = f(x)$ by a neural network model.

Training and test data: The training and test data files are fet_iv_train.dat and fet_iv_test.dat in the NeuroModelerHome\working directory where NeuroModelerHome is the directory on your computer where NeuroModeler is installed. Each file has three columns of data corresponding to V_{DS}, V_{GS}, and I_D. Each row of data represents a sample of (V_{DS}, V_{GS}, I_D). The training and test data are sampled differently.

Step 1: In your Windows NT/95 system, click "Start" and "Programs," then select the NeuroModeler program. After starting NeuroModeler, you are first in the main menu.
Click <New Neural Model>.

Step 2: For neural model structure, select <3 Layer Perceptrons>.
Use the scrollbars to make:
number of input neurons = 2
number of output neurons = 1
Press <OK> to return to main menu.

Step 3: Press the <Train Neural Model> button. Press <Get Training Data> to load the file fet_iv_train.dat;
Press <Start Training> and wait a few seconds (or minutes if the computer is an older one) for the training curve to be plotted. Press <Exit> to return to the main menu.

Step 4: Press the <Test Neural Model> button. Press <Get Test Data> button and select file fet_iv_test.dat;
This file contains (x, y) samples of the original problem and is never used in training. Press <Start Testing>. Press the rest of the buttons to examine the accuracy of the neural model as compared to test data. Press <Exit> to return to the main menu.

Step 5: If model accuracy in Step 4 is not satisfactory, then go to Step 3 for further training.

Step 6: Press the <Display Model Input/Output> button. Press <Start Simulation> to see the FET I-V relationship estimated by your trained neural model. This should approximate the original problem $y = f(x)$. Press <Exit> to return to the main menu.

Step 7: (Optional) If you want to save the trained neural model, then pull down the <File> menu on the top-left corner of the window, and choose <Save As>. A file dialogue will appear; enter a file name WITHOUT extension. A file extension of .prj will be automatically added. The saved file can be loaded back to NeuroModeler in the future.

To exit the program, pull down the <File> menu on the top-left corner of the window and choose <Exit>.

A.4 User Interactions

A.4.1 Minimum User Interactions

The minimum steps to train a neural model in NeuroModeler are:

Step 1: Before starting NeuroModeler, get a data file ready (to be used as training and test data). For file format, see the help topic "Data File Format" in the help menu in NeuroModeler.

Step 2: From the main menu, Press <New Neural Model> and select the correct numbers of input neurons and output neurons corresponding to your problem. <Exit> to the main menu.

Step 3: Press <Train Neural Model> then <Get Training Data>. Press <Start Training>. After the training is finished and the training error curve shown, <Exit> to the main menu.

Step 4: Press <Test Neural Model> and <Start Testing>. Wait until computation finishes and the Average Error and Worst Case Error are shown. If the error is too large, then <Exit> and go to Step 3 for further training; otherwise, <Exit> to the main menu and <Save Neural Model>.

The four steps above are the minimum user-interactions in training a neural model using NeuroModeler, where the user accepts the default values/choices of various options.

A.4.2 Extra User Control

Advanced users who like to have increased user control over the neural model development process can choose differently from the given defaults. For example, in NeuroModeler, users can override the default suggestions by choosing different starting values of weights, different activation functions, different connections between neurons, different neural network structures, different training algorithms, different data scaling schemes, and so on.

A.5 Highlights of the Introductory Version

- **Model Creation and Editing:** Create or modify a neural network structure. The default structure is a three-layer perceptron and the number of hidden neurons is automatically estimated for you. You can also choose from a variety of structure templates or define customer structures of neural models. You can change the number of hidden neurons, or change the activation functions of neurons from a set of predefined activation functions built into the program.
- **Data Processing:** Here you dictate what the neural network should learn from your data. NeuroModeler automatically performs basic data processing for training. It also has a Simulator Driver feature that can help automate the data generation process.
- **Training:** NeuroModeler automatically checks your data and suggests a training technique. A set of back propagation and gradient-based second-order training algorithms are available for you to select. The training process can be controlled through a set of training parameters.
- **Test:** The performance of a neural model can be verified using an independent set of data, either with a set of simple error criterion, or by a variety of detailed plots. You can also evaluate interpolation and extrapolation capabilities of your model.
- **Export:** You can export the trained neural model to a suitable format so that the model can be used in an external user environment. The introductory version allows the trained neural models to be exported to C-source code, Java source code, matlab, and a spreadsheet/Excel format.

A.6 Information on Upgrade to the Standard Version

The standard version of NeuroModeler is currently Version 1.2. In addition to all the capabilities in the Introductory Version, the standard Version 1.2

features several ways of incorporating empirical and equivalent circuit models into neural networks, including KBNN, source difference, PKI, space mapping, and hierarchical neural networks. A visual editor allows users to graphically define, visualize, and edit neural network structures. Also featured are an extended library of neuron activation functions, extended templates of neural network structures, an extended library of training algorithms, and an extended online user help feature. Further information can be obtained from Professor Zhang's Web site at www.doe.carleton.ca/~qjz.

Before Opening This CD-ROM Package, Please Read the Following Agreement. If You Do Not Agree With the Terms of the Agreement, Please Return the CD, Unopened, to Artech House.

The NeuroModeler introductory version (the software) attached to this book is provided purely as a complimentary program. All rights are reserved by the author(s) of the software. The purchase of the book does not imply the purchase of a license of the software. You are only given a complimentary license to install and use the software. No technical support is provided for complimentary licenses.

The software may be installed and used on one machine only. Except for backup, the software may not be copied for any other purpose.

The software is provided as is without warranty of any kind, either express or implied. This complimentary software is for evaluation and trial purposes only. There is no guarantee that the software is free of errors. You install and use the software entirely at your own risk. The author(s) of NeuroModeler assume no liability of any alleged or actual damages arising from the installation or use of (or inability to use) the software.

About the Authors

Q. J. Zhang

Professor Qi-jun Zhang received his B.E. degree from East China Engineering Institute, Nanjing, China, in 1982. He earned his Ph.D. in electrical engineering from McMaster University in Ontario, Canada, in 1987.

Professor Zhang worked at the Systems Engineering Institute at Tianjin University in China between 1982 and 1983. From 1988 to 1990, he worked with Optimization Systems Associates, Inc. (OSA) in Canada, where he helped develop advanced microwave optimization software. He joined the faculty of the Department of Electronics at Carleton University in Ottawa, Ontario in 1990.

Professor Zhang's research interests center around neural network and optimization methods for high-speed/high-frequency circuit design. He has authored over 130 publications in this area. He was co-editor of *Modeling and Simulation of High-Speed VLSI Interconnects* (Boston: Kluwer, 1994), and a contributor to *Analog Methods for Computer-Aided Analysis and Diagnosis* (New York: Marcel Dekker, 1988). He was a guest co-editor of a special issue of the *International Journal of Analog Integrated Circuits and Signal Processing* on high-speed VLSI interconnects, and a guest editor of a special issue of the *International Journal of RF and Microwave CAE,* on applications of ANN to RF and microwave design. He also developed NeuroModeler, the first software program designed specifically for neural-based RF and microwave modeling. He is a senior member of the IEEE and a member of the Professional Engineers of Ontario.

K. C. Gupta

Professor K. C. Gupta received his B.E. and M.E. degrees in electrical communication engineering from the Indian Institute of Science, Bangalore, India, in 1961 and 1962, respectively. He went on to earn his Ph.D. from the Birla Institute of Technology and Science in Pilani, India, in 1969. From there, Professor Gupta moved to the Indian Institute of Technology in Kanpur, where he taught electrical engineering until 1984. He was the coordinator for the Phased Array Radar Group of the Advanced Center for Electronics Systems at the Indian Institute of Technology from 1971 to 1979. He has been a visiting professor at the University of Waterloo, Canada; the Ecole Polytechnique Federale de Lausanne, Switzerland; the Technical University of Denmark; the Indian Institute of Science; and the University of Kansas in the United States. He has been a professor at the University of Colorado since 1984. He currently serves as the associate director for the NSF I/UCR Center for the Advanced Manufacturing and Packaging of Microwave, Optical, and Digital Electronics (CAMPmode) at the University of Colorado.

Professor Gupta's current research interests are in the area of computer-aided design techniques for microwave and millimeter-wave integrated circuits and antennas, RF MEMS, and reconfigurable antennas. He is the author or co-author of seven books: *Microwave Integrated Circuits* (New York: John Wiley and Sons, 1974); *Microstrip Line and Slotlines* (Norwood, MA: Artech House, 1979); *Microwaves* (New York: John Wiley and Sons, 1979); *CAD of Microwave Circuits* (Norwood, MA: Artech House, 1983); *Microstrip Antenna Design* (Norwood, MA: Artech House, 1988); *Analysis and Design of Planar Microwave Components* (New York: IEEE Press, 1994); and *Analysis and Design of Integrated Circuit-Antenna Modules* (New York: John Wiley and Sons, 1999). He has also contributed chapters to the *Handbook of Microstrip Antennas* (London: IEE, 1989); the *Handbook of Microwave and Optical Components, vol. 1* (New York: John Wiley and Sons, 1989); *Microwave Solid State Circuit Design* (New York: John Wiley and Sons, 1988); *Numerical Techniques for Microwave and Millimeter Wave Passive Structures* (New York: John Wiley and Sons, 1989); and the *Encyclopedia of Electrical and Electronics Engineering* (New York: John Wiley and Sons, 1999). He has published close to 200 research papers and holds three patents in the microwave area.

Professor Gupta is a Fellow of both the IEEE and the Institute of Electronics and Telecommunications Engineers (India), and a member of both URSI (Commission D, United States) and the Electromagnetics Academy at the Massachusetts Institute of Technology. He is also a member of the ADCOM for the MTT Society of IEEE, the IEEE Technical Committee on Microwave Field Theory (MTT-15), and the IEEE-EAB Committee on Continuing Education. He serves as chair of the IEEE MTT-S Standing Committee on Education and is the former co-chair of the IEEE MTT-S Technical Committee on CAD. He is a recipient of the IEEE Third Millennium Medal. He is also the founding editor of the *International Journal of Microwave and Millimeter-Wave Computer-Aided Engineering*, and an Associate Editor for *IEEE Microwave Magazine*. He currently serves on the editorial boards of *IEEE Transactions on Microwave Theory and Techniques*, *Microwave and Optical Technology Letters*, and the *International Journal of Numerical Modeling*. He is listed in *Who's Who in America*, *Who's Who in the World*, *Who's Who in Engineering*, and *Who's Who in American Education*.

Index

8-bit digital bus, 207–9
   cross-sectional view, 208
   neural network model, 209
   on PCB, 208

Activation functions, 66–68
   arc-tangent, 67
   for boundary layer neurons, 299
   hyperbolic-tangent, 68, 69
   input neuron, 68
   linear, 68
   MLP, 81
   output neuron, 68
   RBF network, 81
   sigmoid, 66–67
Active component models, 7, 227–46
   direct modeling approach, 228–39
   discussion, 245–46
   indirect modeling approach, 239–45
   introduction to, 227–28
   large signal, 233–37
   small signal, 230–33
   time-varying Volterra kernel-based, 237–39
   transistor DC, 229–30
   See also Models
ANN models, 2, 4
   CAD, 21
   circuit optimization with, 251–55
   CPW patch feed line, 268–69
   design and optimization with, 249–50
   filter design advantages, 262
   hybrid structures, 342
   insertion in design tools, 340
   integration, with circuit simulator, 249
   knowledge-based, 7, 56, 283–334
   linked to design software, 340–41
   macromodels, 56
   multilayer circuit design and optimization with, 255–65
   See also Artificial neural networks (ANNs); EM-ANN models; Models
Antennas
   CPW patch, 188, 265–72, 339
   microstrip patch, 39–49, 187–88
   network characteristics, 51
   slot, 49–55
Arbitrary structures, 61, 62, 88–90
   features, 88
   feedforward, 90
   mapping, into weight matrix, 89
   structural optimization, 90
   See also Neural network structures
Arc-tangent function, 67
Artificial neural networks (ANNs), 3–4
   architectural and training methodologies, 342, 343
   characteristics, 4
   impact, on RF/microwave design, 340–42

modules, 345
possible applications, 344–45
in RF and microwave CAD, 55–57
role of, 344, 345
structures, 5
training, 5
use of, 3
See also ANN models; EM-ANN models
Average test error, 126

Back propagation (BP), 75–76, 129
algorithm, 133–37
comparison, 149
convergence, 134
defined, 75, 133
error, 76–77, 136
training process, 75
update formulae, 133, 134
weight oscillation, 134
See also Training
Broyden-Fletcher-Goldfarb-Shanno (BFGS) formula, 140

Cavity model, 41–46
defined, 41–42
fringing fields, 45
input impedance, 45
limitations, 46
loss tangent, 45
magnetic current source, 46
modal fields, 43
resonant wave numbers, 42, 43, 44
two-dimensional resonator, 46
See also Microstrip patches
CD-ROM, this book, 347
Central composite distribution, 111–12
defined, 111–12
design patterns, 112
See also Distribution
Chamfered 90° bend, 169–70, 171
EM-ANN modeling of, 169–70
optimization, 250–51
structure illustration, 171
See also CPW bends
Circuit analysis, 7
block diagram, 23
defined, 22

Circuit functions
active power, 27
computation of, 26–28
gain, 27
input impedance, 28
insertion loss, 27
phase, 27
reflection coefficient, 28
transfer function, 27
Circuit waveform modeling example, 290–92
accuracy comparison, 292, 293
testing results, 291
See also Knowledge-based neural networks (KBNN)
Clustering
algorithms, 91–94
filter responses, 92–94
K-means algorithm, 94
network, 92
problem, 91–94
Compensated CPW bend, 173
Computer-aided analysis, 21–36
computation of circuit functions, 26–28
defined, 21–22
harmonic balance method, 35–36
linear/nonlinear subnetworks, 34–35
nodal admittance matrix method, 22–26
nonlinear circuits analysis, 34
scattering-matrix analysis, 28–34
Computer-aided design (CAD), 15–17, 19–57
aim of, 15
analysis techniques, 21–36
ANNs role in, 55–57
circuit optimization, 36–38
defined, 15
illustrated, 16
implementation, 17, 19
modeling of printed patches/slots, 39–51, 51–55
printed RF and microwave antennas, 38–55
RF and microwave circuit, 19–38
segments, 15–16
See also Computer-aided design (CAD)

Conjugate gradient training method, 137–39
  application issues, 139
  comparison, 149
  conjugate direction, 138
  defined, 137
  Fletcher-Reeves formula, 138
  See also Training
Connection-scattering matrix, 32–33
Conventional design, 13–15
  defined, 13–14
  difficulty in using, 14–15
  illustrated, 14
  See also Design process
Coplanar waveguide (CPW)
  circuits, 1, 55, 168
  component models, 168–78
  feed, 51
  open-end effects, 267
  resonators, 265
  variable parameter ranges, 169
  See also CPW bends; CPW open circuits; CPW short circuits; CPW step-in-width; CPW symmetric T-junction; CPW transmission lines
Correlation coefficient, 127, 279
CPW bends, 56, 169–74, 339
  90°, 169–70, 171
  chamfered, 169–70
  comparisons, 173–74
  compensated, 173
  modeling of, 169–74
  optimally chamfered conventional, 170–73
CPW folded double-stub filter
  design, 251–52
  geometry, 251–52
  for optimized EM-ANN circuit, 253–54
CPW open circuits, 174–75
  development, 174
  EM-ANN models for, 174–75
  error results, 178
  geometry, 176
  S-parameter response, 177
CPW patch antennas, 188, 265–72, 339
  capacitance model, 267–68

design layout, 269
design optimization with EM-ANN models, 269–72
design using EM-ANN models, 269
EM-ANN model, 266–67, 269–72
geometry, 266
improved bandwidth, 273
layout, 273
modified geometry, 271
patch feed line, 268–69
patch length effects, 272
radiation conductance model, 267–68
response effects, 271
return loss comparison, 270, 274
transmission line model for, 265–69
CPW power divider, 252–55
  design, 252–55
  EM simulation time, 254
  geometry, 255
  optimizable parameter, 252
  for optimized EM-ANN circuit, 256–57
CPW short circuits, 174–75
  EM-ANN models for, 174–75
  error results, 178
  geometry, 176
  S-parameter response, 177
CPW step-in-width, 175–76
  error results, 181
  geometry, 179
  S-parameter response, 180
CPW symmetric T-junctions, 176–78
  error results, 183
  geometry, 181
  S-parameter response, 182
CPW transmission lines, 56, 169, 266, 339

Data scaling, 116–19
  linear, 116–17
  log arithmetic, 117–18
  no, 119
  two-sided log arithmetic, 118–19
  See also Scaling
Data splitting, 114–16
  extreme case, 115
  intermediate case, 115
Davidon-Fletcher-Powell (DFP) formula, 140

Decomposed optimization, 146–47
Delta-bar-delta rule, 135
Design
    central composite distribution, 112
    CPW folded double-stub filter, 251–52
    CPW patch antenna, 265–72
    CPW power divider, 252–55
    design-in-the-large, 13
    design-in-the-small, 13
    detailed, 12
    EM simulation model, 48
    of experiments (DOE), 111–12
    initial, 12
    multilayer circuit, 255–65
    multilayer coupled line filters, 258–61
    See also Microwave design; RF design
Design process, 11–18
    anatomy, 11–13
    CAD, 15–17
    conventional, 13–15
    KAD, 17–18, 56–57, 345
    philosophies, 13
    step sequence, 12
Detailed design step, 12
Direct modeling, 228–39
    large-signal models, 233–37
    small-signal models, 230–33
    time-varying Volterra kernel-based
        model, 237–39
    transistor DC model, 229–30
    See also Active component models
Direct search optimization, 38
Distribution, 108–14
    central composite, 111–12
    discussion, 114
    nonuniform grid, 110–11
    random, 113–14
    star, 112–13
    uniform grid, 108–10
    See also Neural model development
Electromagnetically trained ANN
    (EM-ANN). See EM-ANN
    models
EM-ANN models, 4, 338
    comparison, 167
    for CPW bends, 169–74
    for CPW components, 168–78
    for CPW open-end effects, 267
    for CPW opens and shorts, 174–75
    for CPW patch antenna design
        optimization, 269–72
    for CPW patch antennas, 269
    for CPW step-in width, 175–76
    for CPW symmetric T-junctions,
        176–78
    for CPW transmission lines, 169, 266
    defined, 4
    integration, with circuit simulators,
        157–58
    integration, with network simulator,
        166–68
EM simulation models, 47–49
    design illustration, 48
    finite-difference time-domain
        simulation, 48–49, 52
    integral-equation-based full-wave
        analysis, 47–48
EM simulators, 341
Equivalent circuit model, 239–45
Error back propagation, 76–77
Error mean, 127

Feedforward
    arbitrary structure, 90
    MLP, 70–71
    RBF network, 79–80
    training examples, 148–51
    wavelet network, 84–86
FET neural model, 62–63
Finite-difference time-domain (FDTD)
        simulation, 48–49, 52, 245
    computational domain limitation, 52
    formulation, 48–49
Fletcher-Reeves formula, 138

GaAs microstrip bias, 160–63
    one-port, 160–62
    two-port, 162–63
    variable input parameters, 162
Gain, 27
Galerkin's method, 21
Gauss-Newton training method, 140–41
Generalization ability, 120
Generalized delta rule, 134
Genetic algorithms, 143–45
    comparison, 149
    defined, 143

steps, 143–44
uses, 145
See also Training
Global minimum, 133
Global optimization methods, 143–46
   genetic algorithms, 143–45
   SA algorithms, 145–46
   See also Training
Golden Section Method, 131–32
Good learning, 121–22
Gradient-based optimization, 38
Gradient-based training methods, 130–31
   algorithms, 137–41
   conjugate gradient, 137–39
   Gauss-Newton, 140–41
   Levenberg-Marquardt, 140–41
   Quasi-Newton, 139–40
   See also Training

Harmonic balance method (HBM), 35–36, 233
   comparison, 36
   defined, 35–36
   use of, 36
   See also Computer-aided analysis
HBT neural model, 230–33
   illustrated, 231
   S-parameters, 230, 231
   Y-parameters, 231, 233
   See also Small signal models
Hidden neurons
   activation function, 66
   with bias parameter, 69
   defined, 61
   quality of learning vs. number of, 125
   RBF network, 79, 80
   sigmoid activation functions, 73
   See also Neurons
Hierarchical neural networks, 315–33
   base models, 316–18
   examples, 323–33
   finished model, 321
   index function, 319
   knowledge hubs, 319
   library of MESFET models example, 329–33
   library of stripline models example, 323–29
   low-level neural modules, 319–20

   structure, 316
   structure illustration, 318
   training, 318–21
High electron-mobility transistors (HEMTs)
   indirect neural modeling for, 242–43
   parameters, 243
High-speed IC interconnects, 7
Hopefield network, 98–100
   activation function input, 99
   defined, 98
   dynamics, 100
   output, 99
   See also Recurrent network
HP-MDS simulator, 157, 252, 260, 264
HSPICE simulation, 210, 291
Hyperbolic-tangent function, 67, 68

Index function, 319
Indirect modeling, 239–45, 246
   example, 242–43
   illustrated, 241
   See also Active component models
Initial design step, 12
Input impedance, 28
Input neurons, 68
   activation functions, 68
   defined, 61
   See also Neurons
Insertion loss, 27
Integral-equation-based full-wave analysis, 47–48
   optimization techniques, 47–48
   steps, 47
Inverse Mexican-hat function, 85
Inverse modeling problem, 184

Jacobian matrix, 289

Kirchoff's current law, 23–24
K-means algorithm, 94
Knowledge-aided design (KAD), 17–18, 345
   features, 17
   implementation, 56–57, 342
   PCM, 17–18
Knowledge-based neural networks (KBNN), 7, 56, 285–301, 317–18, 339

accuracy comparison, 292, 295, 296, 297, 298
circuit waveform, 290–92
defined, 285–86
examples, 290–301
finished model, 289–90
introduction to, 283–85
MESFET, 292–96
microwave-oriented, 285
MLP vs., 292, 295, 296, 297, 298
model structure, 286–89
neuron arrangement, 287
output, 290
performance stability, 301
rule-based, 284–85
size of, 291
structure illustration, 288
superiority, 295, 301
training, 289
transmission line, 296–301
weight parameters, 289
Knowledge hubs
defined, 319
stripline library, 327

Large-signal models, 233–37, 344
incorporating, into harmonic balance circuit simulator, 235
MESFET, 233–35
physics-based MESFET, 235–37
Layers
MLP, 74
multiple, 55
RBF network, 78
Learning
curves, 122
good, 121–22
overlearning, 120, 124
quality of, 124–25
rate, 134
training and validation error curves, 123
underlearning, 120–21
See also Training
Least-mean square error (LMS) algorithm, 147
Levenberg-Marquardt training method, 140–41, 148
comparison, 149
with diagonal matrix, 141

Libraries, 315–33
accuracy and development cost comparison, 329
MESFET models, 329–33
stripline models, 323–29
See also Models
Library development, 315–16
algorithm, 322–23
with hierarchical neural network approach, 317
MLP approach, 322
Linear scaling, 116–17
illustrated, 117
of inputs/outputs, 117
See also Data scaling
Linear subnetworks, 34–35
Linecalc, 170
Line minimization, 131–32
Golden Section Method, 131–32
quadratic interpolation method, 132
Links
defined, 61
weights, 88
Local minimum, 132–33
Log arithmetic scaling, 117–18
descaling, 118
illustrated, 118
two-sided, 118–19
See also Scaling
Lumped equivalent circuit, 239

MESFET
DC neural model, 229–30
with different gate-length values, 330
intrinsic structure, 330
large signal model, 233–35
physics-based large-signal model, 235–37
MESFET modeling example, 292–96
accuracy comparison, 295, 296, 297, 298
empirical data, 294
extrapolation data, 294
input parameters, 294
I-V curves, 300
training data, 294
See also Knowledge-based neural networks (KBNN)

MESFET model library example, 329–33
   base model selections, 330–31
   defined, 329–30
   hierarchical neural model, 332
   model accuracy comparison, 332
   number of training samples, 334
   overall library accuracy and
      development cost comparison,
      333
   =2 library model, 331
   =5 library model, 331–33
   See also Libraries
Microstrip patches, 39–49
   analysis of, 51–55
   cavity model, 41–46
   EM simulation-based numerical
      models, 47–49
   finite-difference time-domain
      simulation, 48–49
   integral-equation-based full-wave
      analysis, 47–48
   models, 39–51, 187–88
   multiport network model (MNM), 46
   network models for, 39–46
   schematic, 187
   transmission line model, 39–41
   See also Antennas
Microstrip-to-microstrip multilayer
      interconnect, 165–66
   illustrated, 165
   simulation, 166
   variable parameters, 166
   See also Multilayer interconnects
Microstrip transmission line model,
      159–60
   geometry cross-section, 159
   input parameters, 160
Microwave antennas, 38–55
Microwave circuits
   CAD design, 19–38
   computer-aided analysis, 26
   modeling technique comparison, 286
   multiple layer, 55
   performance predication accuracy, 19
Microwave design, 1–3
   component models, 6–7
   neural net impact on, 340–42
   See also Design

Modeling procedures, 155–58
Models, 155–89
   active component, 7, 227–46
   ANN, 2, 4, 7, 21, 56, 155–56
   base, 316–18
   capacitance, 267–68
   EM-ANN, 4, 157–58, 168–78,
      174–78, 269–72, 338
   equivalent circuit, 239–45
   error measures, 157
   hierarchical neural, 318–21
   high-speed IC interconnects, 195–223
   input/output selection, 156
   large-signal, 233–37
   libraries, 315–33
   microstrip patch antenna, 187–88
   microstrip transmission line, 159–60
   multiconductor transmission lines,
      179–87
   for multilayer interconnects, 163–66
   passive component, 178–89
   polynomial, 199
   radiation conductance, 267–68
   small-signal, 230–33
   spiral inductors, 179
   training data, 63, 156–57
   for vias, 160–63
   waveguide filter components, 188–89
Modified nodal admittance (MNA) stamp,
      203
Modular neural network (MNN)
      architectures, 343
Moment-matching techniques (MMT),
      200
Monte Carlo analysis, 211
Multiconductor transmission lines, 179–87
   analysis ANN model for, 184
   geometries, 184
   synthesis ANN model for, 185
   synthesis procedure, 186
Multilayer coupled line filter, 257, 258–65
   center frequency and bandwidth
      parameters, 264
   conventional method physical
      dimensions, 263
   design example, 258–61
   design method comparison, 261–65
   design times, 264

geometry, 257
modeled response, 260, 261
optimized physical dimensions, 262
optimized responses, 262
physical dimensions, 260
response comparison, 261, 263
specifications, 260
Multilayer interconnects, 163–66
   microstrip-to-microstrip, 165–66
   stripline-to-stripline, 163–65
Multilayer perceptrons (MLP), 61, 62, 64–74, 100, 338
   3-layer, 74
   4-layer, 74
   accuracy comparison, 82
   activation functions, 66–68, 81
   back propagation (BP), 75–77
   defined, 64
   effect of bias, 68–69
   elements, 64
   illustrated, 65
   information processing by neurons, 65–66
   I-V curves, 300
   KBNN vs., 292, 295, 296, 297, 298
   library development approach, 322
   neural network feedforward, 70–71
   number of layers, 74
   number of neurons, 73–74
   one-dimensional function modeled by, 72
   RBF network comparison, 81–82
   structure, 64–65
   universal approximation theorem, 71–73
   See also Neural network structures
Multiport network model (MNM), 46

Near-field to far-field transformation, 52, 53
Neural model development, 106–28
   comparison, 109
   data generation by measurement, 107
   data generation by simulation, 108
   data scaling, 116–19
   data splitting, 114–16
   overlearning and underlearning, 120–26
   problem statement, 105–6
   quality measures, 126–28
   range/distribution of samples, 108–14
   weight parameter initialization, 119–20
Neural-network-based global modeling, 243–45
   defined, 243–44
   illustrated example, 244
   iterative computation process, 245
Neural network feedforward, 70–71
   computation, 70
   process, 70
Neural Networks for RF and Microwave Design
   CD-ROM, 347
   organization, 4–8
   summary, 337–40
Neural network structures, 61–101
   arbitrary, 88–90
   functions, 62
   generic notation, 62–63
   introduction to, 61–64
   MLP, 64–74
   RBF network, 78–81
   recurrent, 97–100
   SOM, 90–97
   summary, 100–101
   types of, 61, 62
   wavelet, 83–88
   See also Artificial neural networks (ANNs)
NeuroADS, 290
NeuroModeler, 347–51
   data processing, 350
   defined, 347
   export, 350
   extra user control, 349–50
   installing, 347
   introductory version highlights, 350
   minimum user interactions, 349
   model creation and editing, 350
   quick starting, 348–49
   standard version upgrade information, 350–51
   system requirements, 347
   test, 350
   test data, 348
   training, 348, 350
   use agreement, 351
   user interactions, 349–50

Neurons
  boundary, 292
  defined, 61
  hidden, 61, 66, 69, 73, 79, 80, 125
  input, 61, 68
  MLP information processing by, 65–68
  number of, 73–74
  output, 61, 68
  SOM, 94–95
Nodal admittance matrix method, 22–26
  defined, 22–23
  definite admittance matrix, 24, 25
  indefinite admittance matrix, 25
  network example, 24
  reference node, 23, 25
  sparse matrix, 26
  See also Computer-aided analysis
Nongradient-based training, 130, 141–43
  simplex method, 141–43
  See also Training
Nonlinear circuits
  analysis, 34
  linear/nonlinear subnetworks, 34–35
Nonlinear subnetworks, 34–35
Nonuniform grid distribution, 110–11
  defined, 110
  illustrated, 111
  use example, 111
  See also Distribution
Normal mode parameters (NMPs), 180–81, 258
  altering, by changing design parameters, 185
  calculation, 185
Numerical inversion of Laplace transform (NILT)
  off-line data generation, 215
  simulation, 211, 212

Optimally chamfered CPW bend, 170–73
  error results, 172
  return loss, 174
  See also CPW bends
Optimization, 249–50
  chamfered CPW 90º bend, 250–51
  circuit, 36–38, 251–55
  component structure, 250–51
  CPW patch antenna, 265–72
  decomposed, 146–47

  direct search method, 38
  gradient method, 38
  multilayer circuit, 255–65
  role, 37
  signal integrity, 197, 210–13
  strategies, efficient, 56, 341–42
  stripline-to-stripline multilayer interconnect, 250
  of three-stage MMIC amplifier, 272–78
  yield, 272–78
Organization, this book, 4–8
Output neurons
  activation functions, 68
  defined, 61
  See also Neurons
Overlearning
  defined, 120
  detection, 124
  neural network model, 121
  training and validation error curves, 124
  See also Learning

Phase, 27
Polynomial models, 199
Printed circuit boards (PCBs)
  configurations, 213
  eight-bit bus on, 208
  neural networks for interconnects on, 213–16
  tree of four interconnects on, 220
Prior knowledge input (PKI) method, 310–11
  defined, 309
  finished model, 310–11
  neural network training, 310
  preprocessing, 309–10
  structure, 309
  structure illustration, 310
Product design specification (PDS), 11
Propose-critique-modify (PCM) approach, 17–18
  elements, 18
  illustrated, 18

Quality
  learning, 124–25
  trained neural model, 126–28

Quasi-Newton training method, 139–40, 148
  comparison, 149
  weight updating, 139

Radial basis function (RBF) networks, 61, 62, 77–81
  accuracy comparison, 82
  activation function, 81
  applications, 77–78
  defined, 77
  feedforward computation, 79–80
  Gaussian function, 78, 79, 80
  hidden neurons, 79, 80
  illustrated, 78, 80
  layers, 78
  MLP comparison, 81–82
  multiquadradic function, 78, 79
  one-dimensional function to be modeled by, 80
  structure, 78–79
  two-step training, 81
  universal approximation theorem, 81
  weight parameters, 80
  See also Neural network structures
Radial function, 83
Radio frequencies. See RF design
Random distribution, 113–14
Recurrent networks, 61, 62, 97–100
  defined, 97
  with feedback of delayed output, 98
  Hopefield, 98–100
  illustrated, 98, 99
  with MLP module, 99
  uses, 98
  See also Neural network structures
Reflection coefficient, 28
RF design, 1–3
  antennas, 38–55
  application modeling techniques, 286
  CAD, 19–38
  component models, 6–7
  neural net impact on, 340–42
  See also Design
Rule-based knowledge networks, 284–85

Sample-by-sample training, 125, 134
Scaling
  input/output, 119

  linear, 116–17
  log arithmetic, 117–18
  two-sided log arithmetic, 118–19
Scattering-matrix analysis, 28–34
  application, 28
  block diagonal matrix, 30
  component characterization, 31
  connection-scattering matrix, 32–33
  illustrated, 29
  implementation, 29
  See also Computer-aided analysis
Search Then Converge (STC) scheme, 135
Segmentation and Boundary Element Method (SBEM), 258
Self-organizing maps (SOM), 61, 62, 94–97, 338
  defined, 94
  neural network model using, 97
  neurons, 94–95
  purpose, 91
  topology ordering, 95
  trained, using, 96–97
  training, 95–96
  See also Neural network structures
Sigmoid function, 66–67
  illustrated, 67
  property, 66
Signal integrity
  analysis approaches, 199
  characteristics modeling, 197
  defined, 195
  optimization, 197, 210–13
Simplex training method, 141–43
  comparison, 149
  error function, 141
  new points generated in, 142
  optimization, 141
  training error, 141
  uses, 143
  See also Non-gradient based training; Training
Simulated annealing (SA) algorithms, 145–46
  comparison, 149
  defined, 145
  steps, 145–46
  uses, 146
  See also Training

Simulators
    circuit, 157–58, 249
    EM, 341
    HP-MDS, 157, 252, 260, 264
    microwave, 265
    network, 166–68
Slot antennas, 49–55
    analysis of, 51–55
    equivalent transmission line model, 50
    evolution, 49
    inductors, 50
    modeling of, 51
    numerical modeling, 51
    simulating, 52
    transmission line resonator model, 49
    See also Antennas
Small signal models, 230–33
    FET, 242
    HBT, 230–33
    use of, 230
Source difference method, 301–9
    defined, 301–2
    finished model, 308–9
    model structure, 303
    neural network training, 308
    preprocessing, 303, 308
    structure illustration, 308
Space-mapped (SM) neural networks, 312–15
    concept, 313–14
    defined, 312
    frequency, 314–15
    neuromodeling, 314, 315
    structure, 312
    structure illustration, 312
Space-mapped neuromodeling (SMN), 313
    frequency (FSMN), 314
    frequency dependent (FD), 314
Spectral domain approach (SDA), 345
Spiral inductors, 179
Star distribution, 112–13
    defined, 112
    illustrated, 113
    number of samples, 113
    See also Distribution
Stripline model library example, 323–29
    all library models, 328–29
    base model selections, 323–25

defined, 323
hierarchical neural model, 327
illustrated, 324
input/output parameters, 324, 325
knowledge hubs, 327
library accuracy and development cost comparison, 329
low-level modules, 327
model accuracy comparison, 328
number of training samples, 329
q=1 library model, 325–26
q=3 library model, 326–28
See also Libraries
Stripline-to-stripline multilayer interconnect, 163–65
    optimization, 250
    simulation, 165
    structure, 164
    variable parameters, 164
    See also Multilayer interconnects
Symmetric T-junctions, 176–78

Three parallel coupled interconnects example, 203–6
Three-stage MMIC amplifier, 272–78
    circuit diagram, 274
    CUP comparison summary, 281
    distributions for statistical variables, 277
    gain and input VSWR, 276–77
    MESFET parameters correlation coefficients, 279
    Monte Carlo sweeps, 278, 280
    variables for nominal design, 276
    yield optimization, 272–78
    yield optimization design variables, 280
Time-varying Volterra kernel-based model, 237–39
    defined, 237–38
    example, 238–39
    transistor models to be used in, 239
    See also Direct modeling
Topology ordering, 95
Training, 128–33
    ANN, 5, 342, 343
    automated, algorithms, 343
    back propagation (BP), 75–76, 77, 129
    base model, 316–18
    batch-mode, 125

concept illustration, 287
conjugate gradient method, 137–39
error, 115–16
feedforward examples, 148–51
Gauss-Newton method, 140–41
genetic algorithms, 143–45
global minimum, 133
with global optimization methods, 143–46
Golden Section Method, 131–32
gradient-based methods, 130–31
hierarchical neural model, 318–21
KBNN, 289
learning and, 120–26
Levenberg-Marquardt method, 140–41
line minimization, 131–32
local minimum, 132–33
NeuroModeler, 350
nongradient-based, 130, 141–43
objective, 116
PKI method, 310
process types, 125–26
quadratic interpolation method, 132
Quasi-Newton method, 139–40
RBF network, 81
sample-by-sample, 125, 134
SOM, 95–96
source difference method, 308
technique categories, 129–30
technique comparisons, 147–48
wavelet network, 87
Training data
defined, 63
generation, 156–57, 343
MESFET modeling example, 294
quality of learning vs., 124–25
scaling, 116–19
splitting, 114–16
transmission line modeling example, 301
Transfer function, 27
Transistor DC model, 229–30
Transmission line model, 39–41
configurations, 41
for CPW antennas, 265–69
defined, 39
illustrated, 40
limitations, 41
for two-port rectangular patches, 41, 42
See also Microstrip patches
Transmission line modeling example, 296–301
accuracy comparison, 302, 303, 304, 305
average testing error, 304
MLOP vs., 302, 303, 304, 305
scattering plot of mutual inductance, 306
testing data in extrapolation region, 303
testing data in same region, 302
testing error histograms, 307
training data, 301
worst-case testing error, 305
See also Knowledge-based neural networks (KBNN)
Trends and challenges, 342–46
Two asymmetric interconnects example, 206–7
Two-sided log arithmetic scaling, 118–19

Underlearning
defined, 120–21
neural network model, 121
See also Learning
Uniform grid distribution, 108–10
defined, 108
illustrated, 110
number of samples and, 110
See also Distribution
Universal approximation theorem
defined, 71
MLP, 81
RBF network, 81

Vias, 160–63
broadband GaAs one-port microstrip, 160–62
broadband GaAs two-port microstrip, 162–63
simulation, 163
substrate thickness, 163
VLSI interconnects
application examples, 203–16
circuit level modeling, 202–3

circuit with nonlinear terminations, 209–10
conclusions, 222–23
coupled 4-conductor transmission line, 201
eight-bit digital bus configuration, 207–9
end-to-end signal, 222
hierarchy, 198
high-speed, 197–203, 212
illustrated, 291
input-output parameters, 201
introduction to, 195–97
modeling, 195–223
neural network approach, 200–203
neural networks, on PCB, 213–16
number of, 196
performance evaluation, 218–22
physical/EM level modeling, 201–2
run-time comparison, 216–18
signal delay histogram, 217
signal delivery example, 199
signal integrity optimization, 210–13
three parallel coupled, 203–6
traditional modeling techniques, 197–200
tree, 202
two asymmetric, 206–7

VSWR
frequency variations, 20
input, 19–20

Waveguide filters
component models, 188–89
E-plane metal insert, 188, 189
fifth order, 188
model development procedures, 189
Wavelet function, 83
Wavelet neural networks, 61, 62, 83–88
with directional feedforward, 86–87
feedforward computation, 84–86
illustrated, 85
one-dimensional function to be modeled by, 86–87
radial function, 83
training, 87
wavelet function, 83
wavelet initialization, 87–88
wavelet transform, 83–84
See also Neural network structures
Wavelet transform, 83–84
Weight parameters
initialization of, 119–20
KBNN, 289
RBF network, 80
Worst-case error, 127

Recent Titles in the Artech House Microwave Library

Advanced Automated Smith Chart Software and User's Manual, Version 3.0, Leonard M. Schwab

Behavioral Modeling of Nonlinear RF and Microwave Devices, Thomas R. Turlington

Computer-Aided Analysis of Nonlinear Microwave Circuits, Paulo J. C. Rodrigues

Design of FET Frequency Multipliers and Harmonic Oscillators, Edmar Camargo

Design of RF and Microwave Amplifiers and Oscillators, Pieter L. D. Abrie

EPFIL: Waveguide E-plane Filter Design, Software and User's Manual, Djuradj Budimir

Feedforward Linear Power Amplifiers, Nick Pothecary

Generalized Filter Design by Computer Optimization, Djuradj Budimir

GSPICE for Windows, Sigcad Ltd.

Introduction to Microelectromechanical (MEM) Microwave Systems, Hector J. De Los Santos

Microwave Engineers' Handbook, Two Volumes, Theodore Saad, editor

Microwave Filters, Impedance-Matching Networks, and Coupling Structures, George L. Matthaei, Leo Young, and E.M.T. Jones

Microwave Mixers, Second Edition, Stephen Maas

Microwave Radio Transmission Design Guide, Trevor Manning

Microwaves and Wireless Simplified, Thomas S. Laverghetta

Neural Networks for RF and Microwave Design, Q. J. Zhang and K. C. Gupta

PACAD: RF Linear Power Amplifier Design, Software and User's Manual, Ramin Fardi, Keyvan Haghighat, and Afshin Fardi

RF Design Guide: Systems, Circuits, and Equations, Peter Vizmuller

The RF and Microwave Circuit Design Handbook, Stephen A. Maas

RF and Microwave Coupled-Line Circuits, Rajesh Mongia, Inder Bahl, and Prakash Bhartia

RF Power Amplifiers for Wireless Communications, Steve C. Cripps

RF Systems, Components, and Circuits Handbook, Ferril Losee

TRAVIS Pro: Transmission Line Visualization Software and User's Manual, Professional Version, Robert G. Kaires and Barton T. Hickman

Understanding Microwave Heating Cavities, Tse V. Chow Ting Chan and Howard C. Reader

For further information on these and other Artech House titles, including previously considered out-of-print books now available through our In-Print-Forever® (IPF®) program, contact:

Artech House
685 Canton Street
Norwood, MA 02062
Phone: 781-769-9750
Fax: 781-769-6334
e-mail: artech@artechhouse.com

Artech House
46 Gillingham Street
London SW1V 1AH UK
Phone: +44 (0)20 7596-8750
Fax: +44 (0)20 7630 0166
e-mail: artech-uk@artechhouse.com

Find us on the World Wide Web at:
www.artechhouse.com